GLOBAL POPULATION

Columbia Studies in International and Global History

COLUMBIA STUDIES IN INTERNATIONAL AND GLOBAL HISTORY

Matthew Connelly and Adam McKeown, Series Editors

The idea of "globalization" has become a commonplace, but we lack good histories that can explain the transnational and global processes that have shaped the contemporary world. Columbia Studies in International and Global History will encourage serious scholarship on international and global history with an eye to explaining the origins of the contemporary era. Grounded in empirical research, the titles in the series will also transcend the usual area boundaries and will address questions of how history can help us understand contemporary problems, including poverty, inequality, power, political violence, and accountability beyond the nation-state.

Cemil Aydin, *The Politics of Anti-Westernism in Asia: Visions of World Order in Pan-Islamic and Pan-Asian Thought*

Adam M. McKeown, *Melancholy Order: Asian Migration and the Globalization of Borders*

Patrick Manning, *The African Diaspora: A History Through Culture*

James Rodger Fleming, *Fixing the Sky: The Checkered History of Weather and Climate Control*

Steven Bryan, *The Gold Standard at the Turn of the Twentieth Century: Rising Powers, Global Money, and the Age of Empire*

Heonik Kwon, *The Other Cold War*

Samuel Moyn and Andrew Sartori, eds., *Global Intellectual History*

Global Population

HISTORY, GEOPOLITICS, AND LIFE ON EARTH

Alison Bashford

Columbia University Press
New York

Columbia University Press
Publishers Since 1893
New York Chichester, West Sussex
cup.columbia.edu

Library of Congress Cataloging-in-Publication Data
Bashford, Alison, 1963–
Global population : history, geopolitics, and life on earth / Alison Bashford.
pages cm. — (Columbia studies in international and global history)
Includes bibliographical references and index.
ISBN 978-0-231-14766-8 (cloth : alk. paper)
ISBN 978-0-231-51952-6 (e-book)
1. Population—Social aspects. 2. Population—Economic aspects. 3. Population—History. I. Title.

HB849.44.B37 2013
304.6—dc23
2013013776

Columbia University Press books are printed on permanent and durable acid-free paper.
This book is printed on paper with recycled content.
Printed in the United States of America
c 10 9 8 7 6 5 4 3 2

COVER IMAGE: David Malan © Getty Images
COVER DESIGN: Milenda Nan Ok Lee

For O. and T.

And for N, M, B, and A

Contents

Acknowledgments

Big projects can sometimes start with a single folio, in a single archive box, in an instant. It seems twenty years ago, but I think it was only ten, when I opened one of the Eugenics Society Papers boxes at the Wellcome Library, London. "Narrow patriotism must go and one must become 'planet-conscious,'" one eugenics leader had written to another in 1954. The planet trumping the nation? And for one of the twentieth century's most fatally nationalist endeavors? It made little sense to me at the time, and if one cannot satisfactorily explain such a statement, especially if it is directly in one's own field, chances are the next project has arrived. Fiction writers talk about characters "taking over"; historians sometimes have no real choice either.

This eugenicist's "planet consciousness" came out of the archive box at a rich moment for historians. This was just when imperial and colonial history was meeting world history, global history, and environmental history. The scholarly context made planet-level talk immediately intriguing. All the more so given my pre-existing questions about population derived from medical history and feminist history: the intellectual allure was irresistible. On an early research trip to Geneva, when the League of Nations card records were still arranged by the League's original organizational sections, I searched for "population" initially under the "Health" section (I am originally a medical historian, after all). I quickly learned to search under

"Economics." This was my first major conceptual lesson. The second in this long research journey was perhaps less obvious: the conceptual return of "ecology" to "economics," a process that restored my background in history of science to the inquiry, but in a new way. Add an awareness of fertility—as much of soil as of women—and all the base historical and historiographical elements were in place. It took some time for this particular historian-alchemist to transmute them, and what follows is far from gold. I do now know, however, just how much was behind, and before, that mid-1950s statement about planet consciousness.

It has taken not just research but a thousand conversations to get it all in order. For incisive commentary on earlier written and spoken versions of chapters herein, my sincere thanks to friends and colleagues. On one continent, thanks to Tomoko Akami, Robert Aldrich, Warwick Anderson, Virginia Brunton, Barbara Caine, David Christian, Ann Curthoys, Graeme Davison, Helen Dunstan, Marco Duranti, Andrew Fitzmaurice, John Gagné, Chris Hilliard, Marilyn Lake, Stuart Macintyre, Jim Masselos, Iain McCalman, Dirk Moses, Soumyen Mukherjee, Hans Pols, Libby Robin, Tim Rowse, Deryck Schreuder, Carolyn Strange, and Patrick Wolfe. With my election to the Vere Harmsworth Chair of Imperial and Naval History at the University of Cambridge, I will shortly be leaving longstanding colleagues in Sydney and elsewhere in Australia, and am only too aware of what an extraordinarily rich community of historians it is. On a second continent, my thanks to many scholarly friends, some of whom shall soon be much-valued colleagues: Sunil Amrith, Chris Bayly, Lesley Hall, Phil Howell, Sarah Hodges, David Livingstone, Hilary Marland, Maria Sophia Quine, Dan Stone, Simon Szreter, and Megan Vaughan. And on a third, to Janet Browne, Joyce Chaplin, Mary Fissell, Karl Ittmann, Paul Kramer, Erez Manela, John McNeill, Randall Packard, Diane Paul, Harriet Ritvo, Charles Rosenberg, Simon Shapin, Kavita Sivaramakrishnan, Alexandra Stern, and Sarah Tracy. Matthew Connelly and Adam McKeown commissioned this book for their series, and I am grateful in particular to Matthew for his vast and generously shared expertise on population. Readers for Columbia University Press clearly put much time and expertise into their reviews: my thanks. And I am grateful to editors at the press for their interest and care.

One has so many conversations about a project like this, in so many locations. A few stand out. I had the privilege of presenting an early version in Canberra, with commentary from J. C. Caldwell, the eminent demographer who has the rare distinction herein of being equally significant as a primary and a secondary "source." It was the lebensraum argument in which Jack

Caldwell was most interested, and I was spurred on by his response. It was slightly later conversations with Sarah Hodges about "waste" and her intellectual and organizational ventures into that idea that made me look again at population from that perspective. I am grateful, especially since she is not necessarily going to like my conclusions. But that is the stuff of academic friendships, and my thanks to all at Warwick University history of medicine, for a welcoming and productive fellowship. Warwick Anderson has generously shared his ideas about ecology. Long may Sydney history of science conversations be inspired by the harborside Oyster Bar. With David Armitage and Joyce Chaplin, I have benefited from conversations on cosmopolitanism and planetary population, respectively, and fine dinners collectively. A year at Harvard University arrived at the perfect time to pursue a new angle on Malthus. I was offered warm hospitality. Finally, this book has been delayed by several other projects that have insistently jumped the queue. The good news is that in the process I learned a great deal from my much-valued coeditors: on different books, Philippa Levine, Stuart Macintyre, and David Armitage. Many have grounds to appreciate the work of these fine scholars; few can have enjoyed working with them as much as I.

A troupe of graduate students and research assistants have helped along the way. In its final stages, Chris Holdridge and Tiarne Barratt offered exemplary assistance. I am grateful also for the research work of Christian O'Brien, Catie Gilchrist, Annie Briggs, Matthew Oram, Émilie Paquin, and Meg Parsons. Kenny Chumbley was the perfect last-minute addition to the team, assisting with permissions for artwork.

Some work in this book has appeared in earlier short versions, and I acknowledge kind permissions: "Nation, Empire, Globe: The Spaces of Population Debate in the Interwar Years," *Comparative Studies in Society and History* 49, no. 1 (2007): 170–201. Copyright © 2007 Society for Comparative Study of Society and History, reprinted with the permission of Cambridge University Press; "World Population and Australian Land: Demography and Sovereignty in the Twentieth Century," *Australian Historical Studies* 38 (2007): 211–27 (Taylor & Francis); "Population, Geopolitics and International Organizations in the Mid Twentieth Century," *Journal of World History* 19 (2008): 327–47 (University of Hawai'i Press); "Anti-Colonial Climates: Physiology, Ecology, and Global Population, 1920s–50s," *Bulletin of the History of Medicine* 86 (2012): 596–626 (Johns Hopkins University Press); "Fertility Control: Eugenics, Neo-Malthusianism, and Feminism," "Internationalism, Cosmopolitanism, and Eugenics," and "Epilogue: Where Did Eugenics Go?" from *The Oxford Handbook of the History of Eugenics*, edited by Alison Bashford and Philippa Levine (New York: Oxford University Press, 2010), by permission of Oxford University Press, USA. I am grateful for the permission of the coauthor of "Fertility Control," Susanne Klausen, to reproduce some material from that chapter. I acknowledge the Galton Institute for use of the Eugenics Society Papers. I have cited the unpublished writings of J. M. Keynes, copyright the Provost and Scholars of Kings College, Cambridge, and am grateful for permission to do so. I acknowledge permissions from the Wellcome Library, the London School of Economics Library, Glasgow Library, and the National Library of Scotland. In Paris, my thanks for the interest and skills of archivists at UNESCO, and in Geneva at the World Health Organization and the Palais des Nations. In the United States, I am grateful for use of papers and collections from the Houghton Library, Harvard University; the Library of Congress; the University of Toledo Library; the Rauner Special Collections Library, Dartmouth College; the Stanley G. Rudd Library, Princeton University; the Rockefeller Archives Center; the American Philosophical Society; the Fondren Library, Rice University, Houston, Texas; and the Milton S. Eisenhower Library, Johns Hopkins University. In Australia, the National Archives in Canberra and archivists at the Australian Academy of Sciences have been obliging. The relevant collections at these repositories are indicated before the notes section. I am grateful also to the Norwegian Polar Institute, Tromsø, for sending me a copy of Charles Elton's Spitsbergen account. This research has been enabled by grants from the Australian Research Council: DP0557329, FT100100077, DP0984518. Kind permissions granted to reproduce artwork are noted with relevant figures. All efforts have been made to locate copyright holders for uncredited artwork.

For restoring walks and talks in our beautiful part of the world, thanks to Barbara and Kylie. And to Nerida, who has been there forever and a day, as have Marjorie and Keith Bashford. In their different ways, Oscar and Tessa, Barbara and Tess, Cat, and Nicholas have made up a wonderful extended family over these years. My deep thanks to each of them.

AB
Manly, Sydney
May 2013

GLOBAL POPULATION

Introduction

Life and Earth

John Maynard Keynes stood to speak. "Ladies and gentlemen, I ask you to raise your glasses and to drink—in silence—*in piam memoriam*."[1] It was late July 1927, at the Holborn Restaurant in London, and Keynes was toasting Thomas Robert Malthus. The Malthusian League was celebrating its first fifty years, an anniversary function chaired by the famous Cambridge economist. A second toast was raised to "the Pioneers," and the most significant living pioneer of all, elderly Annie Besant, responded. She had provoked the league's founding back in 1877, and she was still provoking those around her: guests were asked kindly to withhold wine and cigarettes in deference to her age, star status, and philosophically based abstention.[2] For once in his life upstaged by a rival provocateur, H. G. Wells—essayist, novelist, and neo-Malthusian—deferred. And then he saluted the lot of them, living and dead: "To the Malthusian League."[3]

Malthus, Besant, Keynes, Wells: a curious foursome, in their various conditions of earthly absence and presence in late-1920s London. Besant had long left Malthusianism behind for theosophy and the Indian National Congress. Keynes would later retreat from Malthusian economics but stay interested in population. Cosmopolitan Wells, in love with another guest, the American birth control lobbyist Margaret Sanger, was also enamored of ideas about world government, microbial life, and interplanetary travel.

He was taking time out from writing his massive ecological synthesis, *The Science of Life*, with Julian Huxley, yet another Malthusian raising his self-congratulatory glass that night.[4]

Many of this group were to meet again just a month later in the world city, Geneva, at the first World Population Conference. They were joined by North and South Americans; eastern, western, and southern Europeans; Scandinavians; East and South Asians; and Australians. In the spirit of Geneva, and of the times, some came as citizens of the world.[5] What was the population problem that brought them all together? What was the "world" that they were imagining and actively constructing in relation to it? And how did this pertain to the ideas of the Anglophone Malthusians? Everyone gathering in London, and then in Geneva, knew about the phenomenal acceleration of the world's population growth across the nineteenth century. There had been a doubling between the publication of Malthus's *Essay on the Principle of Population* in 1798 and the first World Population Conference in 1927, from an estimated 1 to 2 billion people. This would double again, faster than most in Geneva predicted, to about 4 billion by 1974, the United Nations' World Population Year. And they all knew that there were two other curious population trends at work: first, in certain global locations, there had been a sharp decline in fertility rates, although no one really knew why or even how, precisely, many people were limiting their families; and second, a place-specific decline in infant mortality rates, though just how that related to fertility, was another intriguing mystery. These factors—fertility and mortality—have been core business for intellectual and social historians of population for decades, governed in part by the need to explain or explode the influential mid-twentieth-century theory of demographic transition.[6] For the players in the early twentieth century, however, changes in fertility and mortality alone did not explain why the world population problem was big and getting bigger. Or even why it was a problem in the first place. There was a further question at hand: How was that aggregate population distributed over the globe? "The earth, and every geographical division of it, is strictly limited in size and in ability to support human populations."[7] Life in limited space—life on Earth—is what framed and announced the first World Population Conference.

Global Population seeks to explain and untangle the great knot of ideas, politics, and public discussion that constituted this world problem: inherited from the nineteenth century; announced in Geneva in 1927; watched warily by the League of Nations; acted upon by imperial states, decolonizing states, and neocolonial states; and after World War II, acknowledged by the early United Nations (UN). It was a phenomenon that came deeply to shape the very idea of "development," the demographically defined three worlds, and for some, an aspirational "one world." It is hardly surprising that so many experts from so many different disciplines and traditions turned to think about the great changes in population trends across time and space. But how was it comprehended and created? "Population" is often taken to be a sexual and reproductive issue in the first instance. Yet it was a spatial and economic issue too, a question of land cultivation and food production. More than that, the population question persistently raised territorial matters: colonization, migration, and ultimately sovereignty. And what was at stake? Commentators at the time would have said unequivocally: war and peace. Density in relation to cultivable land—the crowded and the empty parts of the planet—was the problem of the era, linked to war, even causally many of them thought. It was the spatial context combined with the biological phenomenon of population growth—rarely just the latter—that created a sense of global crisis after World War I. This was a proposition that simultaneously held prospects for global division and for global singularity.

The world population problem as formulated from the 1920s onward, then, was as much about geopolitics as it was about biopolitics. This book traces the twentieth-century story of how a geopolitical problem about sovereignty over land gradually morphed into a biopolitical solution, entailing sovereignty over one's person. It does so through investigation of the creators and the keepers of population knowledge. *Global Population* is only partly about the experts on human reproduction and sexuality—medical doctors, human physiologists, birth control lobbyists. And it is only partly about "demography," the discipline devoted to population that emerged over the early twentieth century. Indeed, of the long list of distinguished participants at the first World Population Conference, only two called themselves "demographers," although they were perhaps not as idiosyncratic as the other outliers: two self-identified "Explorers," plus a humble "Author," Wells.[8] They were all outnumbered by those whose expertise lay with land, food, and territory: plant geneticists, agriculturalists, soil scientists, scholars of international law and international relations, geographers, and economists.[9] After all, those troubled by population growth and distribution were concerned with the fertility of soil as least as much as the fertility of women.

This book analyzes the scholarly and political conversation that energized a transnational Anglophone group who met in and around the 1927 World Population Conference, but whose influence was evident long after. Researchers, writers, and lobbyists, they internationalized the very idea of population from their bases in Britain, the United States, India, and Australia. Born around the 1880s, their mature lives and ideas crossed, time and again from the 1920s to the 1960s. This was the generation for whom world population met world wars. It was the generation who engaged with population thought from the Great War to the Cold War.

The group of American experts I investigate includes biologist Raymond Pearl (1879–1940), geneticist Edward M. East (1879–1938), birth control lobbyist Margaret Sanger (1879–1966), agricultural economist Oliver Edwin Baker (1883–1949), and demographer Warren S. Thompson (1887–1973). The British group includes medical nutritionist John Boyd Orr (1880–1971), zoologist-turned-social scientist Alexander Carr-Saunders (1886–1966), economist John Maynard Keynes (1883–1946), biologist Julian Huxley (1887–1975), and psychiatrist Carlos Paton Blacker (1887–1975). I expand Anglophone expertise beyond the commonly studied transatlantic world to include Indian and Australian experts as well.[10] Lucknow-based economist and human ecologist Radhakamal Mukerjee (1889–1968) brought the disciplines together in a more integrative way than anyone else. And the slightly older, Australian polymath statistician George Knibbs (1858–1929) energized the interwar generation with a series of mathematically sophisticated world warnings, first published in 1917. At the other temporal end of this study, I include an important Indian of a younger generation, Sripati Chandrasekhar (1918–2001). Chandrasekhar's career captured, for the post–World War II world, the linked nationalism, internationalism, and anticolonialism that had in fact shaped the world population problem from the beginning of the century.[11]

Yet this discussion must begin earlier, with Malthus. The *Essay on the Principle of Population* was an intellectual and political touchstone for all these twentieth-century figures. Indeed, it remains so. As influential as it was controversial, the *Essay* has been variously disputed and reinstated, simultaneously mobilized and condemned for a vast suite of political ends. In looking again at Malthus's *Essay*, I isolate and examine its planetary scale, something that was picked up by scholars in the 1920s. I then trace a spatial history of Malthusianism to a high point in the 1920s and 1930s—on which most of this book is focused—and onward to the incorporation of geopolitics into biopolitics that I suggest took place over the 1940s and 1950s. The study ends in a year that is often imagined as a beginning: 1968.

A lot happened in that year, not least the publication of Paul Ehrlich's best seller *The Population Bomb: Population Control or Race to Oblivion*.[12] In tracing Malthusian thought, however, I am uninterested in either celebrating or condemning the intellectual and political history that culminated in *The Population Bomb.* This is not an aligned study, and I remain, for the purposes of this book, agnostic about the Malthusian paradigm, of which more later. Rather, I explore here the entwined intellectual backstories of all kinds of population events in 1968, the multiple strands of political and scientific thought that produced an apparently singular phenomenon, the "world population problem." This book writes together, then, elements of population that have come to be separated out, historiographically speaking: the histories of ecology and political economy; of feminism and birth control; of food and agriculture; of international relations, colonialism, and anticolonialism; and of demography, immigration, and health.

It was the economists and ecologists who probably comprehended population most fully "in the round." Yet even for someone as single issue–driven as Margaret Sanger, "population" was always more than the politics of sex, reproduction, birth control, and women's bodies. It touched on almost everything: international relations; war and peace; food and agriculture; economy and ecology; race and sex; labor, migration, and standards of living. All of this, as Sanger perceived perhaps more readily than many of her colleagues, entailed politics and governance on scales from the intimate to the international. We may or may not agree with what they drew as either problems or solutions, and they, of course, disagreed with each other. Nonetheless, these multidimensional interwar problematizers of population are my object of inquiry, even as the multidimensional problem of population itself was theirs. This book's primary aim, then—conceptually simple, methodologically hard—is to capture these dimensions of the population problem in the early to mid-twentieth century as fully as the players at the time. The keepers of population knowledge made different attempts to organize and integrate their sprawling, fascinating field. I map their complex knowledge differently than they would themselves, through the organizing structure and concepts of Earth and life, geo and bio. It all encompassed biopolitics, geopolitics, and, I suggest, cosmopolitics.

The Politics of Earth

When, in 1927, plans were considered for an international organization of population scientists (what became the International Union for the Scientific Investigation of Population), it was spatial and security questions that dominated discussion. The first item on a proposed agenda concerned the productivity of soil and the second concerned political frontiers. Third on the list was optimal density of population, and fourth was the laws affecting the movements of migrants.[13] All of this I consider part of the "geo" of geopolitics. Earthly and spatial phenomena and knowledge—from soil science to political geography, from agricultural economics to immigration restriction—are explored in part II. But geopolitics in its original usage is also to the point. Indeed, geopolitics is at the center of this book and its title, because population in relation to land was at the center of classical geopolitics.

The *Geopolitiker* who invented and developed the late nineteenth century idea also created the accompanying concept of lebensraum. This expressed the drive of a vital state to expand its "living room." Defined at one point as "the struggle for space and power," lebensraum was argued often enough from propositions about population density.[14] "Overpopulation" manifested not only as a national or imperial pressure, but also as a global pressure that was new to the twentieth century, geopolitical theorists argued. The centerpiece of this system of thought was a newly "closed world," the endpoint of centuries of population growth and geographic expansion out of Europe. With their great multiplying populations and industrializing economies that sought and needed both earthly resources and human markets, vital European and then American powers had claimed all the apparently empty parts of the planet.

The newly closed global epoch was comprehended by these scholars at a very particular moment in international history. The closed-world idea did not belong to German imperial, Weimar, and fascist *Geopolitiker* alone, however. It was widely shared by Anglophone Malthusians, economists, geographers, and the first generation of demographers. While the derivation of German geopolitics from the work of Anglophone geographers is well known,[15] the focus here is on a shared logic. They all agreed that differences of density between nations and global regions was part of population-based international relations. The distinction was the means by which the problem of uneven population distribution would be resolved. *Geopolitiker* imagined that a closed world meant (and justified) war between nation-states, or between nation-states and other peoples with a lesser claim to land. Many Anglophone economists, geographers, and demographers in the 1920s also imagined that such war was likely—indeed, had just occurred on an apocalyptic scale—but that it might be avoided in future by better distributing global populations in relation to an "optimum density."

The most common statement in the thousands of publications on world population that emerged in the 1920s and 1930s was that overpopulation, understood in terms of comparative density, caused war. Some offered this in spurious and unsophisticated terms. Others offered it expediently. There were also population thinkers of great standing in both social and natural science traditions, who worked through this connection slowly and carefully. The whole complex engaged thinkers across the political spectrum. Population and war obviously engaged European fascists. And some Anglophone population commentators were fascist leaning, notably anthropologist G. H. L. F. Pitt-Rivers, founding editor of the journal *Population*. The first issue appeared coincidentally, but nonetheless tellingly, in 1933. Equally, however, the connection between population and war deeply exercised Pitt-Rivers's liberal internationalist, pacifist, and socialist colleagues; in many ways, I am more interested in them. If the management of population density was comprehended by the *Geopolitiker* as a rationale for belligerent territorial expansion, for most Anglophone population experts and most internationalists, the management of population density and population movement was potentially a major instrument in the securing of peace.

The logic of density in a closed world meant that, for a time, the redistribution of people was imagined as a solution to the world population problem, even the first order of business for a future world population policy. From the 1920s, the migration of people from densely populated regions to low-density regions or preferably "empty" parts of the globe, was pressed forward by multiple proponents. Even before the Great War had ended, George Knibbs, for one, was clear that the problem was primarily the spatial distribution of humans: "*migration, aggregation, segregation or wide dispersion*, colonisation, etc., the direction and velocity of movement of populations, the tendency to live in more or less dense groups (large cities or villages), or to spread over the earth."[16] After the war, and the Peace, the stakes were even higher. "Peaceful transfers of population" were imagined, and in some instances materialized. This idea was incorporated in plans from the Roosevelt-sponsored M Project (Where can the Jews of overpopulated Europe go?), to the anticolonial work of Warren S. Thompson ("Where Can the Indian Go?"),[17] to Radhakamal Mukerjee's and Sripati Chandrasekhar's statements about the need for Asian living space. "Asiatic *Lebensraum*" is how Mukerjee put it—in 1946 no less.[18] This particular solution—as contemporaries would have it—is the reason a history of the world population problem in the 1920s and 1930s needs also to be a history of migration and settler colonialism.

Intellectually and politically, admitting a closed world—the limits of the earth—had a number of political implications. The most well known and discussed by historians was the global color line.[19] This was the racial competition for continental space that such limits engendered, and, for some, justified. Since the planet's living space was constrained, the question of the twentieth century was: Who shall inherit the earth? The strident white claiming of land declared by American Nordicists Madison Grant and Lothrop Stoddard through their famous books *The Passing of the Great Race* (1916) and *The Rising Tide of Color Against White World-Supremacy* (1920) has been treated by historians as the representative response to that question.[20] Matthew Connelly has argued that the racialized demographics of white people's "race suicide," on the one hand, and an Asian "yellow peril," on the other, initiated and remained the driving politics of the struggle to control world population over the twentieth century, "from the yellow peril to the population bomb."[21] Yet there is another tradition of population expertise that isolated this nationalist and race-based response as itself a world problem, and sometimes as *the* world problem. American demographer Warren Thompson, for one, would have disagreed with Connelly: "When the question of who is to possess the earth is looked at from a long-time point of view, it is perfectly obvious that people who are no longer 'swarming,' [meaning European populations] who have low birth rates steadily become lower, and who have lost the power of actually taking possession of new lands cannot expect to hold for any great length of time territories which they are not effectively using."[22] Most historians would understand Malthusian "overpopulationists" to be nationalist immigration restrictionists, sometimes the very architects of the race-based immigration restriction acts that proliferated in this period. They would not be wrong, but nor would they be completely right.

Strangely, it was Malthusians—such as Thompson—who presented some of the period's most vocal opposition to the system of race-based immigration restriction. Far from idiosyncratic, this strong and clear critique was articulated by American, British, Australian, and Indian commentators alike. The laws entrenched and even caused deeply problematic regional overpopulation by artificially stemming what would otherwise have been a natural movement of people from low-resource, high-density regions to high-resource, low-density or unsettled global regions. The immigration acts were a problem, not a solution, and they must go.

Others suggested that adequate "living space" might be found by enlarging national areas, not through invasion or war in the fascist model or by old-style colonization, but through peaceful cession of land, one sovereign nation to another, based on need and in the interests of global security.

This is one of the more curious pitches for world peace that emerged in this period—it was a large field—and it was all about population. This was Thompson's pet idea, for example, something he developed in his role as director of the Scripps Foundation for Population Research and as an early theorist of demographic transition. Thompson, especially, but others as well, asked: Can sovereign nations legitimately exclude people who need land (food), and who have the capacity to cultivate it, if that land remains untilled? And even more fundamentally, in a Lockean argument, wherein lay the legitimacy to that claim, if land remained unproductive? In a crowded world, how could that claim be upheld? Population thus became the business of international law.

The sustained discussion about redistribution of people, and the great implications for land use and land ownership, signal not only the need to connect demographic history fully with migration history and the history of geopolitics, but with the history of colonization as well. This becomes a colonial history because of the reliance on "empty space" or "wastelands" in the redistribution model. As socialist director of the International Labour Office (ILO), Albert Thomas, put it in 1927, there was an imperative as never before to redistribute population to "empty" parts of the globe.[23]

The world population problem needs to be understood in terms of the history of internationalism too. When geopolitical theorists were imagining a closed world (indeed *Weltpolitik),* internationalism of all shades was simultaneously emerging. A burgeoning number of international associations and conferences were born at the turn of the century, often trans-European and transatlantic, sometimes pan-American, with East and South Asian presence typically but not exclusively derived from the British and French imperial networks. A series of International neo-Malthusian meetings from 1900 and the derivative World Population Conference of 1927 were just a small part of this modern phenomenon of internationally organized expertise.[24] A liberal internationalism that predated World War I took influential shape in Woodrow Wilson's Treaty of Versailles, and then as a formal body, the new League of Nations. Insofar as population made it onto the League's official agenda, it was conceptualized economically in explicit avoidance of the sexual and health dimensions. Another version of internationalism emerging from Marxism and Leninism rejected the very basis of anything called "overpopulation," following Marx and Engels's critique of Malthus. Finally, an internationalism more strictly described as cosmopolitanism also came to drive and define the population problem. Wells fits this version, but he was by no means unique. Perhaps most surprisingly (because

Malthusian politics is so often assessed as gross epistemic and institutional colonizing),[25] an anticolonialism is also evident. Not infrequently, Malthusian anticolonialism manifested itself as arguments for cosmopolitan world government, a position that the nationalist politics of the immigration acts served to focus. One of the lines that this book traces and explains, then, is the closest of political and intellectual connections between Malthusianism, liberal internationalism, and an anticolonial nationalism that assumed, often enough, an explicitly pacifist politics. This was the case especially, but not only, for the Indian scholars. A deep antinationalism characterized the ideas of Australian statistician George Knibbs, for example, whose work on world population, while not entirely unknown, fails to register historiographically anywhere near as centrally as it did historically. For Knibbs, the population problem was not merely a mathematico-physical one: "It involves our whole conception of life." Citing Keynes, he declared the problem of population "the greatest of all political questions."[26]

It is clear from Knibbs's work, and from that of a whole line of successors, that it was not just an international public sphere that was at stake, but something larger, on a different scale altogether. Indeed, the world population problem engaged matters simultaneously intimate and universal, and in some versions, microscopic and cosmic. The need to think across scales is evident here, from earth as soil to Earth as planet. The work of historical geographers on modern and ancient meanings of "Earth," "world," and "globe" has proved useful in capturing the dimensions of population. Denis Cosgrove caught much of this in *Apollo's Eye*: "Earth," *terra*, is elemental and agrarian; "world" has a more social meaning; "globe" is spatial and connecting, though not necessarily unifying.[27] Yet *terra* was always territory too, especially in the first half of the twentieth century, suggesting the need to fold international history into Cosgrove's schema. At the same time, the international and global history of the twentieth century is often comprehended through a peculiar reuse and politicization of the antique cardinal directions of the planisphere—the north and the south, the east and the west. There is a population and demographic history lying within these conventions too. The twentieth-century "East" and "West," and latterly the global "North" and "South" come to us filtered through population thought; economic regions defined by high and low birthrates, high and low death rates. This is the enlarged intellectual history of a world demographic transition. And one explanatory line of inquiry throughout this book serves to unpack the multiple elements of demographic transition, the massively influential idea presented by sociologist Kingsley Davis in the pivotal global year, 1945.

The "world" of world population is thus part of an intertwined international history and global history. But even the "globe" does not quite encompass the scope and scale of the world population problem as Knibbs and others formulated it. They had more than a tendency to look beyond the globe, and often specialist training and knowledge to do so. Theirs was more like a planetary vision, a scaling up from global history to planetary history that some scholars have recently pursued.[28] Still, there was perhaps nothing new to the kind of planetary consciousness that so many population commentators deployed. In fact, Knibbs's capacity to think about a globe in literal space was already a fairly late recapitulation of a long-established tradition that comprehended population in terms of a singular and limited Earth, even of spaceship Earth, importantly an ecological concept.[29]

We need look no further than Malthus's *Essay*, the scope of which was planetary, and the very premise of which was geopolitical. The "struggle for room and food," as Malthus put it, was permanently in operation, and the *Essay* turned on multiple actual and hypothetical instances of spatial limit, at scales from the very local to the intercontinental and planetary. Malthus never quite formulated this "struggle for room and food" as the need for "living room"—lebensraum. But he might have. And some of his major intellectual descendants did. The Malthusian crossover with lebensraum, in other words, is neither incidental nor coincidental, and the link was an idea that Charles Darwin and Alfred Russel Wallace published simultaneously in 1859.

The Politics of Life

Famously, it was Malthus's formulation of the human struggle for room and food that prompted natural historians to think about the behavior of organisms of all and any kinds, in limited spaces. Wallace's and Darwin's theory of evolution by natural selection was thus born, and it was this Malthusian–Darwinian idea that made its way first into German geopolitics and then into early-twentieth-century plant and animal ecology. It returned to human population studies, demography, and human ecology via many of the biologists studied here.

Chapters in part III, explore the politics and sciences of life that constantly intertwined with the politics and sciences of Earth in the formation of a world population problem. "Biopolitics," after Michel Foucault, typically signals modern techniques for the maximization of human life as individuals and as aggregate national populations—the governance of health, longevity, reproduction, and sex—beginning with seventeenth-century political arithmetic and never really ending. Health and reproductive conduct are perhaps exemplary modern projects of biopolitically aligned state and self-governance.[30] I certainly address this standard account of human biopolitics, but not before I consider the politics of life in less familiar ways.

I begin with ecology, the new method of inquiry that had organisms, populations, and environment at its epistemic center. Unlike many intellectual and political histories of demography, this book seeks to foreground the crossover and conversions between "population" in the natural sciences and "population" in the social and economic sciences. The intellectual traffic between natural history and human history, between zoology and demography, between ecology and economy always ran two ways. Malthus might have offered Darwin the "struggle for room and food," but it was Darwin who offered "standing room only" to generations of writers on global carrying capacity.[31]

What drove it all was the idea generally tracked to Malthus, though one clearly older: too many organisms are always reproduced, and some must always die soon after birth to enable the life of the rest (the fitter, Darwin added in a later edition of *The Origin of Species*). The Victorian neo-Malthusians and their twentieth-century successors understood their own intervention with regard to human organisms as deliberately exchanging this harsh, natural system that saw the constant waste of human life with their preventive systems: life that was only doomed to die (infant mortality because of want of room and food) would be prevented from existence in the first place; this was "birth control." This is why neo-Malthusians always understood contraception as humanitarian, and why they were often incredulous at the critique that they received. This is also why Kingsley Davis, in his original, influential statement of world demographic transition, understood the increasingly widespread limitation of fertility in ecological terms; the transition was from a wasteful system to an energy-efficient one.[32]

The life and death of organisms were always comprehended spatially, in "universes with definite limits," as biologist Raymond Pearl put it.[33] As I show in chapter 3, "density" was not just an economic and geographic concept that gave rise to "carrying capacity," but a biological concept that was actively pursued in cutting-edge research. Indeed, the keynote address of the first World Population Conference was Pearl's on the biology of density. Biologists' recent entrée into the study of human populations drew considerable comment at the time, especially from political economists.[34]

This was a measure of the extent to which population up to that point had been comprehended as economy in the first instance. All kinds of natural scientists who had had nothing to do with human biology, let alone political economy, began to establish themselves first as commentators, then as experts, on human population problems. They typically did so through and because of an explicit politics of human biology in space.

One question in this biological–spatial conception of population was this: In local, national, and global human ecology, which humans died (or would ideally be prevented from being conceived in the first place) to make room for the rest? This turned on the evaluation of human sameness and difference, including, but not only, racial difference. The question was answered both descriptively and prescriptively. Sometimes historians have mistaken scientists' and demographers' accounts of what *did* happen (who *was* dying) for what *ought* to happen (who *should* die, or as formulated in most cases, whose life/reproduction should be prevented). In other cases, biologists and economists were unashamedly prescriptive about the benefits ("racial hygiene," "national hygiene") of certain humans preventing their own reproduction or having it prevented for them. The field of eugenics came to own much of this intellectual territory, though most of the experts studied here crossed fairly seamlessly between eugenic-oriented study of population quality and neo-Malthusian orientation to quantity. Indeed, one of my suggestions is that accounts of eugenics have largely ignored its Malthusian conceptual foundation. In chapter 9, I am concerned with just how eugenics negotiated an international public sphere and the global/planetary domain that Malthusianism brought to it.

There is no doubt that experts on the human population problem engaged deeply with the "quality" dimension, as it was put at the time. There were, of course, all kinds of social and political drivers for this. In part, putting quantity and quality together was prompted by the conceptual revolution that Mendelian genetics offered. Just how or indeed whether the Mendelian laws of inheritance applied to humans as to peas fascinated a whole new generation of natural scientists quite suddenly after 1900. This is a further reason for grounding "world population" in the early twentieth century. It came when the old political economy of population combined with (1) the new geopolitics of a closed world, (2) post-Darwinian focus on interdependent organisms in space, and (3) the new biology of generational breeding.

Animal and plant geneticists who were engaged with the life, death, and quality of fruit flies, chickens, or hybrid corn—occasionally all three—were turning their attention and their research to humans, and with an increasing urgency, as a result of their wartime experience. Indeed, a surprising number of key animal and plant biologists were straightforwardly concerned, in a Malthusian manner, with limited global food supply for humans. Harvard geneticist Edward M. East, for example, saw before him the need for active intervention to bring about "a just peace between the two basic instincts of mankind, nutrition and reproduction."[35] The new world crisis was a "crossroads for mankind," as he put it, one that demanded a political as well as a scientific solution. It was for this reason that he joined Sanger's American (and later international) birth control campaigns and became deeply involved in neo-Malthusianism, as well as in eugenics, whose international organizational dimensions flourished in the 1920s. East framed the crisis in bio-health terms, "nutrition and reproduction." But neither humans nor health really fell within his scientific expertise. He specialized in plant genetics and had been instrumental to augmenting American agricultural yields through hybrid corns, initially in the context of World War I food shortages. For East, and for so many others who appear through this book, food was far and away the most problematic, urgent, engaging, and politicized factor of the whole population conundrum: it was this element that circled back to the geopolitical imperatives about land. Constantly under discussion were issues such as patterns of (food) consumption, the possibilities of organic or synthetic fertilizers, competition between land use for humans and land use for animals, and increasingly the imperative to "conserve" land.

Economic historians have largely focused on two explanatory phenomena as key to understanding Western growth, beginning with Britain: the shift from an organic to a mineral source of energy (the massive significance of coal) and the benefits of New World land that dramatically relaxed Britain's own land constraint.[36] Those who worried about population growth in the early to mid-twentieth century did so largely through an economic prism, but as a rule they remained strongly focused on agrarian possibilities and limitations rather than industrial production of wealth, because "population" always required the production of food from land. Nineteenth-century economic growth, many still thought, was a one-off world phenomenon, a "very exceptional event in world history," as Julian Huxley put it in 1927.[37]

These were John Maynard Keynes's thoughts precisely. When a young Keynes wrote of the economic significance of the inelasticity of land, one limit was area, but another was the capacity of soil to yield, and its diminishing returns. By the post–World War II period, it came to be recognized that innovations in production and cultivation were continuing and might even be continual. By the time E. A. Wrigley wrote *Population and History* in 1969, he could say that industrialized economies had bypassed "the bottleneck caused by the problems of expanding organic raw material supply." Ultimately, inorganic materials have come to substitute for organic materials in all kinds of processes and productions. As Wrigley explains: "Each such change removes another sector of industrial production from dependence upon the productivity of the soil."[38]

The geopolitics and colonialism of a closed world here met Malthus in a most explicit way. Over and over, those such as East would pitch the problem as a recent agricultural closing of Malthus's apparently open world of the late eighteenth century. The colonial and environmental histories of continental settlement, displacement, and land use, especially but not only the United States "closing of the frontier" famously announced in 1893,[39] were both implicit and explicit parts of the world population problem. All the resources of earth sciences and agricultural sciences were brought to bear on this, manifesting as agricultural and environmental politics, as well as innovations: continental grasslands had turned into wheat farms and then into dust bowls.[40] Some commentators thought that the closing of global frontiers could be offset by agricultural innovation (East was not one of them). Others looked to intensification of yields through new fertilizers, new cropping techniques, and new plant breeds. In this sense, this book is in part a backstory to the green revolution, and in part the backstory to environmentalism, which do not have separate histories at all.

Analysis of food, agriculture, and especially the intriguing politics and sciences of soil are placed very deliberately within part III of this book, because I seek to understand how all this Earth business became part of the politics of life. I focus on soil, food, and agriculture in the 1920s and 1930s in chapter 7, and in the 1940s and 1950s in chapter 10. Ultimately, a much larger energy question was at stake that began with human fuel—a new awareness of food consumption patterns geographically and trends temporally. This scaled up to questions about population growth in a fossil-fueled, solar-fueled, or atomic-fueled world. Food, then, was part of the energy economy that enabled human life and reproduction. Although we think of calorie counting as a late-modern phenomenon of personal governance in the West, one aspect of the genealogy of calorie counting was global energy bookkeeping.[41] Many of the agricultural economists and physiologists studied here were deeply engaged in reckoning the daily requirements of the world's population, pitching that against possibilities of producing and distributing total energy needs. Food crises and world food plans were thus one axis around which the world population problem turned. Indeed, I argue that it was via food rather than via sex that population gained what traction it did in early UN agencies, and any traction at all within the earlier League of Nations. It is necessary, then, to tie the international history of food and food security firmly to international histories of birth control and population.

Humans are transformers of energy, and were increasingly comprehended as such: organisms who eat and excrete in cycles of use and waste. Food is thus a pivotal point between "earth" and "life," between agriculture and nutrition. More so, soil. While it seems counterintuitive to analyze the politics of food security and soil science as part of the bio, not the geo, chapters of this book, I do so to draw attention to the extent to which soil was comprehended as itself living, as organic, part of the newly minted "biosphere," another term of the period. Charles Darwin's earthworms are not incidental here, nor is the century of agricultural chemistry that sought to understand the link between the geosphere and the atmosphere through that tricky element of life, nitrogen. When the bacteria on the nodules of legumes were discovered to fix atmospheric nitrogen, soil literally came to life; "active agencies in the organization of life," as the director of Britain's famous Rothamsted Agricultural Research Station put it, in a study of food, health, and population.[42] The world population problem, because it was so centrally a food security problem, was both about the new ecology of (nitrogen) cycles, and the seemingly opposed economy of global fertilization use and resource depletion. Either way, time and again, when "fertility" was mentioned with respect to world population in this period, it was not the fertility of women that was being referenced.

This is why soil erosion came to be such an explosive issue in this context. Especially in the years immediately following World War II, as I show in chapter 10, soil erosion was the object of extraordinary global catastrophic projection, much like climate change now.[43] World populations pressed on global soil, as contemporaries would put it, with ever-increasing demands for food. Aldous Huxley, for one, thought the consequences more devastating than atomic war, a process by which humans had irredeemably changed the capacity of the Earth to sustain life.[44] The soil–food–population cadence explains why American ecologist William Vogt became simultaneously a leading figure in the Planned Parenthood Federation of America, on the one hand, and the Conservation Foundation on the other. Soil conservation is the missing explanatory link, part of the history of population and birth control. But no mistake should be made about priorities here. The fertility of soil was Vogt's prime consideration, the ends to which managing the fertility of women—and the mortality of infants for that matter—was only ever an expedient means.

Thinking about soil as biopolitical also suggests the "life" that was at the core of classical geopolitics. To connect soil science to lebensraum is not my interpretive move. It was a connection the historical actors themselves made. Lebensraum was comprehended less as space in which to live than as space that was itself organic. Well-known political ecologists such as Fairfield Osborn, William Vogt, Julian Huxley, and Paul Sears picked up this idea when they wrote of "living landscapes" and "living Earth." Sears, for example, comprehended "soil as a process—a logical extension of Darwin's greater premises with regard to the unity and continuity of nature."[45] German theorists of lebensraum, left liberal internationalists, American ecologists, and even anticolonial Indian cosmopolitans, as I show, each had an investment in the world-scale geopolitics and biopolitics of soil.

This was global political ecology, what I reconfigure in chapter 6 as the "cosmopolitics of population." The world population problem became the substantive topic over which the holism that underwrote ecology in a very technical sense connected with the holism inherent in political cosmopolitanism: Julian Huxley's "One World," Radhakamal Mukerjee's "Oneness of Mankind," or John Boyd Orr's more sinister "One World or None." Malthusianism, then, turns out to have been a "cosmopolitan thought zone."[46] Many nineteenth-century Malthusians already imagined themselves as citizens of the world and wrote often about federated political structures (after Kant).[47] The population problem early became part of the long, modern trajectory of an aspirational one world—biological, political, territorial, social, and environmental. These were experiments in globality, the imaginative activity through which the "daily work of human beings"[48] was linked to the idea of a global polity.

Political cosmopolitanism was everywhere in integrative population talk, especially favored among ecologists. This all seemed to come together in 1948, a year when multiple books in English popularized the new global political ecology, a year between the world war and the Cold War, when the catastrophic sensibility derived from the bomb and the Holocaust lingered, and when people were rightly worried about global futures. Historians sometimes write about these works—Osborn's *Limits of the Earth*, Sears's *Living Landscape*, Vogt's *Road to Survival* (and more)—as precursors to, or originating texts for, the more famous popularizing that went on in the 1960s, especially Paul Ehrlich's *Population Bomb* (a book about food security).[49] In this study, though, I am interested in the post–World War II political ecology of population as a continuation of post–World War I ideas, and indeed of the nineteenth-century neo-Malthusianism that I discuss in chapter 1. Planet-level overpopulation was a widely discussed problem generations before anything called the population bomb.

There are all kinds of intriguing crossover points between earth and life, geo and bio. If I analyze "soil" within the rubric of the politics of life, I analyze the management of fertility as territorial in chapter 8. Just how did sexual relations connect to international relations—to the geopolitics explored in part II? Many in the 1920s and 1930s, including Margaret Sanger, advocated birth control on the basis of an international security argument: minimizing density differences across the globe, securing food, averting war. That was what birth control would do, and that was precisely how they put it. The investment of well-known feminists in this international argument is one point of interest. Another is the extent to which so many influential men in the 1920s and 1930s were vehement and vocal proponents of birth control. Yet for them, it was not really about sex at all, but about food, land, and war. In other words, it was possible to advocate birth control—fiercely—without having any interest in, or even particular awareness of, the always-gendered politics of reproduction, let alone feminist arguments about health and reproductive rights. Further, it was this "food security" rationale for birth control that was more common, arguably more influential, in successfully normalizing birth control politics, precisely because it was put forward by such authoritative men. Ultimately, for many (even most) population thinkers, food insecurity was the problem for which fertility management was the solution. I suggest also that population's eventual location within world health in the form of birth control (or population control) was a post–World War II surrender to the intractable nature of the geopolitical solutions that the interwar generation had so loudly proposed. Population transfers were too hard; managing fertility at least kept people in their global place.

Interventions

The geopolitical dimensions of population have not often been integrated into sex and reproduction oriented histories of birth control, some of which focus on demographic changes in terms of mortality and fertility, and others on the twentieth-century story of individualized sexual practice.[50] Neo-Malthusianism is historically analyzed rather more often within the framework of reproductive rights and its links to feminism, than to political economy or biological sciences.[51] Much scholarship on population, especially in the history of health and medicine and in feminist studies, simply presumes the question to be one of reproductive sex: advocacy of, theorization on, or opposition to various techniques of fertility control on the one hand, or fertility increase on the other.[52] Put another way, the analytic of sex has consistently trumped the analytic of space. This is a legacy of the success of Sanger-style argument about women's health, of the joint influence of late-twentieth-century feminist and Foucauldian scholarship on the history of sexuality and governance, and on a tradition of critique of reproductive politics and of gender in medical history. In the international history tradition, too—most recently in Matthew Connelly's magisterial study—the analytic and argumentative center is population as reproduction; the critique of international campaigns to limit population growth through reproductive and health practices.[53] As a material *problem* this is absolutely correct: it is impossible to have population without reproduction; improbable to have reproduction without heterosexual sex; and unlikely to have heterosexual sex without a gendered sexual politics between women and men. All this is embodied and experienced by women differently from men, as pregnancy, childbirth, abortion, miscarriage, breast-feeding. Each has attendant health, social, and economic consequences that sometimes escape economic and ecological histories of population, and even historical demographies.[54] And yet, in terms of how "population" has been *problematized*, folding the field retrospectively into a feminist bid for reproductive health is simply incomplete. In other words, to expect the history of the issue of population to be primarily or even, at times, solely about reproduction and health is to miss altogether other lines of thought within which world population came to be problematized. Experts in the 1920s would have puzzled over such a narrow comprehension of "population." Economists and geographers (then, as now) would certainly think this strange. And even biologists, the newcomers to the population problem, always integrated a spatial element to the life/death equation. As Raymond Pearl insisted in 1927, there are three primary variables in the study of population, "natality, mortality, migration"—birth, death, and space.[55]

Several interpretive orthodoxies have flowed from the general trumping of geo by bio in histories of population and have to some extent settled. I question each of them: one substantive, one about periodization, and one about the geography of the problem. These are linked, of course, but I will take them separately. First, this book unsettles a certain presumption—scholarly and popular—that "population" is and was a problem of health in the first instance. In fact, it fell under the institutional brief of international health-care systems quite late in the twentieth century.[56] Only by the late 1960s would the management of fertility be formally recognized as part of primary health care by the World Health Organization. This was bedded down as individualized "women's health" and "family planning" by feminist activism and scholarship of the 1970s, and reproductive health thoroughly displaced population control as the discourse of choice by the time of the Cairo Conference on Population and Development in 1994.[57] Looked at one way, this was all a remarkable success story of feminist attachment of a rights agenda to reproduction, to be ensured and managed through health systems, at least in theory. It did so to such an extent and so thoroughly over the later twentieth century that the United Nations Population Fund (UNFPA) came to describe itself as "an international development agency that promotes the right of every woman, man and child to enjoy a life of health and equal opportunity. UNFPA supports countries in using population data for policies and programmes to reduce poverty and to ensure that every pregnancy is wanted, every birth is safe, every young person is free of HIV/AIDS, and every girl and woman is treated with dignity and respect."[58] There is but a trace here of earlier economic theory, remnant in the aim "to reduce poverty."

This latter-day UN uptake of population through individual rights and health discourse is strikingly different to the squarely geopolitical placement of population within the economics–food security nexus, evident earlier in the century. Within the UN's earliest history, and certainly within its institutional precursor the League of Nations, these emphases would not just have been reversed, the reproductive rights/health argument was formally absent. It was not the League's Health Organization, but its Economics Section, along with the ILO that put population on agendas in the 1920s and 1930s. When the League's Assembly and agencies *did* deal with population as health, it was certainly not reproductive rights that were invoked, rather something more along the lines of the right to minimal food requirement; what would soon be articulated as "freedom from want" by Franklin Delano Roosevelt. World food plans and individual food needs were the major, even the only, way in which world population growth could enter the agenda of early UN agencies.

This book interrogates, then, rather than takes for granted, the checkered emergence of a population problem within international health discourses. This was not a natural coupling, but something that was worked out over the century. Individualized reproductive health and reproductive rights *became* internationally viable, not into a discursive void, but in, around, and through all kinds of other expert investments in, and constructions of, population.

There are, then, parallel historiographical treatments of twentieth-century population. Feminist and medical historians, on the one hand, tend to presume "population" to have been a health issue all along, when it was so much else besides. Economic historians, on the other hand, presume population to be their territory (rightly) but in doing so often separate out the history of population politics and sexual politics. When geopolitics and population are brought together—for instance, by John Perkins in his full and otherwise complex history of the green revolution[59]—sexual politics seems to disappear, as though war, food, economics, and peace never entered the feminist lexicon in their bids for birth control. Nothing could be further from the case.

Second, periodization. There is a generally agreed chronology that it was after World War II that experts and statesmen suddenly comprehended population growth as a problem.[60] Many scholars accordingly focus on the Cold War period as key to the international history of population and foreign policy, analyzing the institutionalization of demographic transition theory, foreign aid, and the idea of "development." The most important element in this periodization of population's link to international history is taken to be the formal uptake of population policies by the U.S. government in its explicit bid to stem Communism.[61] So, for example, we are told that "Neo-Malthusianism had thus arrived on the world agenda" with the U.S. president's incorporation of aid funding for population control and the World Bank's statement that population growth was a major obstacle to social and economic development.[62] World population as crisis certainly belonged to the Cold War era. There is no question about this. However, it is equally plain that population, war, and peace had long been connected by the time population control and aid landed on the Oval Office desk. One of my core aims is to reperiodize the problematizing of world population growth to an earlier moment. A different history is revealed when one thinks not backward from the Cold War, but forward from World War I: less the Club of Rome's *Mankind at the Turning Point* (1974) than Edward M. East's *Mankind at the Crossroads* (1923); less Karl Sax's *Standing Room Only* (1960) than E. A. Ross's *Standing Room Only?* (1927); less James Lovelock's *Gaia* (1979) than George Knibbs's *Shadow of the World's Future* (1928).[63]

What was going on after 1919? Most historians writing about the early twentieth century would isolate the well-known national anxieties about fertility decline that prevailed in Europe, North America, and Australasia, summed up as "race suicide." A 1960s "overpopulation" problem is typically seen to displace an earlier fertility decline problem. And yet, population growth and overpopulation is everywhere in 1920s Anglophone sources, sitting alongside the degeneration and depopulation tomes that historians know so well.[64] Notwithstanding the historiographical insistence on race-suicide anxieties, commentators at the time spoke of a marked Malthusian revival in which birth control was heralded as necessary to offset global growth in a closed world and to help avert the security issues that arose.

Indeed, there were so many books of this sort in 1925 alone that when geographer Marcel Aurousseau was asked to review them, he announced that he was bored. After a few had been absorbed it was all pretty monotonous, he wrote: "They have a habit . . . of telling you all about Malthus, his critics and predecessors; after which they present the state of affairs today and then launch into genetics, to finish with the advocation [*sic*] of birth control."[65] Notwithstanding Aurousseau's ennui, this is just the wave of post–World War I expertise in which I am most interested. It was noted at the time as an outpouring of "economic thrillers."[66] This great wave of postwar scholarship is occasionally noted by those tuned in to Keynes.[67] William Petersen, for example, clearly saw a post–World War I Malthusian revival while writing in the 1950s on Keynes. Overpopulation was everywhere, he observed correctly.[68] And yet so much work on the successive "returns" of Malthus simply bypasses the critical years after the 1919 peace.[69] In other words, the return of Malthus that Keynes himself inspired is often overlooked.

If birth control historians often fail to see neo-Malthusianism as economic history, the reverse is also the case. Economic historians fail to grasp just how centrally, if problematically, an economic agenda governed both the Victorian neo-Malthusians and those of the 1920s and 1930s, in the process also presuming that this "Malthusianism" simply meant "contraception," and calling the connection in one instance, "embarrassing."[70] No doubt Keynes thought many in the Malthusian League embarrassing and peculiar too. But he was one of them. It was Keynes himself, after all, who proposed that toast, looking back to Malthus from his 1927 present, and forward with poignant hope to a world without war.

If there is a received chronology for the twentieth century that I unsettle, there is also a linked geography that underwrites much scholarship, as well as popular discussion on population. Once the world population problem is re-periodized from the Cold War to World War I (and even earlier), a world population problem more directly linked to European population growth over the nineteenth century is brought into view. Third, then, I question a largely presumed and far too simplistically reproduced formulation that the European/Western problem was fertility decline and the Asian/non-Western problem was population growth. In fact, Europe as a region registered again and again as the original and present problem-space when it came to population growth and density; it was rare, indeed, for any of the early to mid-twentieth-century scholars that I examine here not to recognize this. By contrast, the more recent postcolonial critique of a "Third World population bomb" has the unexpected effect of presuming and entrenching South and East Asia and sometimes Latin America to be the sole sites of explosive population growth. Europe is sometimes written out almost entirely. In other words, by insistently analyzing the neocolonialism of a postwar Western discourse about Asia's "population bomb," historians have sometimes unwittingly ignored the European "population bomb." Not a little ironically, recent postcolonial critics and scholars have proven less willing than some of their Malthusian historical actors to provincialize Europe.[71]

There is a long-standing critique of Malthusian population policies that understands post–World War II global population control in the new Third World to be a cloaked neocolonial extension of earlier eugenics. The claim that mid-twentieth-century "world population control" was eugenics in new clothes is common.[72] I agree that these were strongly linked projects, connected not least through the practice of sterilization. But there is another problem of periodization here: "population control" was not a mid-twentieth-century meeting of eugenics and Malthusianism, but something that had driven eugenics from its inception.

Feminist, Marxist, and latterly postcolonial traditions of critique of "population control" dominate both scholarly and popular assessments of Malthusianism,[73] although there is also a conservative, antifeminist, and neoliberal critique that is less familiar.[74] Often the connection with eugenics is deployed to firm up the critical position. In all cases, critique is the purpose of the scholarship, including Connelly's *Fatal Misconception*, a study whose positioned argument is captured in the title.[75] *Global Population*, however, leaves a certain critical orthodoxy behind, partly because it is so familiar. In doing so, I seek less to undermine the politics of prior studies than to open up what else the long history of Malthusianism (or even anti-Malthusian thought) was about. By no means my standard approach to history writing, my agnostic pursuit of Malthusian thought in this project has enabled me to take contemporary scholars rather more at their word, at least in the first instance, the better to see what is revealed.

For example, it is typical, and has been politically important, for feminist historians to be critical of the aggregating and dehumanizing politics of "population control" that often enough targeted the least powerful. Historian Sanjam Ahluwalia, to take just one example, is more aware than many of the history of population in economic theory; historians of India often are. Yet this is used almost solely for the purpose of critique. She is scathing about the likes of Radhakamal Mukerjee, opposing and exposing his "Malthusian slurs" on subaltern groups who (she quotes him) are "breeding like field rats, rabbits, and fruit flies."[76] Mukerjee freely participated in the Indian Malthusian–eugenic class and caste discourse of "excess humans." Yet as I understand Mukerjee, he was also, as an ecologist, quite literally and genuinely interested in how humans breed "like rabbits." This is not just a conceit. Mukerjee's learned head was filled with the new animal ecology and the biology of density: Charles Elton's lemmings, Raymond Pearl's fruit flies, not to mention Charles Darwin's rabbits. They were all thinking about animal ecology. This is the less familiar, certainly less popular intellectual history on which I focus. Moreover, it is often the unexpected and counterintuitive politics that most interest me, precisely because the tradition of critique has very little political room for it. To pursue this particular example, the usual criticism of Mukerjee begs explanation of how he came to be a eugenicist population planner at the same time as he was an anticolonial, antiracist, and ecological opponent of the global color line. All these positions formed his vision of world population growth.

Antinationalism and anticolonialism are both evident in the history of Malthusian thought, connecting demographers from vastly different colonial, national, and racial locations, even if they were all, as experts, elite. This is often overlooked, or actively set aside, because of the bad favor in which Malthusianism is held, as politically dangerous, as intellectually unfeasible, or as economically outdated. But one does not need to agree or disagree with the politics, economics, or even implications of Malthusian ideas, in order to analyze them historically. Malthusianism turns out to be a most interesting, if unlikely, intellectual space in which the political and scientific question of human difference and identity was struggled over and was connected to place. In this intellectual space, not just colonialism and racism were produced, as in the standard account, but also various forms of anticolonialism and antiracism as well.

"Cosmopolitan" is what economist Keynes called the population problem.[77] The single greatest problem facing "this little terraqueous globe," is how Edward East put it in his address "Food and Population," delivered in

Geneva in 1927.[78] But as all at the first World Population Conference knew, this little terraqueous globe was not just a biosphere of life on Earth, a newly globalized set of interconnected continents and economies. It was also a deeply politicized world in a very precarious peace. The globe *could* become the one world of cosmopolitans' dreams, but in the late 1920s it was a world as divided as it had ever been. Not far from Geneva, peacemakers in Paris had recently reterritorialized much of the planet. The "global color line," a long time coming, was at its high point. And Malthus's "struggle for room and food" had become in one version, lebensraum, poised to make a terrible comeback. This was the political context of the world population problem between the world wars.

Part I

The Long Nineteenth Century

1

Confined in Room

A Spatial History of Malthusianism

Where is the fresh land to turn up?

THOMAS ROBERT MALTHUS,
AN ESSAY ON THE PRINCIPLE OF POPULATION (1798)

Malthus liked writing about islands. They illustrated nicely how humans were always, one way or another, "confined in room."[1] The multiple scenarios he offered about the great human predicament were often geographical spaces with limits: the Islands of the South Sea, for instance, or perhaps more pressingly for most of his contemporary readers, the British Isles. "Let us now take any spot of earth, this Island for instance," he invited. If all restraints on population growth were absent, he speculated that in a few centuries, "every acre of land in the Island [would be] like a garden."[2] But even this would be inadequate, because the population would have increased far beyond the capacity of the total island-garden to produce. Malthus hypothesized at a larger scale, a global scale, about "the whole earth, instead of one spot." Indeed, not just one world but "millions of worlds" could theoretically be filled with fast-reproducing life, were it not for the restraining necessity for food and for land in which to grow it.[3] From his rooms in Jesus College, Cambridge, from his garret in London, and later from his chair in political economy at the East India Company College, London, Malthus indulged a planetary imagination: "The whole earth is in this respect like an island."[4]

The *Essay* has been scrutinized many ways,[5] but it still invites a geographical reading as well as further inquiry within the intellectual history of ecology. In short, this famous English text, in its multiple editions, was all about space—"room" Malthus tended to term it—in which animals, plants, and humans struggled to live. "Want of room and nourishment" was a condition that affected all organisms in the animal and vegetable kingdoms. "The contest was a struggle for existence," he wrote in the pages that immediately followed his famous postulata about sex and food in the first edition.[6] Through Malthus's core idea that more organisms reproduce than can survive, political economy and natural history came to be interwoven throughout the nineteenth century. As Charles Darwin and scores of others put it, Malthus was describing the "œconomy of nature."[7] Unsurprisingly, late nineteenth-century Malthusians (and Darwinians)—the "neo-Malthusians"—constantly drew attention to the conceptual link between political economy and natural history. They, too, saw the struggle for room and food as foundational, the economy of nature that became, in one version, ecology.[8]

Looking at the Malthusian line of thought spatially, the extent to which a planetary imaginary was in operation becomes apparent. From Malthus onward, indeed before him as we shall see, this economy of nature was comprehended as checking and balancing itself within the limits of the earth, literally the spherical globe. This was so even for those figures in the neo-Malthusian story who are rarely placed within the history of economic or ecological thought, those have come to most represent the "sex" not the "space" dimension of the population question: Annie Besant, for instance. As the nineteenth century progressed, a capacity to conceptualize a supranational globe was accompanied by an internationalist politics that was explicit among many of Malthus's intellectual successors. Late-Victorian neo-Malthusians, as well as Malthusian economists on the eve of World War I, tended toward cosmopolitan politics and even espoused a grandiose global pacifism: they considered that their own methods would bring about world peace, before there was even such a thing as a world war.

Space and the Principle of Population

Malthus's *Essay on the Principle of Population* was first published in 1798. Too often remembered, defended, dismissed, or reduced to its stark claims about geometrically increasing population and arithmetically increasing food, the *Essay* held a thousand ideas within it, and more with each edition. It was written at a point when the political and economic relationship between the Old World and the New World had re-formed; when British political economy was emerging with expansive and hotly contested ideas about land, people, wealth, and labor; when the fallout from the French Revolution and the radical reconceptualizations of state, citizen, and liberty were taking strange turns, not least war on several continents; and when the Pacific world—the South Sea—was being explored and assertively colonized by the British and the French. All this, one way or another, is in Malthus's *Essay*. Yet perhaps the most tantalizing feature only became apparent retrospectively; the first *Essay* was written on the cusp of a changing economy and of unprecedented population growth in Malthus's own backyard, but without much awareness on his part of these imminent great changes in the modern world.[9] In 1798, Malthus wrote about an old demographic regime, what he saw as centuries of more or less stationary balance between European births and deaths. He thought the increase of his own nation's population since the Glorious Revolution was, if anything, very slow.[10] In fact the population of England and Wales was already increasing, from about 5.7 million in 1750 to 8.6 million in 1800 to 16.5 million in 1850.[11] For the twentieth-century demographers analyzed in later chapters, it was the massive increase in domestic British population, even greater over the later nineteenth century, that was the accelerant in the global population story. But for Malthus, writing at the turn of the eighteenth century, the intriguing point was just how populations were kept as low and as stationary as they seemed to be.

Geopolitics and biopolitics—land, labor, food, and sex—were conceptually central from the Malthusian beginning. In the first anonymous edition, Malthus stated that basic needs underwrote all his arguments: "First, that food is necessary to the existence of man. Secondly, that the passion between the sexes is necessary and will remain nearly in its present state."[12] The power of the earth to produce food, and the power of humans to reproduce, were unequal, however. Thus, the effects of these powers must be (and he claimed were) continually "kept equal" by various natural and human interventions. In later editions, he considered that the relative effects of food and sex might be kept more balanced by "moral restraint," that is, by delaying both marriage and sex.[13] So-called preventive checks were practices that limited fertility, and the Victorian Malthusians later advocated contraceptive practices or devices as the preferred means to check fertility; preferable, that is, over the celibacy theoretically required of "moral restraint" and the prostitution that they thought would result in practice, whatever Malthus himself thought. The balance between population and subsistence, was also kept equal over time by "positive checks" that increased mortality: starvation, disease, and war, as well as infanticide. Malthus's assessment of the capacity of various restraints to ameliorate the conditions of those near subsistence levels differed over successive editions. His first edition doomed humans of the lower classes in various "stages of civilization" to perpetual suffering. By his sixth edition, this was largely modified, and early twentieth-century interpreters of Malthus understood him to argue that restraints could mitigate this suffering significantly.[14]

Malthus's famous "checks" were bodily matters in the first instance, the stuff of biopolitics: sexual conduct, birth, health, illness, and death. But population determinants were also crucially about land and space. This made population, for Malthus, not just a domestic but also an intercontinental matter. Land available in the New World was important and he detailed its use, briefly in the first edition, extensively thereafter: for the Spanish in Mexico and Peru; for the Portuguese in Brazil; for the Dutch and the French in their multiple respective colonies. All these, despite variously questionable governance, he wrote, currently have plenty of room and food and therefore "have constantly increased with astonishing rapidity in their population." But none had increased so quickly as the North American colonies, "now the powerful People of the United States of America." It was the thirteen colonies and the new United States that were the important cases. There, in addition to a healthy measure of liberty and equality, the key factors were "plenty of good land" and social/legal systems that ensured its maximized cultivation. With all this in place, the population in the thirteen colonies had doubled in twenty-five years.[15]

Initially unwittingly, Malthus was reiterating Benjamin Franklin's observations, though the latter had in fact claimed that American populations doubled by natural increase in twenty years.[16] Malthus derived this fact from Richard Price's *Observations on Reversionary Payments* (1771) via Yale president Ezra Stiles's *Discourse on the Christian Union* (1761), whose own original source was Benjamin Franklin's 1751 pamphlet, "Observations Concerning the Increase of Mankind." From his second edition (1803) onward, Malthus cited what was to become Franklin's enduring instance of fennel, the single species that might overrun all else, if the circumstances were right: "were the face of the earth . . . vacant of other plants." And Malthus quoted Franklin on humans: "were it empty of other inhabitants, it might in a few ages be replenished from one nation only; as, for instance with Englishmen."[17]

From Franklin, through Malthus, and eventually to Charles Darwin, as historian Joyce Chaplin has elegantly shown, the doubling of human populations in twenty-five years as a maximum rate of growth was a critical reference point. It came to be a Malthusian mantra for Victorians such as John Stuart Mill and Annie Besant, for twentieth-century moderns like John Maynard Keynes, and well beyond.[18] But critically, for Malthus, the capacity of a population to increase was dependent on geography—"fresh land"—that diminished the need for or effect of preventive checks. It was the availability of land in North America that encouraged people to marry younger and therefore to have more children. In this original moment of the political economy of population lay an intricate and even causal relationship, between sex and space, life and earth, bio and geo.

Over time, all kinds of demographic, geopolitical, and ecological arguments were to be made about the global significance of apparently empty land in North America, its seeming limitlessness, its opening and eventual closing, as Frederick Jackson Turner was famously to frame this geography and history in 1893.[19] For Malthus almost a century earlier, the whole point of introducing the American case was to argue that limits theoretically applied even there. Ultimately, even the vast spaces of the North American continent ended, the famously fruitful soil could only yield so much. Even if the United States of America was "almost entirely vegetable," he wrote, beyond that hypothetical state "where is the fresh land to turn up?"[20] For Malthus, the industry, happiness, and population of the Americans depended on their great plenty of land and their "superior degree of civil liberty." But even civil liberty, he finished, "all powerful as it is, will not create fresh land."[21]

Franklin's and Malthus's interest in numbers of people in the eighteenth century was a product of the more general project of political arithmetic. Beginning in the seventeenth century, modern states sought to reckon population trends with new statistical techniques and with new objectives. Malthus read and cited one of the foundational texts of this tradition, William Petty's *Several Essays on Political Arithmetick* (1699).[22] Enumerating a population, maximizing its aggregate health and longevity, and devising ways systematically to know and predict birthrates, mortality, and morbidity was the business of the modern state. It became "statistics" ("the science of the state") or later "vital statistics," with ever-growing links to insurance industries that traded in probability.[23] For Michel Foucault and subsequent scholars, this was the beginning of biopolitics.[24] Early political arithmetic also included occasional attempts to calculate and project total world populations, for example Gregory King's 1682 projection of 630 million in 1695 that would reach 780 million in 2050.[25]

It was received wisdom that populousness was an index of national wealth and strength, and, as David Hume put it, the happiness, virtue, and wisdom of a nation's institutions.[26] But for Malthus and other late eighteenth-century political economists, the issue was not populousness per se, but the nature of the relation between land and people. Progressive wealth, wrote Adam Smith, was "in proportion to the improvement and cultivation of the territory or country."[27] Theirs was a preindustrial world still based on organic production and conversion of energy. But there were limits. As Wrigley has put it: "Malthus, Smith, and Ricardo . . . shared the conviction that economic growth must be limited because the land (in a literal and narrow sense) was a necessary factor in almost all forms of material production, and the supply of land was virtually fixed."[28] The *Essay* was thus concerned at various points not just with the availability of fresh land, but with the specifics of land use and land reform: the enclosure of commons, the reclamation of land, the use of fertilizers, the economic impact of growing food for animals rather than humans. Malthus thought, for example, that turning corn-growing land into pastureland to feed animals resulted in a smaller quantity of food for human subsistence. Elsewhere he noted that the increasing demand for meat, and the number of horses kept for leisure, together tended to diminish the total yield of food for human consumption.[29] This is why Malthus repeatedly turned to look at Chinese agricultural techniques: he was astonished by the estimates of population density there. Pointedly, he accounted for it differently than Montesquieu, who thought simply that the climate favored women's fertility. The fertility of soil, rather, was what Malthus noted: "the excellence of the natural soil, and its advantageous position in the warmest parts of the temperate zone," combined with "incredible fatigues in cultivating the earth . . . cultivating land, by dunging, tilling, and watering it." Put simply, the Chinese did not waste land: "Little land is taken up for roads, which are few and narrow, the chief communication being by water. There are no commons, or lands suffered to lie waste by the neglect, or the caprice, or for the sport, of great proprietors."[30] The remarkable fact of two harvests per year fascinated him, as it would later Anglophone agricultural economists and demographers, and indeed current historians of an economic divergence between China and Britain.[31]

Precisely how land produced food for humans was critical for Malthus, for his opponents, and as much of this book shows, for his twentieth-century intellectual descendants. His calculations about the most efficient use of land, in terms of food/energy dividends for labor/energy investment, is the same broad paradigm that early to mid-twentieth-century economists, demographers, and agriculturalists deployed, even though they were analyzing a completely different post-industrial economic world. Calculating total (possible) world population on the basis of the habitability of land was a standard technique of the 1920s,[32] but it derived from much earlier attempts to do so. In a response to Malthus, William Godwin in the 1820s calculated a maximum world population of 9 billion. Godwin arrived at this figure by multiplying the population density in China, which represented for him a maximum, by the habitable areas of the globe.[33] Malthus had taken his original earthly terms from Godwin, citing in order to dismiss the latter's claims in *Political Justice*: "Three-fourths of the habitable globe is now uncultivated. The parts already cultivated are capable of immeasurable improvement. Myriads of centuries of still increasing population may pass away. And the earth be still found sufficient for the subsistence of its inhabitants."[34]

Writing slightly later than Godwin, economist Nassau Senior pursued lines of argument about number of acres per person (which for him meant per family) that would result from hypothetical increases in population. Assessing the British situation in 1829, he continued the Franklin–Malthus tradition of speculating on populations that doubled:

> Supposing our population to have increased, as would be the case by the beginning of the next century, to one hundred millions, about an acre and a half would be allotted to each family; and, as I before observed, I think that allotment might be sufficient. But it can scarcely be supposed, that three roods would be enough, which would be their allotment in twenty-five years more, or granting that to be enough, it cannot be supposed that at the end of a further term of doubling a family of four persons could live on the produce of a rood and a half.[35]

The long line of nineteenth-century socialist and Marxist critics also pursued the question of land area, availability, and yield. One aspect of Friedrich Engels's mid-nineteenth-century challenge—and this was an increasingly standard response—was that as population increased, so did scientific and technological capacity to augment food production. Take the great advances of agricultural chemistry, Engels exclaimed, even the work of two scientists alone, Sir Humphry Davy and Justus Liebig, and there is evidence of capacity to vastly increase produce. In any case, Engels continued in 1844, it is "ridiculous" to speak of overpopulation while only one-third of the earth was under cultivation and the productivity of this third could be increased sixfold and more merely by applying improvements that were already known.[36] The Mississippi Valley alone could accommodate the whole population of Europe in its wasteland, he said, dismissively.

Engels, too, was thinking globally, or at least bicontinentally. By that time, British domestic population had increased beyond anything Malthus could have imagined, partly enabled by the clearing of global lands. The British and British-descended settler-colonists were certainly "replenishing the earth."[37] Frontiers were pushed back, grasslands were turned into grain lands, millions of acres were cultivated. New machines and settler populations yielded unprecedented agricultural returns. But the race was also on to defer the eventual diminution of those returns. For Malthusian-inspired economists like John Stuart Mill, this was a trend toward increasingly "laborious and scantily remunerative cultivation."[38]

The process of colonization was, among other things, a process by which land occupied by indigenous people was rendered officially "waste"—unused—so it could be legally reoccupied. The Imperial Waste Lands Occupation Act of 1846, for example, enabled the cultivation and population of New South Wales by English, Welsh, Scottish, and Irish colonists and convicts. At the same time, in Ireland itself, "waste lands" were being reclaimed. In economist William Thomas Thornton's *Over-Population and Its Remedy* (1846), the occupation of Irish wastelands manifested as the "draining and cultivating and rescuing [of] the marsh from the water."[39] The problem there became one of land ownership, tenure, and leasing: the high imperial politics of Irish land in the 1840s. Even as they wrote in the midst of phenomenal economic and population growth, the mid-Victorian Malthusians thought that the reward for cultivating effort would steadily diminish, a scenario that for some made indigenous claims to land an irrelevance, if they bothered to think about it at all. Empty lands, wasted lands, unclaimed land, and reclaimed land: this was the colonial geopolitics of population growth and expansion.

Struggling for Room and Food: Natural History and Political Economy

When Charles Darwin and the younger natural historian Alfred Russel Wallace read about "a perpetual struggle for room and food," they were both startled into intellectual ferment. The result was the theory of evolution by natural selection; in a way, they were each adding their own law of population and space to Malthus's original.[40] What Malthus brought to the question of life for Darwin was the power of reproduction, the power of death, and the "struggle" produced by their relation, in limited spaces with limited resources. As Malthus had put it, writing of early human society, "the contest was a struggle for existence . . . death was the punishment of defeat and life the prize of victory. . . . The prodigious waste of human life occasioned by this perpetual struggle for room and food was more than supplied by the mighty power of population."[41] Using Malthus's phrase, Darwin postulated that a "struggle for existence inevitably follows from the high rate at which all organic beings tend to increase." This was, as a score of neo-Malthusians (and historians of Darwinism) were to repeat "the doctrine of Malthus applied with manifold force to the whole animal and vegetable kingdoms."[42] Over time, Malthus and Darwin came to represent the intellectual exchange between, respectively, the political economy and the natural history of population. The Malthus-inspired section of *On the Origin of Species* was, and is, often quoted, and became foundational to Victorian Malthusian thought: "There is no exception to the rule that every organic being naturally increases at so high a rate that, if not destroyed, the earth would soon be covered by the progeny of a single pair."[43]

By 1859, however, this particular statement was already more repetitious than revolutionary. John Stuart Mill, to take one of the more influential examples, worked over the same Franklin–Malthus fennel-inspired insight in *Principles of Political Economy* (1848): "There is no one species of vegetable or animal, which, if the earth were entirely abandoned to it, and to the things on which it feeds, would not in a small number of years overspread every region of the globe, of which the climate was compatible with its existence." The power of living species to reproduce, and the limits of that reproduction, were basic principles of political economy for Mill. Including humans: "to this property of organized beings, the human species forms no exception."[44] While human populations could possibly double in twenty-five years, limit to this growth was determined by the factor of land, limited in quantity and also in productiveness. By the middle of the nineteenth century—a century after Franklin's original—the North American case was still being cited as the society growing most rapidly by natural increase. It was seen to be faster even than industrializing Britain, which doubled only over forty-three years, according to French statistician Moreau de Jonnès, or fifty-three years, according to neo-Malthusian George Drysdale.[45]

Charles Darwin himself offered a range of curious instances of this theoretically infinite power of increase in the *Origin,* some hypothetical, some actual: a plant that produced only two seeds would become a million plants in twenty years (Linnaeus's calculation); a pair of elephants ("the slowest breeder of all known animals") would become 15 million by the end of five centuries; rabbits introduced into Australia, with nothing to check them, had multiplied rapidly; the tall thistle introduced into the plains of La Plata had come to dominate, "almost to the exclusion of all other plants." And what about humans? "Even slow-breeding man," Darwin wrote, casting back to Franklin, "has doubled in twenty-five years."[46] Of course for all of these writers on population—political economists and natural historians alike—the whole point of reiterating the potential for growth was to substantiate the actual functioning of the "checks." Population does not in fact continue to increase at such a rate, or "the earth would, long ere now, have become unable to support her offspring."[47] It is absurd, declared another Malthusian in 1861: if Great Britain had 21 million people, left unchecked in three hundred years the population would be 1.3 billion, "more than the total population of the globe, which is estimated at about 1,000 millions."[48]

The publication of Darwin's *On the Origin of Species* (1859) and *The Descent of Man* (1871) held massive implications for precisely how humans were to be comprehended. But the generic inclusion of humans within natural history was one of the more standard statements made within political economy of population. "Nature's economy," in other words, already included humans, and conversely, political economy already considered "all organic life." Far from Darwin (or Wallace) extrapolating Malthus's idea about human population and applying it to other organisms or even all organisms, Malthus himself had already made those connections. His law of population operated "through the animal and vegetable kingdoms," literally a law of nature that functioned among "the race of plants, and the race of animals. . . . And the race of man."[49] It was nonetheless Darwin's and Wallace's great insight to link this to variation, adaptation, and the origin of separate species. Versions of this exchange between the natural history of population and the political economy of population repeated through the rest of the nineteenth century. It was basic to the work of late-Victorian neo-Malthusians.

The Law of Population and the Victorian Malthusians

Founding father of the neo-Malthusians, George Drysdale was a medical doctor trained at Edinburgh University. While practicing alongside his brother, Charles Robert, he published anonymously (and elsewhere under

"GR") the cryptically titled *Physical, Sexual and Natural Religion* in 1855.[50] The first edition focused on reproductive development and physiology, venereal disease, and the "evils of abstinence," all interestingly allowing for the "separate points of view" of men and women.[51] By the third edition, however, Drysdale added a much-expanded section on social science. Malthus and Mill, he suggested, underwrote the new science of society with their law of population, "the most important and terrible subject for the contemplation of mankind."[52] In the retitled *Elements of Social Science*, Drysdale underscored for his readers the apparently incontrovertible truth that "it is only *by checking still further the reproductive powers of our species,* that it is possible to remedy poverty and to raise wages." In so doing, Drysdale rehearsed Malthus's original linking of natural history with political economy: "the capacity of increase in the human race, as in all organised beings, is, in fact, boundless and immeasurable," Drysdale wrote. And he spent pages discussing a particular demographic fact: populations in North America doubled by natural increase in twenty-five years.[53]

If Thomas Robert Malthus produced the late eighteenth-century credo, it was perhaps less the Drysdales than the extraordinary Annie Besant who was the modern prophet for many Malthusians (and their historians) in the twentieth century.[54] *The Law of Population, Its Consequences and Its Bearing upon Human Conduct and Morals* was written at the height of Besant's involvement with neo-Malthusianism.[55] Like Drysdale's before it, this 1877 pamphlet was at least as much about the natural history of political economy, as it was about women's health, feminism, and contraception, in which context it is almost always cited. Like most neo-Malthusians, Besant spent considerable time and effort reiterating the cadence of political economists and natural historians: Franklin, Malthus, Mill, Drysdale, and now Darwin. This she established on page one. The law of population has become axiomatic for the political economist and for the naturalist, she wrote, now it should become so for the masses. Besant unmistakably extended the spatial history of Malthusianism into a post-Darwin paradigm. But before she discussed Darwin, she presented Franklin's observation: "there is no bound to the prolific nature of plants or animals but what is made by their crowding and interfering with each other's means of subsistence." Nature scatters the seeds of life everywhere, but also provides comparatively little room and nourishment for them.[56]

Annie Besant's life was immeasurably larger than Malthusianism and the Malthusian League, with which she became so closely associated. As one biographer put it, Besant's first five lives included secularism, socialism, and feminism, and her last four included theosophy, Hinduism, Gandhian pacifism, and a term as president of the Indian National Congress.[57] All of the twentieth-century manifestations of these social, intellectual, and spiritual movements went on to claim Besant, and each with good reason. However, one twentieth-century intellectual and latterly political movement with which she is almost never connected is ecology, and yet that is what she anticipated in linking Franklin, Malthus, and Darwin.[58] As historian of ecological ideas Donald Worster suggests, "Darwin's reading of Malthus can make good claim to being the single most important event in the history of Anglo-American ecological thought."[59] The term "ecology" was invented by a German admirer of Darwin, Ernst Haeckel, in 1866, and it is useful to read Besant herself as part of the intellectual conversation that propelled the consideration of humans and the world population problem into early ecology.

Besant recounted in her *Law of Population* the basic Malthusian insight that drove Darwin as he observed the life and death of plants on the small scale familiar to the earliest botanical ecologists. Darwin, she told her readers, carefully sowed 357 seedlings and watched 295 of them die for want of room and nourishment. "Of the many individuals of any species which are periodically born," she quoted him, "but a small number can survive." This, she announced perhaps more grandly than Darwin himself ever would, is a law beyond controversy, a law as fixed as "the sphericity of the earth." Besant meant this in other senses too, as the literal limits of the earth. As she summed up in 1877: "Since the size of the globe inexorably limits the amount of vegetable produced . . . [this] limits the amount of animal life which can be sustained."[60] Besant placed humanity in nature and as subject to natural laws: not just the theory of natural selection, but the law of population. This was the Malthus–Darwin connection. The law of population "runs through the animal and vegetable worlds," Besant repeated yet again, including man "the highest in the animal kingdom, not a creature apart from it." Indeed, reiterating these enduring ideas was really the whole point of her pamphlet. A slightly later Malthusian author put it thus: "Darwin's great discovery . . . [was] that man was part of the order of Nature, and, in regard to his organs and powers of reproduction, simply a superior animal."[61] For the late nineteenth-century generation, Malthus was authorized via Darwin.

Besant catapulted the question of population into a concerted phase of activism about poverty and connected it to a newly public set of rationalist and feminist politics. More than anything else, perhaps, she was driven by the politics of secular free thought. In 1876–1877 she was famously tried, with secularist Charles Bradlaugh, for their Freethought Publishing Company's reproduction of an old American "physiology" book that advised on contraception. This marked the beginning of the Malthusian League in Britain. Less thoughtful retrospective observers even claimed the famous trial as causative of the slowing of the birthrate, soon to be statistically observable.

The Malthusian League in Britain, the small organization with the big ideas, instrumentally involved secularists Besant and Bradlaugh. They were critical but passing figures, both moving on to multiple other causes. By contrast, the remarkable Drysdale medical family was dynastic. Medical man George Drysdale, his brother the homeopath Charles Robert Drysdale, the latter's wife Alice Vickery (Drysdale), their son Charles Vickery Drysdale, and his wife Bessie Ingham Drysdale provided a continuous link between the Victorian Malthusians and the 1920s.[62] Never large, though internationally represented, the Malthusian League was motivated and driven in the first instance by what they saw as the horrific consequences of the law of population for humans. They perceived their work to be ameliorative. Drysdale's mid-Victorian book was "dedicated to the poor and the suffering," and later Malthusians always perceived their mission—however erroneously, though rarely disingenuously—to be the alleviation of poverty: minimizing birthrates meant wages would rise. Initially at least, birth control—the new "preventive check"—was the means to this end, not an end in itself.

The structure of Besant's *The Law of Population* underscores this. The first two chapters established the law of population from Malthus, Darwin, and Mill, and explained the consequences of this law with respect to humans. It was only then that she explained why contraception within marriage was preferable to celibacy as a human check to population growth. And the final chapter anticipated, considered, and answered various objections. Humans may be subject to the natural law of population, and therefore the inevitability of "checks" on growth, as Besant put it, but such checks caused great suffering, manifesting as the "killing of human beings, either slowly or rapidly." This life-destroying process—"anti-human, brutal, irrational"—could and should be substituted by more humane and effective processes of active prevention of the production of life; life that might otherwise be born only to die. Substituting the "life-destroying" checks—famine, starvation, infanticide, abortion—with the new "prudential checks" was the means by which humans could, in a more civilized manner, acknowledge the constant operation of the law of population and pursue the Malthusian "crusade against poverty" in a way that was about life not death.[63] That was how neo-Malthusians saw the situation. Accordingly, they constructed their work as "humanitarian" in the nineteenth century and "progressive" in the twentieth century.

A derivative benefit began to enter the Malthusian social vision as well: preventive checks helped to realize women's liberty, a term that in this context instantly signaled John Stuart Mill's *On Liberty*, cited throughout Besant's work.[64] While advocacy of contraception was far from coterminous with a mid-Victorian women's movement either in the United States or in Britain, nonetheless an important new feminist rationale for birth control was being articulated. Malthusianism was gathering up an explicitly gendered politics, the so-called woman question that was well and truly on the public agenda by 1877. If Malthus had focused on men's sole control over sex and marriage (posing problems, incidentally, for Darwin's vision of female choice in sexual selection as he struggled toward *The Descent of Man*),[65] some Malthusians began to talk and write about women's capacity—even right—to exercise control over their own reproduction and over their households. Alice Vickery in particular grafted a late-Victorian feminism onto British economic thought on population, arguing for a sexual politics based on the rights to bodily integrity of an individual woman. Keynes later identified her as someone who always linked neo-Malthusianism to the emancipation of women.[66] She elaborated, through a Woman's Malthusian League, a "Programme of Women's Emancipation" that included the right to vote, the right to sit in Parliament, the right to education, the right of married women to property, and to the "Limitation of Offspring—Children should be made a scarcity article rather than a drug in the market, so that all born should be regarded as precious treasures and future citizens, whose lives and well-being should receive the greatest attention and care."[67]

Although most nineteenth-century feminists eschewed birth control and steered well clear of the Malthusian League, a number of issues were just then turning intimate politics into public and national politics in Britain: the agitation against compulsory screening and treatment of prostitutes galvanized by the controversial Contagious Diseases Acts; the public raising of marital rape and the law; and of changing divorce laws on the basis of cruelty. All these were about women's control over their own bodies, deeply troubling the public and private spheres of classical liberalism, but at the same time emerging from that liberalism.[68] The next generation turned this into international politics. For such feminist Malthusians as Vickery, the exercise of various forms of birth control was an individual "right" that neatly produced simultaneously individual and social benefits, especially, it was argued, for working-class women and their families. Their message was that fewer children—spacing births—secured individual and familial health and happiness and was also the key means by which poverty would be addressed at larger scales, with the effect of raising wages. This aspect of neo-Malthusianism was an important part of the intertwining histories of birth control, medicine, feminism, gender, and sexuality.[69] But it should be assessed, too, as the grafting of a newly public politics of sex, gender, and liberal rights onto the prior political economy of population that was also the economy of nature. These were not separate but deeply integrated intellectual histories.

If Besant supported, indeed relied upon Darwin to explain and sell her *Law of Population*, the reverse was not the case: Darwin declined Charles Bradlaugh's request to be a witness in Bradlaugh and Besant's trial in 1877.[70] He believed that artificial checks to the natural rate of human increase were undesirable and morally problematic, in that they would threaten chastity and ultimately the family.[71] Darwin's opposition to family limitation is often elided with an apparent opposition to neo-Malthusianism. At one level correct, such an assessment depends on an attenuated understanding of neo-Malthusianism-as-contraception and fails to comprehend the repetition of Malthusian truisms within Darwin's own work. For their part, the Victorian neo-Malthusians always saw that Darwin, like John Stuart Mill, had incorporated Malthus on the "struggle for room and food," and the vision-cum-mission of the economic/ecological law of population was what the new Malthusians constantly pressed forward. Rather taking liberties, they would even speak of an interchangeably "Malthusian or Darwinian law of population." When Darwin's supporter Thomas Henry Huxley died in 1895, for example, he was claimed as a "pronounced Malthusian," and *The Malthusian* cited his recent essay "The Struggle for Existence": "So long as unlimited reproduction goes on, no social organisation which has ever been devised, no fiddle-faddling with the distribution of wealth, will deliver society from the tendency to be destroyed by the reproduction within itself, in the intensest [*sic*] form of that struggle for existence the limitation of which is the object of society."[72] A century after Malthus's *Essay*, both "want of room and nourishment" and the resulting "struggle for existence" seemed conceptually to belong equally to Malthus and Darwin, to political economy and natural history. And certainly to the neo-Malthusians.

"The Size of the Globe": International Neo-Malthusians Before World War I

The English Malthusians went international with the new century. The first international meeting was in Paris (1900), followed by meetings in Liège

(1905), The Hague (1910), and Dresden (1911), part of that year's International Hygiene Exhibition. The first decade of the twentieth century was the high era of international meetings, and like so many other groups, "international" for the neo-Malthusians simply indicated a gathering of delegates from different nations:[73] Britain, The Netherlands, France, the United States, India, Belgium, and Spain.[74] The first object of the International Neo-Malthusian Bureau established at The Hague in 1910 was straightforwardly organizational: "To bring Neo-Malthusian propagandists of various countries into touch with one another."[75] Quickly, however, and possibly more importantly for the problematizing of world population, internationalism itself gained currency and began to govern the neo-Malthusians' discussion substantively. Charles Vickery Drysdale declared to those gathered at Liège in 1905 that population was "the fundamental question which underlies every phase of international politics." He explained that population problems and neo-Malthusian solutions were international, because so few countries produced all the food they consumed, and there was national struggle, because total world food production "for even a good year" was not adequate. "Hence we have the race for new possessions and colonies, protective and retaliatory tariffs, and national jealousies, ever tending to misunderstandings and war."[76] What would become an enduring connection of neo-Malthusianism, pacifism, and internationalism started here, and the Victorian Malthusian dedication to the domestic "poor and suffering," as George Drysdale had put it, was elevated by the next generation of Drysdales to nothing less than the "brotherhood of the human race."[77]

There were many shades of internationalism at this point, many kinds of brotherhood, and not least an international sisterhood.[78] Although often understood retrospectively as something produced by World War I, pacifist internationalism was already a force in the first decade of the century, as Jay Winter has shown. It was displayed, for example, at the first Hague Convention in 1900 on the peaceful resolution of international conflict.[79] Earlier, nineteenth-century British liberalism had produced one version of internationalism based on the principle of the peaceful benefits of free trade. This found some expression as a solution to population growth, linked often to the need for free global movement. Another liberal internationalism, characterized by William Gladstone's ideas, favored free trade tempered by appropriate interventions, as a means by which liberty would be defended and ensured.[80] At the same time, a political cosmopolitanism was current, drawn from Kant's *Perpetual Peace*, itself written at the moment of Malthus's original *Essay*.[81]

Neo-Malthusians engaged with the idea of cosmopolitan federations with some regularity, a politics rarely ascribed to them. Besant, for example, spoke of a union that would "bind together every land in one great commonwealth . . . one vast Parliament where all should make their voices heard." This was the idea of an ever-expanding Greater Britain: "the Parliament of that English commonwealth which spreads over every part of the habitable globe."[82] Malthusian H. G. Wells was also writing at the turn of the century about population, world government, and world peace, influencing a great number of people, not least John Maynard Keynes.[83] A rarely cited late publication of George Drysdale's dealt with an unlikely proposal for the federal union of France and England as the first step to the federation of the world. "It seems to me that one of the grandest aims ever conceived—indeed, next to the removal of poverty and the other population evils, the very greatest reform that could be effected in human affairs—is to get rid gradually of the present system of independent sovereign states." The "Federation of Mankind" was Drysdale's ambition, in which the "less civilized races" would gradually be "done away with as the backward populations grew in enlightenment." He meant "improvement" and even "assimilation"; the extermination of difference, perhaps, but not of people. Eventually all constituent parts of a federated whole would be placed on equal political footing.[84] Here, Drysdale was both rehearsing the problem of economic difference constructed as civilizational difference that Malthus had grappled with and anticipating the dilemma of development economics.

Prewar internationalism and pacifism often intersected with Malthus's and Darwin's ideas in general and the population problem in particular. As Gregg Mitman has shown, revisions of Darwinism put forward by U.S. biologists were deployed to drive a liberal pacifism well before the war, but already in response to German militaristic interpretations of "struggle."[85] Marxist critics of Malthusian economics held internationalist aspirations of a very different kind, sharpened after the turn of the century through Lenin's writing. For them, Malthusian bids for "brotherhood" were both conceptually shallow and politically sinister.[86] Indeed, competing claims to internationalism became part of Marxism and Malthusianism's own struggle over the long twentieth century.

Internationalism came to shape the very language within which the new internationalist-pacifist Malthusianism was presented. The proceedings of the 1910 conference in The Hague were published in Esperanto and in its recently streamlined form, Ido. Charles Vickery Drysdale not infrequently presented his material in the world language that had been developed over the 1870s and 1880s and around which a world congress gathered in 1905. Drysdale insisted for a time that key articles in *The Malthusian* appear simultaneously in Esperanto, English, and, rather more usefully, French.[87] Malthusians wanted to speak an international language, because as far as they were concerned, population was an international issue.

Malthusians drew from the long-standing global visions and calculations established by Malthus and set to calculating global resources as well as people. It was Frenchman Gabriel Giroud in *Population et subsistances* (1904) who drew up an inventory of food for every country of the world for which he could gather reasonable data. In addition to Europe, he audited Asia (Siberia, China, India, Japan, and Java), Africa (Algeria, Egypt, Tunis, and Cape Colony); the Americas (Canada, the United States, Mexico, Chile, Argentina, Uruguay, and Paraguay); and the Pacific (Australia and New Zealand). Of the total cereals produced, and subtracting the amount used for alcohol, for seed, and for feeding "other animals," the amount of wheat, maize, and rice available was set against world population and its needs. In "The Semi-Starvation of the Human Race," *The Malthusian* editors gushed that this world survey corroborated Malthus's thesis that "the earth has been unable to feed a population which has been abandoned to the energy of its natural fecundity."[88] The hypothetical global scenarios of calculations of food and people that Malthus and others of his generation had imagined were becoming actual scenarios. The limits of the earth were approaching as the globe seemed to shrink in time and space. The planet was hauntingly pictured as "the raft to which we cling in the boundless ocean of space," in an early issue of *The Malthusian*, "hopelessly and for ever beyond the reach of any neighbor raft, and alone throughout the whole course of its existence."[89] Such images not just anticipated, but informed, both Wellsian science fiction and the later twentieth-century vision of spaceship Earth.

In Britain, geographer E. G. Ravenstein presented his own world survey in 1890, a formulation of global carrying capacity in a significantly titled article for the Royal Geographical Society, "Lands of the Globe Still Available for European Settlement."[90] Troubled by the paucity of world-level information, he gathered his own data on population "and its probable increase"; on the total area available for cultivation; and finally, "the total number of people whom these lands would be able to maintain." The puzzling decline in European fertility rate was just registering in 1890. Ravenstein did not assess this as a problem, however, but as an interesting possible solution to the limits of the "Lands of the Globe": "Are we to look upon these checks to an increase of the population as a natural law, the operation of which would, without violence, war, or pestilence, prevent the over-population of this world? Are we permitted to suppose that a time will come when mankind shall be so civilized and provident as to submit to checks to over-population, which at the present time are submitted to only by a few?"[91] He thought the trend was a good one and should become global.

Statesmen's anxieties about the fall in birthrates in this period have been extensively analyzed: "race suicide" and its corollary, pronatalism, have dominated historical discussion in many national histories.[92] Less recognized is the extent to which commentators like Ravenstein were heartened by the figures of fertility decline. Demographers and statesmen alike should welcome the trend toward the restriction of births, "as progress in human prudence," the international Malthusians told each other.[93] In 1906, Charles Vickery Drysdale identified those limiting their families as "individual neo-Malthusians." They should be distinguished from national Malthusians, and those like himself and his good colleagues, the international neo-Malthusians.[94] In this context, United States president Theodore Roosevelt's personal and national anxiety about a decline in the birthrate was becoming a thorn in the Malthusian's side. "President Roosevelt is evidently unteachable on the question of birth-restraint," they worried.[95] Together, the neo-Malthusians of England, France, Germany, and the Netherlands, now united as the Fédération Universelle de la Régénération humaine, protested. It was obvious to them that Roosevelt had given no attention whatsoever "to the economic aspect of the population question." And they suggested he read Malthus and Mill, to start with.[96]

Population and Economics, Circa 1900

By 1900, a whole library of economic scholarship existed on population, deriving in one way or another from Malthus and Mill. In prewar economics of the most scholarly kind, Malthus was discussed in the light of the new international networks of exchange. His ideas were, needless to say, dissected and opposed by some, but substantially accepted and refined by other economists of the period, and not a few in Roosevelt's own backyard;[97] American progressive Richard T. Ely, for example, or Harvard's professor of international economics Frank William Taussig.[98] Taussig affirmed for a new generation of Americans that the human species could double its numbers over a period of about twenty-three years, "unless counteracted," and argued that "restraint on the increase of numbers is one essential condition of improvement. Stated in this way the Malthusian position is impregnable."[99] British economist Alfred Marshall also broadly affirmed Malthus's core proposition that the increase of population everywhere would have been rapid unless "checked either by a scarcity of the necessaries of life, or some other cause, that is, by disease, by war, by infanticide, or lastly by voluntary restraint." This element of Malthus's ideas Marshall agreed with, while other premises had become "antiquated" because of "the great development of steam transport by land and sea, which have enabled Englishmen of the present generation to obtain the products of the richest lands of the earth at a comparatively small cost." But Marshall thought impossible the global replication of this local circumstance, unless checks on population were increased.[100] With the international neo-Malthusians and the *Geopolitiker* to be examined in the next chapter, economists tended to agree that the world was closing. "The world is really a very small place," wrote Marshall in 1907, affirming the prospect of diminishing agricultural returns, notwithstanding growth over the nineteenth century. That was, he said, an exceptional place and time, an "age of economic grace," that would run out by the end of the new century.[101]

Economists like Marshall were theorists of population, extending and adjusting Malthusian ideas for an already global and globalized world. But they were generally not "neo-Malthusians," like the men and women meeting in Paris, The Hague, and Liège. There were some important exceptions, however. In Cambridge, a student of Marshall's was pursuing mathematics and economics, linking it to his liberal pacifism and his interest in the changing views of evolution then proliferating among a new generation of biologists. The young John Maynard Keynes was enough persuaded to become an active member of the Malthusian League, and, driven by chronic bibliophilia, to purchase and cherish his own original editions of the *Essay*.[102] He was thinking deeply about population at an intercontinental scale, after the work of Marshall.[103] In spite of the decline of the birthrate, "the population of the world is still increasing fast," Keynes soon lectured to his own Cambridge students.[104] He also considered nineteenth-century growth a one-off phenomenon of expansion that was impossible to replicate, and he regarded the tendency of the birthrate to decline "as one of the most hopeful signs of the times."[105]

"Some day," Keynes mused, "the world will have to determine what is the fit and desirable number of inhabitants for it."[106] He perceived, like all the neo-Malthusians, that the Malthus–Darwin exchange between political economy and natural history was both ongoing and deeply relevant for the new century. Indeed, he drew a direct line in the genealogy of population thought from Malthus to himself, catching Darwin, Mill, and Marshall along the way.[107] Keynes considered population to have become a "cosmopolitan" issue since Malthus's time, in large part because of the intercontinental nature of modern economy, especially the goods and labor moving between Europe and North America, the migration exclusions then building up, and the place of India in an imperial economy. In fact, as John Toye notes, his scope was rather less global than Malthus's had been. But it was in this context that Keynes thought about race politics explicitly, finding the militaristic patriotic responses to the new world problem of population growth and race-based labor restrictions distasteful. He claimed that his own sympathies in seeking solutions lay "with the cosmopolitans."[108]

When Keynes was preparing his lectures on Malthus and population, just before the outbreak of World War I, Benjamin Franklin's eighteenth-century calculations were still, apparently, current and correct. "The increase in USA . . . has about doubled every generation," while the European rate of growth was slower, Keynes noted. But, he lectured to his Cambridge students, "If the population of the world were to increase no faster than that of Europe during the last 25 years, we should at the end of 1000 years be standing shoulder to shoulder over the whole habitable globe."[109] Such statements were repetitions for the new century of the long and intertwined political economy and natural history of population.

Franklin and Malthus had offered Darwin the "doubling population" story and the tradition of whole-earth scenarios that John Stuart Mill, George Drysdale, Annie Besant, and many others passed on to early twentieth-century moderns like Keynes. But in an unwitting return gesture, it was Darwin who offered the powerful "standing room" image to the long modern history of population thought: the rapid multiplication without natural or human checks meant that "there would literally not be standing room for his progeny."[110] Book after book would use the evocative phrase, "standing room only," inflected at every iteration with a particular politics. Considering overpopulation, George Bernard Shaw wrote in his 1889 *Fabian Essays* that there will be "only standing room left."[111] In 1917, statistician George Knibbs, in the tradition of Malthus's *ad absurdum* scenarios, wrote that many times the current surface area of Earth—that is, many Earths—would be necessary for a population growing at the current rate, even when calculated at the "standing room" of 1½ square feet per person. The necessary surface area would be a number "so colossal that it is difficult to appreciate its magnitude."[112] American sociologist and immigration restriction advocate Edward Alsworth Ross published his *Standing Room Only?*—with a question mark—in 1927, a moment when the world population problem was also the problem of the global color line. British biologist turned social scientist Alexander Carr-Saunders was to broadcast for the BBC on "Standing Room Only."[113] This is what Keynes repeated as "standing shoulder to shoulder," and it is why, in his 1914 lecture on population, he advised students not to be too concerned about what the Americans tended to overdramatize as "race-suicide."[114]

In America, a young doctoral student was also busy scrutinizing what he saw as the misplaced scare about the declining birthrates of the native-born population, compared with the immigrant population. Warren Thompson was just then undertaking his Ph.D. in economics and sociology at Columbia University, and he, too, thought through the new century's situation in terms of Malthus. *Population: A Study in Malthusianism* was Warren Thompson's dissertation, published in 1915. It covered Keynes's intellectual territory: an economic study of prices, wages, standards of living, population trends, and agricultural production in Europe and North America, between 1880 and 1900. It was one long study of trends in food production: cereals, meat, poultry, vegetables, fish. Thompson found Malthus to be "essentially correct" and argued that his own data substantiated the law of diminishing returns. He deployed a spatial metaphor to explain Malthus's thesis on population and food, one that derived perhaps from the frontier of Thompson's own Midwest:

> He [Malthus] looked upon population as a great force gathered behind a movable barrier, which was the food supply. As this barrier was moved ahead and left an open space, population proceeded to fill up this open space. If the barrier were moved rapidly the population would increase rapidly, if it were moved slowly population would increase slowly, and always those close to the barrier would be in distress, and fear of being pressed close to the barrier would keep many others from having as large families as they would like to have.[115]

Thompson agreed with Malthus: first, for the great majority of the people of the Western world the pressure upon the means of subsistence was the determining factor in family size; second, Malthus was correct in his argument that misery and suffering were due to overcrowding, that this meant a large number of people were always in want, although fewer people in the Western world than the Eastern world would die for lack of food. "The process of starvation is more refined" in the West, he wrote. And importantly, "population cannot continue to increase at its present rate without being more and more subjected to the actual want of food." A simplified standard of living in the West would help, although he doubted that this would enable ever-increasing proportions of the population to increase their standard of living. Ultimately, "a greater and greater control over the growth of population is essential to a growth of rational social control."[116]

The young Warren Thompson also argued that while the growth of population over the nineteenth century seemed to prove Malthus wrong, it was in fact an exceptional century and such growth could not continue. Why? The geopolitics of room and food. The frontiers of all the continents were closing, he wrote in 1915. "Fertile land is no longer to be had for the asking in the United States and will soon be taken up in the other places where Europeans can thrive."[117] In his view, if the nineteenth century had been open, the new twentieth century was unequivocally closed.

Part II

The Politics of Earth, 1920s and 1930s

2

War and Peace

Population, Territory, and Living Space

. . . a problem which is now menacing the peace of the world.

GEORGE HANDLEY KNIBBS,
THE SHADOW OF THE WORLD'S FUTURE (1928)

Imperial Germany entered World War I loudly proclaiming its need for *Raum*. German foreign policy was strongly influenced by the *Geopolitiker*, scholars who detailed the theory (and indeed the practice) that "vital" nation-states were not fixed in Westphalian agreement, but were necessarily and organically expansive.[1] In this system of thought about people, land, and political territory—this geopolitical worldview—population pressure was both metaphor and quasi-physical law. "Organic-biological *Weltanschauung* of the Geopolitikers," émigré Robert Strausz-Hupé explained, "is the urge to territorial expansion involving the revolutionary use of the population pressure of a growing nation."[2] At one level, the "pressure" argument employed by imperial German leaders, developed by second-generation *Geopolitiker* in Weimar Germany, and implemented as policy by the Nazis over the 1930s,[3] was a nationalist ruse for the acquisition of territory and power and widely understood to be so by the world's statesmen and scholars alike. But at another level, this formative idea of classical geopolitics had much in common with emerging Anglophone geography and demography. In general, they agreed with the geopolitical proposition that the world had become a closed system because of colonial expansion into all continents over the preceding three hundred years: without "empty land" to claim, colonizing nations struggled both with each other and with an emerging

Asian nationalism. This was what made the twentieth century a new epoch, involving a global struggle for power.[4] The proponents of geopolitics and territorial expansion, and the countering antinationalists, pacifists, and internationalists disagreed about what should be done, but they nonetheless employed the language of population and "living space." In short, they shared a nineteenth-century intellectual genealogy rooted in Malthus's "struggle for room and food."

Both the war and the 1919 peace were the key international contexts in which a revival of Malthusian ideas took place. And this was the context within which the League of Nations danced nervously around the population question, to some extent taking up the economic and geopolitical implications of world population growth, yet remaining anxious about its biopolitics: the League steered well clear of either birth control or pronatalist discussion. Other international and purportedly peace-making organizations of the 1920s and 1930s, however, faced population, geopolitics, war, and peace far more squarely, and not just in Europe. It was the strong entry of Japan, and therefore the entire Pacific, into first-tier international relations that unmistakably added another hemisphere to the complex whole, making the "struggle for room and food" global.

Geopolitics: The Struggle for Room and Food

Geopolitics as an idea originated with naturalists, geographers, and historians who were deeply engaged with ideas about the peopling of political territory. Lebensraum—living room—was theorized originally by Friedrich Ratzel, whose zoological training inclined him to consider populations of organisms and adaptation in limited spaces—the Franklin–Malthus idea within Darwin. For Ratzel, population and space were always interrelating factors, and he translated these ideas to human populations in *Anthropogeographie* (1882) and *Politische Geographie* (1897).[5] In these works, Ratzel began to write specifically about Germans needing new land and soil.

With direct reference to Malthus, Ratzel thought often about the limited geographies of islands that prompted either population-limiting practices or population-driven expansion. Both England and Japan greatly interested him in this respect. "While the population of a small country can spread beyond its boundaries as far as the habitable land extends, in islands all habitation ceases at their shores." He nominated the "early statistic maturity" that such geographic circumstances engendered as the condition for emigration, colonization, and commerce.[6] Early statistic maturity signaled a demographic stage, of sorts. When space is limited, attention is directed to the relation of area to population, he argued. "The question arises early, therefore, on islands and in other confined regions. They soon lead either to emigration—voluntary or compulsory—and colonization." This was Malthus's insight as well, and it was no accident, Ratzel wrote, that the *Essay on the Principle of Population* emerged from an island country. In making this claim, he was reading Malthus in the light of massive nineteenth-century population growth and the idea that emigration was an outlet for an oversupply of labor. Without a territorial outlet, Ratzel argued, confined countries could shift from a "wholesome" urge to expand to undesirable forms of population checks, "the evils of a redundant population . . . and especially the fundamental evil, the low value put upon human life." The British Isles were fine examples of the former, the depopulating islands of Polynesia and Melanesia were Ratzel's examples of the latter, where, he claimed, infanticide was rife.[7]

Ratzel's ideas entered the American turn-of-the-century debate largely through the work of his student, collaborator, and translator, the Chicago geographer Ellen Churchill Semple. She explained in a series of scholarly articles that were part translation, part commentary, part coauthorship, that states were organic, biological entities. The state, according to Ratzel, was a "living political organism."[8] Another student of Ratzel's, the Swede Rudolf Kjellén, published *The State as a Living Organism* in 1916, and it was he who coined the term "geopolitics."[9] Strausz-Hupé neatly explained to the next generation of Americans that geopolitics was the struggle for "national survival . . . waged for the redistribution of space."[10]

The state, then, was both territory and it was living.[11] But the politics of life were not just part of the operation or the implementation of geopolitics—expressed as population policies, for example. Rather, "life" was inside the very concept itself: space was biological, and population was spatial. This biopolitics within geopolitics was manifestly evident to its architects, though less so to subsequent historians and analysts. Contemporary critic Hans Weigert perceived that geopolitics "contemplates space and mankind as one inseparable unit."[12] What constituted the state was land and the people who literally grew from it. Thus lebensraum was not simply "living space"—an area to inhabit—but space that was, itself, living.[13] Correspondingly, *weltanschauung* was not just a "worldview," but a biological worldview. Geopolitics was, in short, a kind of human ecology.

This worldview mapped onto the literal globe; for Ratzel, geopolitics raised the relation between political and territorial parts in relation to a very large whole. "Just as the different races are members of the one human family, so countries are parts of this maximum political area . . . the special relation of every geographic phenomenon to the earth as a whole."[14] It was from German geopolitics, then, that one of the more powerful visions and images of a political–territorial globe circulated in the early twentieth century. A true world power, Ratzel considered, had to span the earth, and "since the size of the earth's surface sets limits to this development, the zenith can be reached by only a few states at the same time."[15] Geopolitics was one line of thought in which historical time, geological time, and global space were each comprehended politically.[16]

Such expositions emerged when German expansion was not just an idea but a reality. The new German nation had acquired a range of colonies from 1884: land and people in Africa (Tanganyika, German Southwest Africa, Togoland, and Kamerun) and in the Pacific (German New Guinea, Nauru, German Solomon Islands, Samoa, Bougainville Island, the Marshall Islands, and the Mariana Islands). Imperial Germany was an expansive, vital, political organism, in fact and in theory.[17] When, in 1914, this expansive ambition came up against the Austro-Hungarians and the Ottomans, the British, French, and Americans, it is not surprising that the outcome came to be called the first "world" war, a nomenclature not coincidentally offered by Darwinian "ecologist" Ernst Haeckel.

Notwithstanding obvious Anglophone opposition to an imperial Germany, many of the core principles of "geopolitics" were far from exclusive to the Ratzelian school. Ratzel's ideas were picked up within British geography by the influential Halford Mackinder, who imagined world history in the manner of the *Geopolitiker*, as successive open and closed systems. The closed world of medieval Christendom, he wrote, had become a bicontinental open world in the age of Columbus, between 1500 and 1900. In what Mackinder then nominated as the post-Columbian age (after 1900), a closed world had returned, but now on a global rather than a continental stage.[18] In the United States, Ellen Churchill Semple widely disseminated the geopolitical idea of a closed system.[19] So did the influential geographer, later president of Johns Hopkins University, Isaiah Bowman, whose work on global frontiers, density, and land use became key to both policy and scholarship on war, peace, and population distribution.[20]

Biographer Neil Smith has characterized Isaiah Bowman's own set of ideas as "American *Lebensraum*."[21] And well might he do so, because lebensraum as an idea derived in part from American politico-legal thinking on national territorial possession. Germany was hardly the only nation-empire struggling for room and food. The continent-level concepts and the "pan-region" ideas so central to geopolitical thought owed more than a little to the nineteenth-century Monroe Doctrine and the principle of manifest destiny; the idea that "national peoples" should and would (and did) claim continents.[22] Just when Ratzel was thinking about expanding political organisms, the American historian Frederick Jackson Turner comprehended the "closing of the frontier" as an American vital force playing out, one that energized and shaped U.S. institutions and its "expanding people."[23] The United States was in fact claiming not just a continent, but a hemisphere. While Germany was consolidating its new Pacific colonies in the period before World War I, the United States had grown offshore, annexing Hawaii, and after the Spanish-American War, controlling the Philippines, Guam, Puerto Rico, and temporarily Cuba.[24] Put another way, one might say that U.S. manifest destiny and the Monroe Doctrine together were a nineteenth-century version of lebensraum. Indeed, the ideas and policies were so interchangeable that when a Paris International Studies Conference took on the question of population and peace in 1937, "The Doctrine of Manifest Destiny" was not discussed with reference to the United States at all, but with respect to the German perception of a "natural right [to territorial expansion] of a people endowed with superabundant energies."[25]

Notwithstanding this history, U.S. commentators, alongside the British, tended to see "expansionism" as a German trait. By the outbreak of World War I, the Anglophone world had been well primed to think of geopolitics as substantially German. It was no longer possible for American scholars to present geopolitics in the admiring way Semple had done at the turn of the century. Prussian general von Bernhardi announced *Weltmacht oder Niedergang* (world power or downfall), and his tract *Germany and the Next War* (1911) was available in English in 1912. War was "a biological necessity," von Bernhardi claimed with reference to nature's "struggle."[26] The Washington publication in 1917 of *Conquest and Kultur* (incidentally by historian Wallace Notestein, cousin of Frank Wallace Notestein) brought German imperial geopolitics to a wide U.S. audience. The book was issued by the U.S. secretary of state and its committee on public information, and with seemingly unarguable veracity cited "the Germans' aims in their own words." Arthur Dix's geopolitical words in particular: "The first prerequisites for world power are extensive territory and population, together with a powerful tendency toward expansion. A world power needs extensive territory." Because the German population increased by around 800,000 inhabitants a year, he continued, "they need both room and nourishment for the surplus."[27]

Population pressure was a simple idea that could be mobilized popularly, as much by American leaders to justify entering the war, as their German counterparts to initiate it.

Demographically, the war, postwar influenza, and famine, especially in Russia and eastern Europe, manifested as increased mortality. As significant in demographic terms was the altered age composition of many national populations. Both of these factors compounded pre-existing fertility decline in France, in Britain, in the United States, and in Australasia, and certainly fostered renewed national anxieties and nationalist policies to turn this around. Anglophone Malthusians, however, welcomed and heralded fertility decline, not by any means, but by "civilized means." They tended to see World War I itself as one massive Malthusian check,[28] but there was nothing inevitable or necessary about such a "check," and certainly nothing desirable. The events over 1914 to 1918 were for them an object lesson in the relation between overpopulation and war. The corollary was that "optimum" population—secured through equitable distribution of people over land and through a universalized birth control—might secure peace.

Population and the Peace

The postwar world was ushered in by the Paris Peace Conference. And by its critics: the Germans, the Japanese, and John Maynard Keynes. Advisor to the British prime minister David Lloyd George, Keynes came to disagree sharply with the economic plans for peace and the reparations demanded of Germany. He squared off across the table against geographer Isaiah Bowman, advisor to American president Woodrow Wilson. Where Keynes saw economic consequences, Bowman saw geographical solutions, an opportunity for boundaries to be redrawn in a new political geography. Both thought in terms of population, food, and space, understanding that 1919 represented a new era, a new world.

It was Keynes's massively influential best seller, *Economic Consequences of the Peace* (1919), that brought a Malthusian argument to postwar readers by the millions. "The diminishing yield of Nature to man's effort" had become an actual problem for Europe from about 1900, he argued.[29] A diminishing agricultural return was already a reality, and its effects would be felt with increasing severity. "The earth heaves," he wrote after World War I. In Europe, "there it is not just a matter of extravagance of 'labour troubles'; but of life and death, of starvation and existence."[30] The war had shaken the existing economic system, endangering European life as well as its economy: "A great part of the Continent was sick and dying; its population was greatly in excess of the numbers for which a livelihood was available."[31] Keynes repeated his prewar message that the nineteenth century had been a one-off period of growth. And it had ended.[32] The temporary suspension of the law of diminishing returns brought by new world lands and new means of transport was nonrepeatable. Although the temporal "new world" was being pronounced, not least in Bowman's *New World: Problems in Political Geography* (1921), Keynes was still considering the geographical worlds, the ratio of exchange between Old World manufactured products and New World food and raw materials.[33] This situation, already grave for Europe in the new century warned Keynes, was not in the least helped by the ill-advised agreements and requirements of the Peace.

Keynes's work suggests the extent to which Europe figured as a focus of the world population problem in this economic tradition. Famously, economist and director of the London School of Economics William Beveridge, disagreed, but nonetheless also understood the discussion in terms of Europe. At his presidential address to the British Association in 1923, Beveridge tried to overturn Keynes's view that overpopulation threatened "the civilised world." He thought that Europe was perfectly able to sustain an increase: "Nature's response to human effort in agriculture, on each unit of soil and for each unit of total population in Europe, has increased, not diminished, up to the very eve of the War."[34] It was an uphill argument for Beveridge, however, because so much nationalist politics and policy was based on the premise that pressure on the soil was too great, and territory therefore needed to be enlarged.

A decade after the publication of *Economic Consequences of the Peace,* the political economist A. B. Wolfe still counted Keynes's book as pivotal: "So prominent a place given to the Malthusian spectre in the prologue of a book of this kind, so widely read at so psychological a moment, could not fail to bring home to thousands of readers the fact that the population problem is far more than an academic pastime."[35] There was indeed a significant re-entry of Malthus into Anglophone public discussion in the light of World War I and in the light of Keynes, both in popular and expert fora. A great rush of studies followed, every one of which considered population growth an issue of war, peace, and international security.

Taking his cue from Keynes, editor of the influential *Edinburgh Review,* Harold Cox, was the first to publish, with *The Problem of Population* in 1922. Educated in mathematics and political economy at Malthus's own Jesus

College, Cambridge, Cox had been a Liberal member of Parliament before the war. The population question, he announced, "affects the peace of the whole world." Placing himself within a long intellectual line of biologically literate political economists, he thought that Franklin's statement on fennel, repeated in Malthus's *Essay*, "embodies the whole essence of the matter."[36] Cox rehearsed the core ideas of lebensraum as well, suggesting that international relations were determined by healthy nations increasing in number, expanding their frontiers, and requiring new territory for their surplus populations. In what would turn out to be a surprisingly common response, Cox the Malthusian lifted the particular German circumstance to a common problem of human organization shaped by the shared conundrum of living space: "Their form of expression may perhaps be peculiarly German, but in fairness to these German writers it must be admitted that the truth which they so bluntly express is universal."[37] This is an extraordinary statement made in the shadow of the war and a keen measure of Cox's Malthusian persuasion, all the more so because he had served on the Bryce Committee on Alleged German Outrages in Belgium in 1915.

Also in 1922, Keynes commissioned Harold Wright to write *Population* for his Cambridge Economic Handbook series. Wright crafted a key chapter as "International Population Problems." The tragedy of recent history, he summarized, "is the trans-formation of the spirit of nationalism from one which seeks to unify and resist oppression into a jealous exaggeration of differences and a desire to oppress others."[38] Keynes contributed a sharp preface. The same year, a biologist weighed in. Oxford zoologist Alexander Carr-Saunders turned his attention to human populations, publishing the lengthy and weighty *Problem of Population*. He followed this with the popular précis, *Population* (1925), and a decade later, *World Population* (1936) was written for the Royal Institute for International Affairs. Like Cox and Wright, Carr-Saunders saw the crux of the matter to be population and international relations.[39] The book itself, he said, was an accident of the war, an outcome of his five years' active service, which gave him time to consider such an ambitious project.[40] Carr-Saunders's intellectual and institutional career brought together the political economy and natural history lineages of population thought from his early Oxford work on ecology to his directorship of the London School of Economics, where he was to succeed William Beveridge and oversee major demographic work. From that position, Carr-Saunders was to go on to direct any number of international initiatives, including the League of Nations's cautious forays into the economics of population in the late 1930s. In the early 1920s, though, Carr-Saunders was still a biologist, just beginning to contemplate a crossover into the human and social sciences.

In the United States as well, the war prompted a series of Malthusian studies by scientists communicating in both specialist and popular fora. Plant biologist Edward East published *Mankind at the Crossroads* in 1923 in the light of the new genetics as well as his wartime work with the U.S. Food Administration. Populations of humans needed to be considered within biology as much as economy: "His life, growth, and death are subject to natural laws. His fundamental instincts of preservation and perpetuation are common to the whole organic world."[41] The sharp young agricultural economist Oliver Baker produced a series of studies on population growth and land use that sprang from his work on soils with the U.S. Department of Agriculture. He had also found rich intellectual ground between the human and organic world, undertaking a political science degree at Columbia in 1905, followed by forestry studies at Yale and agricultural studies at the University of Wisconsin–Madison.[42] Sociologist Edward Alsworth Ross, one of the first Americans to develop university courses on population, published the popular *Standing Room Only?* Ross was still looking back to imperial Germany in 1928, citing von Bernhardi's classic geopolitical statement from *Germany and the Next War*: "Since almost every part of the globe is inhabited, new territory must, as a rule, be obtained at the cost of its possessors—that is to say by conquest, which thus becomes a law of necessity."[43] Ross did not disagree in principle. Rather, he disagreed with the means by which necessary "living space" might be acquired. For Malthusians, German expansion was regrettable, because it was belligerent. It was nonetheless an international relations object lesson on the dangers of overpopulation, lessons that extended beyond the specificities of Germany. Warren Thompson, now comfortably inhabiting his directorship of the Scripps-funded population institute at Miami University, Oxford, Ohio, wrapped up the decade with *Danger Spots in World Population* published in 1929, the entire premise and object of which was the prevention of another war.[44]

For some of these population experts, both economists and biologists, the connection between population and war was based on an economic proposition about agricultural subsistence. Population caused war, because it was about land, and it was about land, because it was about food. Harold Cox argued that scarcity prompted the search for new territory. The growth of population "not only creates occasion for war," he argued, "but makes war inevitable."[45] For others, the pressure of population, the "saturation" of national land, was the issue: "their over-flowing numbers threaten the world with war. Sooner or later the whole habitable globe will be full."[46] Always a cautious and careful scholar, Alexander Carr-Saunders thought that over-population was rarely a direct cause of conflict. He nonetheless argued that friction between nations was linked to population trends and changes: the need to secure markets overseas, which at base was about exchanging manufactured goods for food.[47] It was Carr-Saunders the zoologist who was most skeptical about the crude idea that war was somehow biologically determined: war was not biology, it was "a custom."[48] He was already more anthropologist than biologist, and well on his way to becoming one of Britain's key social scientists of population.

Carr-Saunders might have been right about many things, but his mid-1920s political forecasts were overly optimistic on several major counts. He confidently predicted that with respect to Japan and the east Asian mainland "there is happily no immediate likelihood of war." And of Europe: "No one imagines nowadays that war between nations long settled in their territories could end in the annexation of territory belonging to the conquered state, in the driving out of the previous inhabitants and in the settling there of members of the victorious state."[49] He could not have been more wrong.

Peace on Earth: George Knibbs and the Malthusians

What were later called the "demogenic causes of war" also invited "demogenic" solutions for peace.[50] Already strongly present in Malthusian politics before World War I, population, peace, and internationalism became a familiar triad after 1919. From versions of Wilsonian internationalism that retained the significance of national entities,[51] to full-fledged ideas of cosmopolitan world government and world citizenship that would do away with nations, population was one object of inquiry, one social, political, economic, biological, and spatial problem through which the twentieth-century "quest for one world" was pursued.[52] This quest was most alive when the world was most divided.

Australian statistician George Knibbs epitomized this generation's linking of population to pacifism and internationalism. "Economic equity, the abandonment of unscrupulous competitions, and the promotion of a world-concentration on the great issue [of population] is the way of peace."[53] This was a position he had honed before and during the war, and it culminated in his acclaimed 1928 book written the year before he died, *The Shadow of the*

World's Future, or the Earth's Population Possibilities and the Consequences of the Present Rate of Increase of the Earth's Inhabitants. Like those of Cox, Knibbs's arguments were antinationalist before they were internationalist, written in strong response to the belligerence of imperial Germany, what he called the "Clausius-Treitschke-Bernhardi doctrine of frightfulness." To avoid the "indescribable horror" brought by that species of nationalism, the international community and all intergovernmental bodies should cooperate to consider the interests of "the whole human race."[54] And yet he also considered that territorial aspirations based on population pressure might sometimes be legitimate: "Both Italy and Japan must find outlets for their people," he wrote to his fellow statistician, the Englishman Bernard Mallet, recently registrar-general.[55] Securing such "outlets" should not be forced, but with international cooperation, might be a peaceful process in itself, serving to ensure peace for the future. Knibbs argued that the necessary counter to imperial Germany's bid for world power was the normalization of the expectation that each national unit "should recognize its obligations to mankind as a totality,"[56] with particular implications for those who occupied "wastelands," as we shall see. Knibbs summarized his book as "an exposition of the consequences of the limited population-carrying capacity, under various conditions, of our earth."[57] Population, war, and peace were intricately connected for him, as for so many others, in a world comprehended as geopolitically closed and limited. While some sought systems of world governance and world federalism as a solution to such limits, for George Knibbs, the population question ultimately begged a revision not an abandonment of nationalism. As things stood, thought Knibbs in the late 1920s, the world picture looked grim. But "a new liberalism, and a less egoistic regard for the well-being of all races, is being called into existence."[58]

George Knibbs was trained in surveying, geodesy, and astronomy, and he brought an earth science dimension to the cosmopolitics of population. But it was as a statistician that he was most influential. Knibbs's initial Malthusian concern about population growth and war was published in 1917, a text in which his innovations in statistical method also became internationally noted. It was reviewed as "a veritable milestone in statistical theory . . . which for all times will stand as one of the leading works on mathematical statistics of the twentieth century."[59] This might be so, but Knibbs's *Mathematical Theory of Population* had its unlikely origin as the (450-page) appendix to the 1911 Australian Census Report. Knibbs, we might say, was the modern and antipodean embodiment of William Petty and his "Political Arithmetick," the first chief statistician of the new (1901) Commonwealth of Australia.

Charged with the responsibility to implement and analyze the nation's first census, George Knibbs brought political arithmetic to a whole new level. Straying well beyond the brief and genre of a census report, Knibbs left the national(ist) project behind in the wake of a catastrophic world warning:

> The limits of human expansion are much nearer than population opinion imagines; the difficulty of future food supplies will soon be of the gravest character; the exhaustion of sources of energy necessary for any notable increase of population or advance in the standards of living, or both combined is perilously near. Within periods of time, insignificant compared with geologic ages, the multiplying force of living things, man included, must received a tremendous check.

He was quite specific: "only 450 years to exhaust the food required." Schooled in the "geologic ages," for Knibbs the future of humans on Earth was reduced to the population growth "of the last ten decades" and the effect of this "within the next ten."[60] *Mathematical Theory of Population* became widely known and cited in any number of official and expert domains. It was sought by European government statisticians and U.S. officials in Washington, D.C., and was extensively quoted in South African Census Reports, for example.[61] Importantly, the Czech probability theorist Emanuel Czuber translated and repackaged Knibbs's work into a widely cited German edition.[62]

George Knibbs's geodesy, combined with his mathematical, insurance, and census work as a government statistician,[63] gave what he called the "New Malthusianism" a particular flavor; it added expert knowledge to, rather than grated against, the long-standing global and spatial dimension of Malthusian thought. His work was deeply political, representing antinationalist and pacifist expositions of the political economy of life, death, food, and space on planet Earth. Knibbs moved from the application of statistics to the complex of human biological and social life at a national scale, to the problem of human reproduction at a global scale. He thought the planet was time-relative: it had both sped up and had shrunk. The conditions of modern life functioned to connect previously disconnected peoples: "The whole plexus of relations which modern transport and economics, and the intricacies of trade and commerce have established, have really welded the peoples of the earth into a kind of pseudo-solidarity."[64] Some argued that this led to competition. Knibbs's prediction was that greater populations would require of necessity a more unified, economically globalized world. The problems posed by accelerating world population growth produced an interconnectedness that if managed well, might turn into a new kind of cosmopolitan federation and even unity.

In the 1920s, international neo-Malthusians eagerly claimed Knibbs as an elder statesman. In India, *Sir George Knibbs on the Menace of Increasing Population* was distributed by the Madras Neo-Malthusian League. In Britain, C. V. Drysdale sought his contribution to the fifth International Neo-Malthusian and Birth Control Conference in London, at which H. G. Wells and his wife, Amy Catherine Robbins, were to receive and entertain the foreign delegates; Maynard Keynes was to chair the economic and statistical section; and Harold Cox the "international" section. And if that was not enough to encourage Knibbs to steam across the globe to talk about the limits of the planet, on the Saturday of the conference there was promised "an automobile excursion to Dorking, Surrey, in order to visit the birthplace of the Rev. T. R. Malthus."[65]

For all of them, World War I had tragically vindicated the Malthusian position. And it had raised the stakes for the pre-existing Malthusian claims to prevent war and usher in peace. Overpopulation was "now menacing the peace of the world," as Knibbs put it.[66] Drysdale spoke of Malthusianism as necessary to secure a "universal brotherhood of man" that would bring peace on Earth and goodwill to men. Accordingly, the International Neo-Malthusians developed the formal aim: "To remove the international rivalries caused by the pressure of overpopulation, and thus give opportunity for the establishment of international law leading to federation and permanent peace."[67] The reduction of population growth and the redistribution of populations across the globe were to be more or less constantly tied to bids for peace and international security over the next decades. Carlos Paton Blacker, British eugenics leader, psychiatrist, and Oxford student of Carr-Saunders (and Julian Huxley), asked: "From the last war no lesson as to the importance of population control has been learnt. Will another war be necessary to teach us this lesson?"[68] By 1926, when Blacker posed the question, fascism was already on the rise.

Fascism, Population, Land

In Italy, land had been a central policy problem for Benito Mussolini since 1922, captured in the fascist dictum: "redeem the land, with the land the people, and with the people the race."[69] Mussolini established thousands of new farms and agricultural towns on reclaimed land, as part of the so-called battle for land and battle for wheat. What Carl Ipsen has labeled the "realization of totalitarian demography" involved both forced and voluntary internal migration to relieve population pressure and produce more food.[70] Even before fascist rule in Italy, Eritrea had been colonized with this kind of demographic justification, to keep emigrants closer to hand (closer than the United States) as well as keeping them under the law and dominion of Italy. They could be counted, not lost, as citizens. Under Mussolini, the state pursued a policy of encouraging the demographic colonization of Libya and invaded Ethiopia in 1935, announcing in 1936 an Italian Empire, the modern rendition of the Roman Empire.[71]

In his explanatory pamphlet, *Living Space and Population Problems*, the demographer Robert Kuczynski summarized Italian fascist logic: "Our country is over-populated. In the past emigration has relieved population pressure, but we need a bigger population within our State boundaries in order to increase our political power in Europe and in the world. We must therefore expand our population and at the same time expand our territory."[72] Notwithstanding the great international controversies over Italian colonization, and even Mussolini's volte face on Malthus, Anglophone Malthusians were on record endorsing the Italian need for more cultivable territory and doing so remarkably often. George Knibbs thought it imperative that Italy find "outlets."[73] Isaiah Bowman presented the Italian case sympathetically, even in the late 1930s, after the Ethiopian invasion had caused such an international controversy. Italy, he wrote, was "overpopulated at home and on human grounds absolutely required a place in the underdeveloped territories of Africa."[74] And yet, the final message was that population pressure urgently required peaceful rather than belligerent solutions; the situation demanded "normal outlets," not violent expansion.

Meanwhile, from the beginning, German fascists nurtured ambitions to expand both population and land, fully incorporating early geopolitical ideas. After the Versailles Treaty and the Allies' reallocation of German colonies to various powers, including Japan, the doctrine of expansion turned to contiguous land to the European east.[75] The third point of the Twenty-Five Point Program issued by the Nazi Party on February 25, 1920, read thus: "We demand land and territory (colonies) for the nourishment of our people and for settling our superfluous population."[76] Adolf Hitler, needless to say, was strongly engaged with the theorization of political territory and the question of the peopling of space. "The foreign policy of the folkish state," he wrote in *Mein Kampf*, "must safeguard the existence on this planet of the race embodied in the state, by creating a healthy, viable natural relation between the nations' population and growth on the one hand and the quantity and quality of its soil on the other hand."[77] This was the Ratzelian idea that Semple had earlier summarized for Americans: "The state is an organism, indissolubly connected with the earth's surface, deeply rooted in the soil."[78] National socialism sought to eliminate a disproportion between population and area, and as Hitler put it, space, or *Raum*, was understood not just as necessary for food production, but for power.[79] Once in power, the vast Nazi programs for domestic land reclamation, the draining of marshes, and the resettling of ethnic Germans from the east, were constantly underscored by ideas about population, density, and *Raum*.[80]

The idea that people were held (back) in a constrained space held great currency in Germany. The popularity of Hans Grimm's *Volk ohne Raum* (1926)—"people without room"—captured the sentiment through the Weimar period.[81] By that point, Kjellén and Ratzel's ideas had been developed significantly by Karl Haushofer, who taught geography, geopolitics, and military science at Munich's Ludwig Maximilian University, where he established *Zeitschrift für Geopolitik*. Haushofer was explicit about population density forming the core rationale for the need for *Raum*. "The inhabitants per QKM, is most important," he wrote. "It is the necessary basis of almost all geopolitics."[82] Population density led naturally—biologically Haushofer would also have it—to struggle between peoples: "Man has a continuous fight in gaining, preserving, changing and newly distributing 'living space' and power."[83] It was from Haushofer, via Rudolf Hess, that Adolf Hitler derived his own understanding of lebensraum detailed in *Mein Kampf* in 1925 and 1926, although Haushofer always denied direct influence.[84] In that context, Hitler translated the density question into a particular post-Versailles German ressentiment: not least against British and U.S. colonialism. In some nations, Hitler wrote (in this instance meaning the United States with its recent acquisitions of Puerto Rico, Alaska, and Hawaii) there are "not even 15 inhabitants to the square kilometre, while other nations are forced to maintain 140, 150, or even 200 in the same area. But in no case should these fortunate nations further curtail the living-space of those peoples who are already suffering, by robbing them, for example, of their colonies."[85] The "haves and have-nots" mode of analysis that became such a standard part of interwar international debate was argued constantly through and over the question of population density.

The politico-spatial dimension of population was unsurprisingly put by fascists to the various international meetings on population in those decades. The leading Italian statistician, Corrado Gini, argued that population density spreads "the thought of the nation beyond its national frontiers."[86] He complained about the constant assertions of the "limits of the Earth," not because it was invalid, but because such limits were not to trump the needs of a truly vital state. If internationalist Malthusians sought the containment of expansive nations, geopolitical nationalists vigorously pursued a right to land. Gini challenged what he comprehended to be the weak and limiting powers of internationalism itself: cooperation or global unity, the language of international Malthusians, was not part of his political vocabulary. As Gini put it in "The Scientific Basis of Fascism," for Italians, "the concept of organic unity" is perceived in and as the nation, not "larger human society" as liberal internationalists like Knibbs would have it. Gini was especially trenchant about an international polity that the League of Nations purportedly represented. That institution, he considered, "place[s] limits upon the free action of the national organism."[87] The Indian demographer and sociologist Benoy Sarkar was another commentator deeply skeptical of internationalism, labeling it "the great fetish." He defied all internationalists and asserted the "bankruptcy of internationalism as a cult." Sarkar referred to Haushofer, whose "geopolitics teaches the world to remain awake to the one great reality of life, namely, that it is nothing but nationalism that rules mankind and that the eternal problem of today is . . . to study the science and art of *Macht,* i.e. *shakti* or power."[88] All the so-called "international" endeavors were, he thought, limited by this incapacity to transcend the determinations of space. They were permanently *Raumgebunden* (space-bounded).

When Corrado Gini told his colleagues that population density made energetic nationalists think "beyond current frontiers," he was offering both a population theory and a foreign policy of the Italian fascists.[89] More revealing, though, was the fact that New York University's professor of sociology Henry Pratt Fairchild agreed with him: nations would thus be "reestablishing the equilibrium on the basis of a larger circle."[90] The specific "population pressure" argument that drove the lebensraum idea, was neither an invention of nor limited to the German imperial context, or to Italian or German fascism of the 1920s and 1930s. To some considerable extent, lebensraum was not exceptional conceptually, but was part of a shared set of ideas about population, space, and density. Whatever Italian or German (or Japanese) statesmen actually thought about demographic rationales for territorial expansion, Anglophone statisticians and other population experts tended to engage with such claims on their own terms, even into the late 1930s, when events made such claims far from theoretical.

Population and Pacific Relations

The German idea of *Weltmacht oder Niedergang*—world power or downfall—certainly propelled one European vision of global population and war. Events in east Asia rendered the population question global in more substantial terms. The expansion of the Japanese Empire began in the Meiji era, especially from 1875, when islands to the north and south were incorporated. After the Sino-Japanese War in 1895, this extended to the east Asian mainland. With victory over Russia in 1905, the southern half of what was then Saghalien was ceded to the Japanese, Korea became a protectorate, and the Chinese province on Kwantung was leased to Japan.[91] The Paris Peace Conference granted Japan a mandate over the former German islands north of the equator—the Caroline Islands, the Marshall Islands, and the Marianas. The United States also had a clear geopolitical stake in the Pacific, and acted on it. The Washington Conference of 1921 and 1922, shifted local agreements, involving all the Western powers with interests in east Asia, as well as China and Japan.[92] This marked a key turning point toward the prominence of the Pacific in global politics, driven by the desire of the United States to restrain Japan in the region.

Population was everywhere in the talk around these well-known events, prompted not least by Japanese framing of east Asian and Pacific tensions. One Japanese delegate explained to the Institute of Pacific Relations in 1927 that the "energies of the whole nation" were being called forth by new national and global circumstances. "The pressure of population crying for a new orientation of policy, the irresistible force of democracy challenging the old order . . . the changing psychology of the nation after sixty years of occidentalization and the consciousness of the dawn of a new era—the great era of the Pacific."[93] "Energies," in this context, was not a neutral word: the geopolitical links—intellectual and political—between Germany and Japan were strong.

Karl Haushofer had almost single-handedly offered this lexicon to the Japanese. He had conducted his early and formative research work in Japan from 1908 to 1910, liaising between the German Imperial Army, of which he was an officer, and the Japanese Army. His doctoral dissertation and no fewer than four published studies dealt directly with Japan.[94] It was an interest and knowledge that drove his important *Geopolitik des Pazifischen Ozeans* (*Geopolitics of the Pacific Ocean*) originally published in 1925, revised and updated for the Nazi regime in 1936 and 1938. This book was translated and published in Japan, energizing groups like the Japanese Geopolitics Society and the Pacific Society, which sponsored translations of Haushofer's works and Japanese commentary on them.[95] Haushofer's reliance on an idea of biological struggle was widely upheld,[96] and his assessment of the process of Japanese imperial expansion often cited.[97] According to Kjellén, the term "geopolitics" was first used in Japan in 1925.[98] Thereafter, it strongly shaped the "occidentalizing" nation's foreign policy and colonial practice, with population pressure arguments strongly justifying territorial expansion. This involved increasing identification, and ultimately of course alignment, with Germany.

In his 1934 article, the "Fascist Tendencies in Japan," German economist and sociologist Emil Lederer wrote that "probably nothing in a nations' consciousness is more disturbing than the knowledge that population is growing quickly in a small country."[99] A familiar idea after Malthus. And yet, in the Japanese case, the much-pronounced significance of population density that was apparently pressing people outward territorially bore little relation to what was happening on the ground. Very few Japanese in fact migrated anywhere, and by 1928, there were only 203,000 Japanese in Manchuria, the territory that was supposedly the great pressure valve.[100] Nonetheless, when Japan invaded Manchuria in 1931, commentators on population, geopolitics, and international relations immediately proclaimed evidence of "pressure on land."[101] This was the beginning of the Asia-Pacific War of 1931 to 1945. Many who had thought for some time that it was over the ocean named for peace that population and war was especially likely were more or less repeating the official Japanese position. The Manchurian invasion was regularly justified by Japanese statements about geopolitics and "living space." Kaku Mori, secretary to the government cabinet, claimed "Japan's 70,000,000 people are penned in a group of narrow islands, not blessed with natural resources. The vital factor in the existence of a civilized nation is that its people should be able to turn their energies to the increase of national wealth and strength."[102] The League of Nations condemned the 1931 military action, prompting the withdrawal of Japan from the League in 1933. When this invasion escalated to the Sino-Japanese War in July 1937, neo-Malthusians pronounced a tragic vindication: overpopulation had led to war again.[103] Warren Thompson proclaimed the east Asian circumstance as evidence of the impact of population pressure and the need for population management and control measures to ensure that the situation was not compounded. For Thompson, managed geopolitical expansion would secure the world against just this type of aggressive action.[104]

This all made the need for good demographic work on Japan pressing. However, there was no consensus on population trends. One set of data indicated a total population growth of 78 percent between 1872 and 1933 (still slower than that of England and Wales, it was noted).[105] However rapid Japan's population growth was estimated to be, this fact was not perceived to be significant in and of itself, but only in relation to Japan's insular and limited geography. Harold Cox, for example, argued in 1922 that although Japan had a lower birthrate than India, population growth was more serious, because of the density produced by its small area. This, he considered, explained Japanese statesmen's eagerness to find territorial outlets. He was notably sympathetic to the Japanese position: "Western Powers who have built up their dominions by similar methods cannot on moral grounds protest against this Japanese expansion."[106] Edward East had warned in 1923 that Japan was "overpopulated," but he too expressed some sympathy with the move into Manchuria: what else would one reasonably expect of a hungry polity?[107] For some commentators, Japanese militarism was about a specific belligerent national danger. For others, it was about the generic relationship between overpopulation and war, playing out in the east Asian case. "Why has Japan broken out and attacked China?" asked the editor of *The New Generation*, the new name for *The Malthusian*. "Because of her sufferings from overpopulation." And significantly the Malthusian solution linked the management of sex and space: "The only way to bring such doings to an end is universal birth control coupled with a reasonable division of the lands of the world."[108] Demographer Grzegorz Frumkin thought that Japan paralleled industrializing Britain, an increasingly common comparison. But whereas Britain had been able to relieve its pressure through mass emigration, this was not possible for the east Asian island. It "suffocates within its narrow frontiers."[109] Likewise, Warren Thompson insisted that Japan's colonial expansion should be regarded as just as legitimate as Anglo-Saxon demographic expansion.[110] In Anglophone literature and politics, this was occasionally put as a special Japanese "Right to Live," an argument that economic necessity might even justify a violation of existing international law.[111] The connection between overpopulation and war explicitly drove foreign policies, however spuriously related they were to the actual impact of population growth. Anglophone scholars largely bought into this premise, even as alternative responses were sought.

One outcome of the tense new Pacific focus and circumstance was the Institute of Pacific Relations. Established in 1925 to improve "mutual relations" between peoples of the Pacific, it derived from Woodrow Wilson's

Versailles position and carried forward a particularly American vision of liberal internationalism. It was an unofficial institute strongly supported by the Rockefeller and Carnegie foundations, one that engendered international conversation without expectation of consensus. At least thus ran its formal brief. In a new age that had "reduced time and space as factors in international relations," the Pacific needed to become a bridge not a barrier, the institute officially announced. A founding principle was to engage key players in unofficial roundtable discussions based on a new vision of international interdependence that included recognition of "the rights of the weak."[112] Its initial conferences were held in Honolulu in 1925 and 1927, a third in Kyoto in 1929, a fourth in Hangzhou and Shanghai in 1931, after which Japanese delegations withdrew: roundtable discussions of mutual relations between China and Japan proved impossible.[113]

From the beginning, population was a central issue, explicitly discussed in terms of national and international "struggle for room and food." This was understood to have consequences for international relations of the highest order:

> What will be the political, economic, cultural, consequences of the differential rates of population growth among nations bordering the Pacific? What is the most reasonable way of straightening out the difficult situations arising from the above conditions? . . . What elements in the food and population problems of the Pacific are likely to give rise to international difficulties? In what forms are these difficulties likely to take shape in conflicting national policies?[114]

That Japan's land had come up against its physical limits, and almost to its economic limit was up for discussion. For some, this was a fact for which solutions should be found. For others, the proposition itself was debatable. An American expert was reported saying: "This co-existence of unequal economic standards of living in a world which is practically an economic entity is the outstanding international problem of the world as well as of the Pacific. . . . [W]e can hope for peace and fair play among the nations, when, but hardly before, they have each reached a fairly high standard of living."[115] This may well have been John D. Rockefeller III, who attended the Institute of Pacific Relations as "secretary" for the United States group. Deeply interested in Japan, Rockefeller was to become a key player in shaping the global population agenda after 1945. After the end of the Pacific war, Japan, its land, and its people, were in completely different international circumstances.[116]

Unsurprisingly, how and whether limits could be territorially extended was presented strongly by the Japanese delegation at the Institute of Pacific Relations meetings. Manchuria was everything. One Japanese delegate announced that all territorial frontiers were up for question and Western insistence on current national boundaries was an entirely spurious position: "The present national boundaries . . . are mostly the result of occupation or conquest. They have been changing throughout the centuries. Nobody can say that the Versailles Treaty is the last word spoken by God."[117] There were no Germans involved in the Institute of Pacific Relations, but such statements betrayed an aligned German and Japanese *ressentiment*, produced by and directed against the Versailles Treaty and the League of Nations: although there as victors, Japanese proposals for the League to overturn doctrines of racial inequality had been roundly rejected.

The Institute of Pacific Relations heard Professor Royama Masamichi of Tokyo Imperial University justify Japan's presence in Manchuria in an extensive paper that drew in part on Karl Haushofer's ideas. But Haushofer himself was on record as saying that Japan should not be participating in the institute at all. The various pan-Pacific unions, associations, and institutes were neocolonial enterprises cloaked as Wilsonian-inspired internationalism. He thought them simply clever incorporations and diffusions of the new Japanese "living force" that should instead be mobilized to different ends. Japan should oppose all such U.S. neocolonial initiatives and turn instead to Germany, who would defend Japan's bid for Pacific lebensraum against both European colonialism and the American "Monroe Doktrin."[118] Haushofer presented such views on a striking map that depicted the United States itself pushing farther and farther westward across the Pacific (figure 2.1). "*Gegen Genf*"—against Geneva—was the political dynamic that for Haushofer, and for Japanese delegates at the Institute of Pacific Relations meetings, linked Germany and Japan. This was the 1930s context that framed Anglophone work on population and war.

Peaceful Change: The League of Nations and "Demographic Tension"

The Institute of Pacific Relations was hopelessly caught between internationalism and nationalism, and in this sense it was the Pacific shadow of the Geneva-based League of Nations. Unsurprisingly, personnel of this generation of internationalists crossed over both bodies. While unofficial national

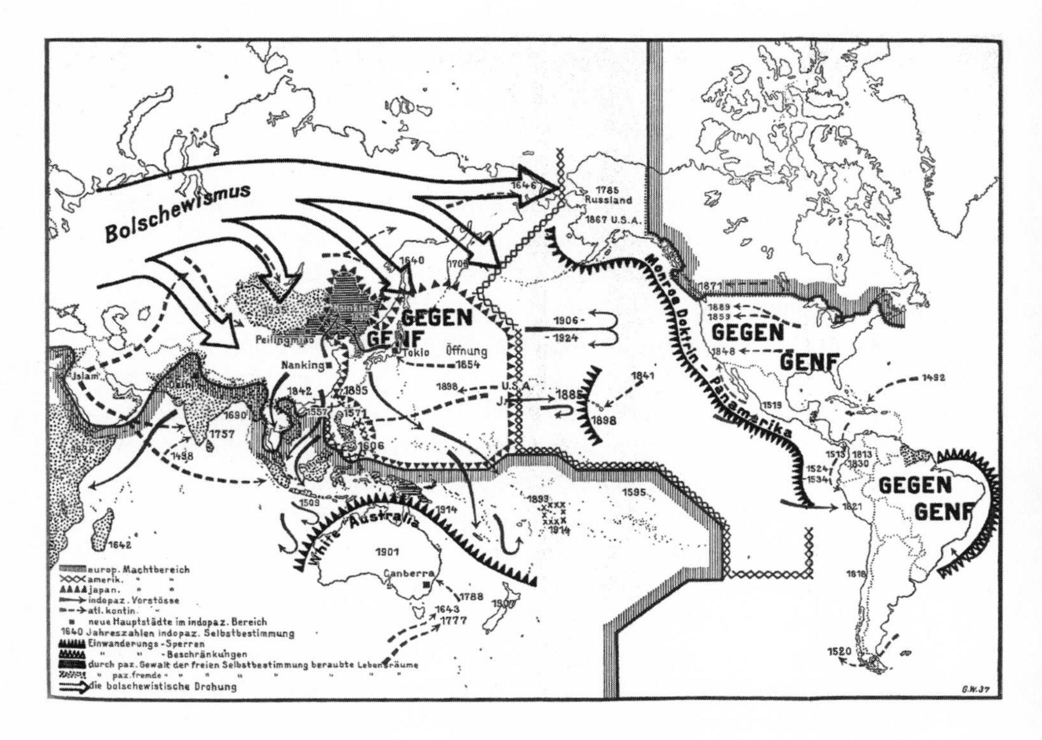

FIGURE 2.1 "Against Geneva." Japanese and German identification in the Pacific context, aligned against the American "Monroe Doktrin." ("Karte der politischen Raumverteilung und Selbstbestimmung im Pazifik," in Karl Haushofer, *Geopolitik des Pazifischen Ozeans*, 3rd ed. [1924; Heidelberg: Vowinckel, 1938], 95.)

delegations formed the core of the Institute of Pacific Relations meetings, international observing delegates were also present, including Louis Varlez from the International Labour Office (ILO) and H. R. Cummings, who represented the League of Nations secretariat. Agricultural economist Nitobe Inazō chaired the Japanese delegations to the Institute of Pacific Relations meetings, having recently served as under-secretary-general of the League of Nations. He was an important interventionist commentator on Japan's place in global perception of, and law on, racial equality. A converted Quaker, Nitobe brought multiple traditions of pacifism to his international work; his internationalism found occasional expression in Esperanto, in his work with the Carnegie Foundation for International Peace, and within the League itself.[119] Like George Knibbs, he was appalled most by German nationalist justifications for war, and by accompanying German race theories.

"I hail all scientific researches," he noted during his time at the League, "but I am doubtful of their hasty application to social politics as was done by 'Politische-Anthropologische Revue' set."[120]

The League of Nations itself had a most ambivalent relation to the world population problem. Any reproductive rights or health-based arguments for birth control were forcefully excluded, under pressure from Catholic national delegations. At the same time, because the League was established to secure the Peace, it is perhaps unsurprising that the secretariat was pressed again and again to consider the population question formally by any number of international groups, not least the International Neo-Malthusians. It was in terms of war and peace that most proponents pitched birth control to the League. At the 1922 International Malthusian conference in London, Baroness Ishimoto spoke as president of the new Birth Control League of Japan, arguing for contraception "as a sure preventive of war, and a certain cure for poverty and starvation."[121] And when Margaret Sanger wrote internationalism into the American Birth Control League's manifesto, her eyes were fixed firmly on Geneva. She urged on "all international bodies organized to promote world peace, the consideration of these aspects of international amity."[122] There was no need to elaborate: by that point, the connection between population control and world peace was perfectly well understood. As the Czech anthropologist and statistician František Jaroslav Netusil stated categorically, the whole point of discussing world population was to move toward international peace. "Otherwise it would have no meaning."[123]

It is a measure of the difficulty of the whole question for the League that it was not until the late 1930s that population growth came to be acknowledged, and it was entered as a problem of density. At the 1937 assembly, the Polish delegate spoke about the need to return to pre-1914 "freedom of movement" as a way to equalize differential population densities that were causing war. This prompted a successful proposal that a League committee "study demographic problems in their international aspects." Wary after German, Italian, and Japanese territorial expansion, the assembly indicated "solutions other than territorial solutions were to be preferred."[124] This was an early expression of the difficulty that geopolitical "solutions" to overpopulation posed. And, although the Catholic-sensitive assembly itself could never say so explicitly, there was a hint there that "other solutions" might just be reproductive.

The relevant point for the moment is that population was not, in the first instance, comprehended as an international health or reproductive question. It was an economic question. The Committee on Demographic Problems was delegated not to the Health Section of the League, but to Alexander Loveday's Economic and Finance Section. It was chaired by his countryman Alexander Carr-Saunders, and the Demographic Committee barely registered birth, sex, or health: it was all about land. How might the condition of agricultural populations be improved? How might new land be found "with a view to settling surplus peasants?" What were the possibilities of emigration "to reduce the effects of demographic tension?" And was industrialization an alternative remedy for "demographic tension?"[125] Informing and surrounding this committee was work on population done not by the League's secretariat, but by its agencies. The ILO, established in 1919, dealt constantly with population data, and as we shall see, the question of global migration. The International Institute of Intellectual Co-operation [IIIC] also extensively discussed world population as a key element of its examination of "international tension." Indeed, international tension and demographic tension were coming to be regarded in such circles as the same thing.

The IIIC was the Paris-based agency established as advisory to the League in 1926, charged with promoting cultural exchange among intellectuals: scientists, researchers, teachers, and artists. It was the forerunner to the United Nations Education, Scientific, and Cultural Organization, and its founding director was Nitobe Inazō. In 1937, a large meeting of population and international relations experts was held in Paris in response to the call for "peaceful change." This was an offshoot of mid-1930s international studies discussions on the prevention of war, a recognition that the idea of collective security had failed. The aim was to gather new intellectual energies toward the settlement of international disputes without violence, disputes understood to have origins in demographic circumstances. Significantly titled, the 1937 Peaceful Change meeting brought 150 scholars together to discuss a question for which a neologism was created: "What, in a word, is the nature of the problem of international 'decrowding?'"[126] Core business was security, peace, and population, an investigation of the increasingly common arguments of various national governments (Italy, Germany, Japan) that "demographic congestion . . . be relieved by distributing this excess population over underpopulated territories." The constituent elements of the population problem at this meeting were the politics of earth more than the politics of life: the peaceful solution for economic and social problems of population was not a diminished birthrate, but the redistribution of people—migration, colonization, population transfers—and perhaps the redistribution of raw materials.

The meeting on peaceful change early decided to differentiate its work from Knibbs's unified global perspective. Referring directly to him, the contributors were not to be concerned with overpopulation "in the world as a whole." True to its League roots and imperatives, the IIIC and the Peaceful Change meeting was far more international than global or planetary. Its brief was in specific relation to countries that claimed changes in the status quo on the overpopulation/living space argument. The Paris meeting of experts heard from and about the "dissatisfied countries"—Italy, Japan, Germany, as well as Poland. It studied the basis of population claims, looking at overpopulation, the doctrine of manifest destiny, and what it called "demographic imperialism." It examined the notion of overpopulation itself and of demographic remedies, which were organized as "emigration," "colonial expansion," and "economic remedies." Its final meetings on "international planning" featured the prospect of the international organization of migration.

The meeting looked forensically at the nature of the basis of the claim that population pressure caused war. Was this a social or even biological fact? Was it sometimes so, but questionable in the particular instances under discussion? Or was it an effective pretext, effective precisely because of the long-standing expertise that authorized it? The very concepts of overpopulation and underpopulation were argued. Given the national claims (and threats) then being made in the international arena on the basis of "overpopulation," clarity over both the concept and the claims themselves was key. Some thought "felt population pressure" was important as an objective economic measure of overpopulation. And indeed, a German participant, Heinrich Rogge, asserted, "We [Germans] understand the concept of 'overpopulation' in the sense of a vital lack implying the co-existence of subjective and objective needs."[127] The League of Nations Economic Committee accepted the hypothesis that overpopulation was at the basis of territorial claims, with reference to Haushofer's "autarky" idea.[128]

A Japanese delegate focused the meeting's attention: "The sole difference between Japan and Western Europe is that, in the case of Europe, the population increased at a time when emigration and foreign commerce were free, likewise colonization, whereas Japan no longer benefits by such conditions."[129] This was essentially the argument that Hitler used vis-à-vis the British Empire and U.S. expansion. Unsurprisingly, the clearest responses to German and Italian arguments came from French demographers. Economist and socialist politician Adolphe Landry, author of *La révolution démographique* (1934) was having none of it, with his eyes fixed on the German delegates; lebensraum was simply a ruse for expansion.[130] Yet even among the French, there were concessions to the power of the idea of overpopulation. André Touzet in *Le Problème colonial et la Paix du monde* wrote that, rightly or wrongly, the more populous nations think themselves over-crowded. "These consider, more or less sincerely, that they are encircled, as if subjected to blockade. And claim that their existence is threatened; hence an extremely dangerous political tension."[131] The consensus of the international meeting in 1937 in fact allowed for the demographic causes of war.[132]

Population was a territorial and geopolitical matter on the eve of World War II, just as it had been on the eve of World War I. One effect of the Peaceful Change meeting was to focus and collate the massive amount of scholarship that had been produced in the 1920s and 1930s on the population question. It prompted its own fresh set of assessments. Carr-Saunders's volume *World Population: Past Growth and Present Trends* (1936) was published under the auspices of the Royal Institute of International Affairs for the Conference.[133] Isaiah Bowman's important and much-cited collection *Limits of Land Settlement* was submitted by the New York–based Council on Foreign Relations. And the Institute of Pacific Relations submitted *The Problem of Peaceful Change in the Pacific Area.*[134] From Hitler's *Mein Kampf* to Landry's *La notion de surpeuplement,* to Grover Clark's *Balance Sheets of Imperialism*, the Peaceful Change meeting, held on the cusp of war, represented the extent to which the population problem was still comprehended as one of international security.

One effect of all this talk was to highlight how minimally the League had involved itself. Imre Ferenczi, in charge of population and migration data and policy at the ILO, was briefed to write one of the meeting's synthesis documents on population and war, and considered that the absence of an international policy on population prevented the League from fulfilling its principal mission—the prevention of conflicts.[135] Well might he have wondered about the League's mission, since in the Paris summer of 1937 when this large group of intellectuals gathered to present their accumulated knowledge on "population and peace," war was under way in east Asia between Japan and China, the Spanish were in civil war, and the Nazis were poised to create "Greater Germany" and were already in pact with Japan and Italy. Population and territory had long been linked in geopolitical theory. Both the war and the peace engendered a new generation of global population thought framed by economics, land, and resources, as peace was secured and then undone, as territorial boundaries were redrawn and people transferred. Events in 1919 and the two decades that followed saw boundaries change, peace and war come and go. The land–food–population nexus was raised again and again in relation to international security and conflict in the era after Wilsonian liberal internationalism.

3

Density

Universes with Definite Limits

All populations of real organisms live in universes with definite limits. The absolute size of the universe may be small, as in the case of the test-tube . . . or it may be as large as the earth.

RAYMOND PEARL, "THE BIOLOGY OF POPULATION GROWTH" (1927)

The first World Population Conference gathered in the high summer of 1927, right in the middle of the decades that separated one world war from the other. Provocatively, to say the least, it was located in Geneva. Socialist conscience of the meeting, Albert Thomas, reminded those assembled just what was a stake: "the question is one of peace or war."[1] And yet this remarkable conference did not start with war or peace, or even, for that matter, with humans. It began with fruit flies. Johns Hopkins biologist Raymond Pearl gave the opening paper, "The Biology of Population Growth," a précis of his recently published work on *Drosophila,* their fertility and morbidity in contained spaces, and their biological response to density.[2] Pearl specialized in test tube–scale universes, though what this taught him he considered applicable to all life, in all spaces, in all times. In theory (though not with respect to his particular theory), many of his colleagues agreed. If geopolitics created a new "biologization of global space,"[3] biology was equally spatial.

The World Population Conference has been most noted for the fact that birth control was formally off its agenda. Yet the very absence of birth control discussion also makes plain the other registers in which population was problematized: geographical, geopolitical, and economic. World population deserved its own high-level conference not just because of an acceleration in the rate of growth or localized declines in the birthrate, but also because of growing concern about population density. Geographers, biologists, and economists all had a specialist investment in population adjustments in relation to space, which was sometimes considered simply as area, sometimes as land with different capacity to yield, sometimes in complex, multifactorial ecological terms.[4] They engaged both separately and collaboratively in the project of calculating the best ratio of people to land, as *Drosophila* to test tube. This constituted an "optimum population," although even the most laboratory-bound biologist well knew that determining an optimal population for human organisms was necessarily deeply political. It is no wonder that even after birth control was ruled out of the meeting, there was so much else to say.

Margaret Sanger's World: The Geneva Conference, 1927

Before World War I, Malthusian Charles Drysdale displayed his internationalist credentials by publishing in Esperanto. After the war, it was Margaret Sanger who spoke a much more effective international language. In plain English and French, she lobbied, politicked, and argued her way into a prime bicontinental neo-Malthusian position. Phenomenally energetic and diplomatically effective, Sanger occupied more of a middle ground in the 1920s than in her early years of socialist activity. She was entirely happy to liaise with and benefit from the influential and the monied, yet was always ready to put herself on the line in pursuit of birth control and women's health and to internationalize both agendas. Any way one approaches the population problem, Margaret Sanger was a key and remarkable figure.

Sanger's work as a nurse in New York City is often recounted as her training ground. She initiated this narrative, autobiographically casting herself as "nurse," and thus accruing a useful suite of traits: caring, feminine, but also knowledgeable and expert. It was all slightly disingenuous, however. Sanger's real training ground was in the labor, socialist, and feminist organizations and activities then gripping the United States and Europe.[5] There, she was deeply schooled in political ideas, as well as in operational diplomacy. Neo-Malthusian thought and socialism of various kinds were not always as conflicted as historians suggest, and Sanger is a case in point. She had been strongly involved in turn-of-the-century socialism and the anarchism, associated with Emma Goldman, who had herself attended the foundational international neo-Malthusian conference in Paris in 1900. In its French manifestation, neo-Malthusianism was thoroughly connected to anarchism,[6] and so it was a clandestine and underground meeting, she recounted. The link with anarchism was one of the reasons that birth control was later outlawed by the French state. Goldman's journal, *Mother Earth* (1906–1917), ran a special issue on birth control in April 1916, a problem, she wrote, that "represents the immediate question of life and death to masses of people."[7] It was just then that Sanger, having jumped bail for the technical obscenities in her own United States journal, *The Woman Rebel*, was visiting the few Dutch and British birth control clinics for working-class women that had recently opened, and along the way she took in various Malthusian meetings and personnel. Returning to the United States, she served time in jail after the opening of a birth control clinic in Brooklyn. In 1921, she established the American Birth Control League, and soon after married for the second time, into serious oil money. Sanger's socialist credentials make it all the more remarkable that she so effectively brought with her, rather than superseded, established British Malthusians like the Drysdales or new economists like Keynes. She effectively facilitated links between these men and a generation of scientifically trained neo-Malthusians based in the United States. Margaret Sanger was a major Anglo-American link.

After the string of prewar meetings, the international neo-Malthusians gathered again at their fifth meeting, held in London in 1922. "Birth Control" entered the formal title of the meeting for the first time, but this was a grafting, not a takeover. The conference certainly retained its political economy roots. Like Annie Besant in the 1880s, organizers were keen to keep a focus on what they saw as the largest issues: "The great principles of Malthus and Darwin stand in the same fundamental relation to sociology as the law of gravitation does to astronomy."[8] They gathered up the long and distinguished tradition in natural history and added birth control to it, knowing that both Malthus and Darwin would have been appalled. Expert discussion of the "struggle for room and food" was ongoing, then, and it was appropriate that Keynes, who now presided over the new school of Cambridge economists, also presided over the economic section of the international neo-Malthusian meeting in 1922: he was deeply interested in the private session on contraception held for the medical men and women.[9]

The sixth International Neo-Malthusian and Birth Control Conference was held in New York City in 1925, driven by Sanger. It was enormously successful and entwined even more strongly the political economy and natural history of Malthusian thought. Havelock Ellis's contribution, "The Evolutionary Meaning of Birth Control," began with the foundational idea that a great many more organisms are born, of any species, than can possibly survive to maturity. His own example was as curious as Darwin's instance of elephants: "A single oyster, if all its progeny survived would speedily accumulate, it is estimated, a heap of shells eight times the size of the world." And he told his audience—as though they needed the connection to be made—that likewise, without any checks it was possible for human populations to double in twenty-five years, citing Darwin as the original, thereby gaining credibility, but losing the eighteenth-century provenance wherein Benjamin Franklin was the source.[10]

After the success of the New York meeting, Sanger immediately planned what was to have been the seventh international neo-Malthusian and birth control meeting with Dr. Clarence Little, a biologist trained with the first generation of Harvard Mendelian geneticists under W. E. Castle. Having worked at Cold Spring Harbor and having assisted Sanger with the establishment of the American Birth Control League in 1921, by 1925, Little was president of the University of Michigan, Ann Arbor, at the age of thirty-six. The early plans between Little and Sanger reveal a conference that looked like its neo-Malthusian predecessors: the initial program was structured with "medical," "ethical," "biological," and "economic" sections. They had major international ambitions: "We should go direct to the heart of things and hold the conference at Geneva,"[11] meaning that they wanted the League of Nations Assembly and secretariat to consider population and birth control. By March 1926, Sanger, Little, and their key administrator, English campaigner for women's suffrage, Edith How-Martyn, approached key men for the organizing committee: Edward East, geneticist at Harvard University and author of the recent *Mankind at the Crossroads*;[12] Raymond Pearl, biologist at Johns Hopkins University; and Adolph Meyer, professor of psychiatry at Johns Hopkins Medical School. Slightly later, they introduced the British connection: Alexander Carr-Saunders, now social science professor at the University of Liverpool; Bernard Mallet, recently registrar-general of England and Wales; and Julian Huxley, the well-known grandson of Darwinian naturalist Thomas Henry Huxley.[13] After a few years at Rice Institute, Texas, and then as demonstrator in zoology at Oxford, Huxley was professor of zoology at King's College, London, from 1925. By the time he attended the Geneva meeting, he still held that affiliation, but was in the process of quitting to write *The Science of Life* with the famous H. G. Wells and his less-famous but more biologically literate son, G. P. Wells.

The seventh International Neo-Malthusian Conference became, at the eleventh hour, the first World Population Conference. Difficult negotiations between Sanger, Little, and Pearl resulted in a merger between the conference idea and Pearl's vague plans for an international union of population scientists. This required, in Pearl's view, "that birth control, or Neo-Malthusianism shall not appear as being the dominant element in the organization or plan,"[14] and references to both were dropped, more or less with Sanger's strategic endorsement. A different kind of conference emerged that included scientists only. Sanger agreed not speak—a major concession—but remained the critical organizing (and fundraising) presence. This meeting, then, was explicitly packaged as a "scientific" meeting, rather than as an applied or lobbying one: "Propaganda of any kind, or for any objective or doctrine whatever, will find no place in the Conference," the call announced.[15] This placed the population question on a "higher plane altogether," according to Bernard Mallet.[16] The gendered divisions that drove the planning of the meeting went on to characterize the two spin-off organizations: the International Union for the Scientific Investigation of Population, chaired by Raymond Pearl; and the Birth Control International Information Centre, with a completely different purpose, chaired by Margaret Sanger.

Given the neo-Malthusian provenance of the conference, the organizers distanced themselves from birth control with some difficulty. One effect of doing so, however, was the incorporation into the meeting of those explicitly opposed to Malthusian theories, as well as politically opposed to birth control; Corrado Gini of the Institute of Statistics, University of Rome, for example. Gini told Bernard Mallet that many in Italy, including those in "high quarters," thought the conference a camouflage for neo-Malthusian propaganda: "I have been asked for explanations." In May 1927, Mussolini had roundly and publicly reversed his position on birth control, announcing that Italian numbers were diminishing and birthrates needed promoting. Mallet diplomatically reassured Gini that the conference was not "Neo-Malthusian."[17] Privately, however, Mallet was far more open-minded, as he himself put it: birth control had always existed, one way or another, repeating the signature argument of his countryman, Alexander Carr-Saunders.[18] If Mallet's public assurances secured Gini's and the Italian presence at the conference, it had the opposite effect on other scientific men: Sanger and Mallet lost, as well as gained, delegates and support. When the Norwegian doctor Otto L. Mohr learned from Julian Huxley that the Geneva meeting had abandoned birth control, he advised the organizers of his regretful withdrawal.[19]

In holding the 1927 Conference at Geneva, the organizers were at once courting and challenging the League of Nations, but the same flurry to distinguish between public and private belief, between sanctioned comment and open secrets about birth control, also marked the response of the League's officeholders. The secretary-general, Sir Eric Drummond (himself Catholic and later British ambassador to Rome) declined any personal or institutional presence, because the League did not endorse birth control. This was because of the illegal status of contraception literature in many countries; because of Catholic opposition then consolidating, especially in Italy, France, and the Irish Free State, and soon to be proclaimed by the Vatican as the *Casti Connubii* (1930); and because the League was bound not to intervene in clearly national issues. Pronatalist policies, such as those materializing in Italy and long implemented in France, were avowedly nationalist. Indeed it was as much pronatalism as birth control advocacy that "roused the strongest national feelings" according to Drummond, thus making official League attendance at the conference potentially controversial. He warned the League's secretariat to keep clear.[20]

In their personal capacities, however, significant officers in the League of Nations were greatly interested in the conference. Dame Rachel Crowdy, chief of the Opium and Social Questions Section, pushed Drummond firmly, seeking permission to attend unofficially.[21] She nonetheless advised Margaret Sanger not to press the League's Assembly for an opinion on birth control: "she would do her cause more harm than good."[22] Dr. Norman White, acting director of the Health Section, attended as "observer," as did Sir Arthur Salter, director of the Economic and Financial Division.[23] Considerably less constrained, Arthur Sweetser, Woodrow Wilson-appointed member of the American Peace Commission at Versailles and founder of the International School in Geneva in 1924, was also reported privately to endorse new contraceptive methods as "an issue as important if not far more important to the future of mankind than nine-tenths of the subjects dealt with by the League . . . a thing that gives man control of his destiny."[24] Yet as a member of the secretariat's Public Information Section, his attendance was difficult. The irony in all this diplomatic care and concern was that birth control was not to be discussed at the conference in any case.

This meeting brought some of the world's leading biologists into the population fold.[25] Most of them supported birth control but thought it a means to an economic and geopolitical end, not of itself the matter at hand. Soil scientists and agricultural scientists, especially experts in plant breeding and the new genetics were actively involved: Wisconsin zoologist and specialist in animal and plant breeding, Leon Cole; distinguished geneticist, F. A. E. Crew, then director of an animal breeding department at Edinburgh University; J. B. S. Haldane, then a member of the "Genetical Department" of the John Innes Horticultural Institution at Merton; and Francis Marshall, reader in agricultural physiology at Cambridge. The population geneticist R. A. Fisher was also present, at that point in his phenomenally significant career employed as a statistician in Britain's Rothamsted Agricultural Research Station. Geographers, integral to the shape of the world population problem over the 1920s and 1930s, contributed as well: Glasgow geographer J. W. Gregory; Liverpool geography professor Percy Roxby; and Sir Charles Close of the International Geographical Congress. Economists and political economists included Mabel Buer of Reading; T. N. Carver of Harvard; Wesley Mitchell of Columbia; Charles Gide of Paris; and Rajani Kanta Das, economist with the International Labour Office, secretariat of the International Labour Organization (ILO). The German Jewish demographer Robert René Kuczynski identified himself as a political economist at this point in his influential career. In 1927, he was still based in Munich, but in 1933 was to flee to Britain, there to contribute strongly to the demography program at the London School of Economics. Corrado Gini identified himself as a political economist too, and brought with him the Italian statistician Livio Livi. Legal scholars, and in particular international lawyers, included the Czech Stanislas Kohn. The papers themselves ranged from the proposition of new population laws, to differential fertility studies, to economic concepts such as "standard of living" and "optimum population." Food supply and international migration were discussed, and in the process, both underpopulation and overpopulation were problematized. Delegates disagreed. The Malthusian League's Charles Vickery Drysdale, for example, sat uncomfortably across the table from August Isaac, the French president of the Federation for Large Families. Some papers were nationally defined or comparatively national, especially when it came to fertility rates. Carr-Saunders delivered a paper on fertility differentials by occupation in the United Kingdom; Paris professor of statistics Lucien March spoke similarly on France; and Berlin's professor Alfred Grotjahn on Germany. Some delegates, not least Sanger herself, conceptualized world population more globally than nationally or internationally, drawing on the long Malthusian tradition.[26] The director of the International Labor Office, Albert Thomas, rose above it all, announcing in the particular vocabulary and political sensibility of the moment that he attended as "a free man and citizen of the world."[27]

Raymond Pearl's Universes with Definite Limits

Density was put on the world population table in the very first paper, the address by Raymond Pearl. The Johns Hopkins mover and shaker represented work that many of the delegates would have already read in his *Biology of Population Growth* (1925), indeed, which some of them had already reviewed. This was Pearl's theory, or as he would have it, law of population growth. After considerable experimental and statistical work, he considered that all communities of organisms grew in number in the same way and that rates of population growth followed a pattern and were therefore predictable. When plotted, this pattern was consistently represented by a particular curve, a version of the logistic equation first developed by early nineteenth-century mathematicians.[28] "In the matter of population growth there not only 'ought to be a law' but six years of research has plainly shown that there is one," Pearl claimed.[29] With Lowell Reed, Pearl demonstrated the applicability of this curve to any number of different organisms and their growth in spatially contained environments: first yeast cells, then *Drosophila*, then poultry. Finally, he ventured into human populations, indicating that the curve described the pattern of population growth of France, the United States, and Sweden.

Casting about for a clinching human case, much of *The Biology of Population Growth* presented complex and confused data on the Algerian population. "In most cases," he lamented, "where Nature has staged a reasonably simple experiment in human biology man has neglected to take adequate records about it." But thanks to the French and their refined census-taking in Algeria from 1886, and their thorough-going colonization of that part of North Africa, Pearl considered he had access to unique data. After a long exposition of their cultural, economic, and sexual behavior, he concluded that Algerians had completed a single cycle of growth over a period of seventy-five years. Pulling widely and wildly from the anthropology and history of sex relations, health, and birth and death customs among Algerians, he felt able to argue that all organic growth, including that of human populations, followed a biological law.[30]

Pearl postulated that there was a correlation between numbers of organisms in a bounded space and fertility and morbidity rates. Fertility, he claimed, was always "markedly affected adversely by small increases in density." And so, density was significant and applicable to all living organisms, from "yeast cells" to "man."[31] As human ecologist Radhakamal Mukerjee later explained it, Pearl posited an automatic control of optimum density: "A limit is imposed simply by the life-process itself on growth in a limited environment." The space factor, he explained, thus had a definite role in shaping population growth rates.[32]

Raymond Pearl was trained in zoology at the University of Michigan and participated in the pioneering Biological Survey of the Great Lakes led by aquatic ecologist Jacob Reighard.[33] Like Carr-Saunders, he had spent a year in London with Karl Pearson, the biometrician and eugenicist, which prompted a shift in his interests not just toward a new statistical method but also toward the scientific consideration of human populations. For several years, Pearl was head of biology at the Maine Agricultural Experiment Station, where he did close work on the breeding of poultry and their population trends. He was subsequently a professor of biometry and vital statistics at the Johns Hopkins University School of Hygiene and Public Health and became Director of the Institute for Biological Research in 1925, founding the journal *Human Biology* in 1929.[34] It was in these years that his international reputation consolidated, largely on the basis of his books, first *Studies in Human Biology* (1924) and then *The Biology of Population Growth* (1925), although he had published widely before this, his output significantly including *The Nation's Food* (1920) and *The Biology of Death* (1922).[35] Seemingly, Pearl had all aspects of population covered.

Raymond Pearl was perhaps surprised, then, at the critique leveled at his population "law." George Knibbs, for one, simply could not credit its mathematics and was intellectually horrified at Pearl's use of the Algerian census data. On the one hand, Knibbs agreed that the world population problem was about the connection between human economic evolution and "the reaction of this upon his numbers and the density of his aggregation."[36] On the other, Pearl's law so was technically and conceptually problematic that he wondered why it was being presented at all, let alone opening the conference. In the lead-up to Geneva, Knibbs wrote to Bernard Mallett (who as registrar-general was certainly mathematically literate) suggesting that Pearl's whole argument was invalid and its assumptions elementary, incomplete, and insufficient. So much so, he thought, that it was "amazing that it should have found acceptance in the external politics of a great nation."[37] Knibbs was referring to the widespread idea that the United States had reached its optimum density, one rationale for its recent series of immigration restriction acts. Knibbs explained his issue with Pearl's theory to his fellow statistician: "The factors that determine the rate of increase of population are so manifold that they cannot hold constant: the utilizable physical resources of a population including its territory; the population's skill in exploiting these resources; the population's response to the reproductive impulse; its death-rates; its emigration rate."[38] Others less qualified mathematically than Knibbs sought similarly to emphasize the vast and unpredictable complexity of human organization, not to mention the constantly shifting spatial limits that humans regularly transgressed, unlike *Drosophila* in test tubes. All this seemed to render Pearl's law spurious. In Geneva, many commentators puzzled over how human interactions were to be considered, let alone investigated, through such a natural law. William Rappard, Geneva-based and Harvard- and Vienna-trained political economist, sniped that "everything is delightfully simplified in the rarefied atmosphere of the laboratory."[39] But even the great biologists of the 1920s who were present at the World Population Conference responded similarly to Pearl's law. The portly Marxist Jack Haldane thought that the recent period of history from which Pearl gathered his data on humans was unique, and its events abnormal; World War I and the Russian Revolution affecting "cyclical growth." Still, he was interested in the political extrapolation of this apparent law: did Pearl think that the reduction in fertility, apparently triggered by a certain density, would come about in human populations without catastrophes such as famine or war? That was a promising idea, to say the least, thought Haldane. If parts of the world were overpopulated, "we must undoubtedly hope that their population will decline in that manner rather than catastrophically."[40]

The controversy surrounding Pearl's curve has largely dominated historiographical discussion of him.[41] But it is less the accuracy or inaccuracy of the logistic curve per se that is important in this context. As the first paper at the first World Population Conference, it foregrounded the spatial question of density. The biology of population was itself about "living room," a premise that held simultaneously biological, geographical, economic, and political significance.[42]

Edward East's "Little Terraqueous Globe"

Raymond Pearl told his Geneva colleagues that "in any and every case, there is ultimately a definite limit to the size of the universe in which any real population lives."[43] It was his biologist colleague Edward East who brought this idea home to the topic at hand: human populations living in what East called "this little terraqueous globe."[44] The land–food nexus was

East's entrée into the population question, and his talk was directly to the point: "Food and Population."

East could make claims about food and agriculture with considerable authority—more than most—since he was an expert in the genetics of crops. East was another of the U.S. plant and animal biologists busy extending their research on the genetics of lower organisms to the politics of human populations. They were doing so not just in terms of eugenics or "quality," but also in terms of quantitative dynamics. Educated at the University of Illinois, originally in chemistry, East worked on key early agricultural research projects. He held a post at Harvard's Bussey Institute from 1909, an agricultural and horticultural research center that later become the Graduate School of Applied Biology. Appointed in plant morphology, his title changed to professor of genetics in 1926, mirroring Pearl's uptake of a title in genetics. And again like Pearl, East was formatively linked to food administration during World War I. His appointment to the Statistical Division of the U.S. Food Administration showed him, he later reflected, "how narrow is the margin between the world's food supply and its ever increasing needs."[45] East read Malthus at this point, and his publications immediately began to include commentary on neo-Malthusian work. After years of writing on plant genetics (he cofounded the journal *Genetics*), East began to address human population questions, "The Agricultural Limits of Our Population," being one example of his output.[46] He reviewed Margaret Sanger's *The Pivot of Civilization*,[47] and his popular book, *Mankind at the Crossroads,* was delivered first as "Civilisation at the Crossways" at a Chicago Birth Control Conference. It was published in *Birth Control Review*, alongside Keynes's "crushing" responses to William Beveridge's attempt to "banish Malthus's devil."[48]

East saw before him a smaller, faster, more connected world and put it all journalistically: "The world has been explored from pole to pole; its resources have been chartered, from aard-varks to zymogens. The seas are dotted with ships; the lands are meshed with railroads. Our hands, our voices stretch from continent to continent. We have become neighbors, whether we care to be neighborly or not."[49] Indeed, they were all imagining this kind of little, terraqueous globe. Knibbs spoke often about a "shrinking world" with more people bringing about closer and quicker economic exchange. "The world is now an economic unit," Harold Wright wrote, foreshadowing a newly globalized century that might see a "growing measure of economic solidarity in the affairs of mankind."[50] Czech statistician František Jaroslav Netusil argued similarly that the evolution and management of human population absolutely required "the organization of the world in one political unit." He also was drawing on the economic links that tightly connected geographically disparate parts of the globe: "Only the world as a unit is really self-supporting. . . . [T]he human population on the globe of the earth needs one organization which would embrace economic life."[51] This might well retrospectively be called globalization, precisely because these commentators were bringing together economic, cultural, and spatial factors. The world was not just small and geopolitically closed; it was economically linked.

East began his paper in Geneva agreeing with Julian Huxley (and Keynes) that population growth was the most important problem confronting the human race. The problem reached back to the deep human past as well as the more recent colonial past: "There was always new land to be brought under the plow. Unexploited reserves of virgin soil lay ready for the coming squatter." But the present epoch, now only a century old, he wrote, was one in which there was no new land.[52] Given space and food limits on this small planet, East posed the question of the day: What were the corresponding population limits? Calculating for his own country, and then for the world, East concluded influentially that each person required an average of 2.5 acres of land to sustain him or her. For the continental United States, this meant an optimal population of 197 million and an "ultimate" population of 331 million: less than usually planned for, he wrote in 1921, quipping that to limit anything in the United States was blasphemous.[53] This figure also determined a maximum global population: 5.2 billion.[54] This was calculated on a projection that 40 percent of the globe could be inhabited and cultivated, and that at the present rate of increase, this number would be reached in just over a century.[55]

East estimated the world population to be 1.85 billion,[56] like others relying heavily on George Knibbs's calculations and on estimates produced by the Institut International de Statistique, the *Statistical Year-Book* published by the League of Nations, and the *International Year-Book of Agricultural Statistics*, published by the International Institute of Agriculture in Rome. In 1930, the consensus was a world population between 1.988 billion and 2.028 billion.[57] Projections were as important. East used Knibbs's estimate that global population doubled every eighty years, anticipating 3.9 billion by 2008. Total possible populations were suggested for the planet. In 1891, British migration geographer Ernst Georg Ravenstein had reckoned this to be 5.994 billion.[58] A quarter-century later, Knibbs thought 5.2 billion a likely maximum, and slightly more with better social organization.

Carrying capacity, he thought, was in part dependent on economic and political cooperation: 9 billion perhaps, with all scientific advances, and with "economic solidarity of the world." And if there was assured peace and "absolute friendliness among all the nations," he projected 11 billion as the ultimate limit. But this, Knibbs advised, would also require a common plan for birth control.[59]

Assessing global carrying capacity required calculations of arable area, yet not all spaces were equally useful, or even relevant when it came to the human population–food equation. "Crude density," argued Henry Pratt Fairchild, professor of sociology at New York University attending the Geneva meeting, "has almost no significance at all."[60] The nature of that land was the point: whether it was habitable and by whom; whether it was cultivated; or whether it was barren—ice-bound, desert, or swamp beyond reclamation.[61] Ravenstein had calculated global cultivable area as 46,350,000 square miles, excluding the polar regions from the point where cereals could not be grown. Of course, he said, these regions yielded game, fish, berries, and even vegetables, "but the few hundreds of thousands of people whom they support at the present time, are of no account at all when dealing with hundreds of millions."[62] Population experts quibbled constantly over whether inland seas and lakes should be counted or not—the Great Lakes in North America, especially—or whether Arctic regions might be included, and to what latitude. Should they be assessed as inhabited or empty? And might the calculation take into consideration area that was potentially habitable and had a latent agricultural capacity but was currently uncultivated? London geography professor C. B. Fawcett estimated the total habitable land area of the world to be 50 million square miles ("if we exclude the uninhabitable ice-covered lands of the Polar Regions").[63] Since world population was about 2 billion, the mean density of inhabited land was about 40 people per square mile.[64] Because of the significance of density, statisticians of human populations found themselves thinking more like geographers or agricultural economists than mathematicians. And so, when George Knibbs, the great census-taker, approached the global population question, he systematically calculated and assessed different kinds of land. For the globe as a whole, he distinguished between meadows and pastureland; natural grasses; woods and forests; marshes and heaths; land that produced crops for industry, for cash, for fertilizer; fallow land; and perhaps most importantly, uncultivated but productive land—land that was wasted.[65]

Estimating the world's carrying capacity thus required great surveys of land and soil as well as of people, a trend that had begun in earnest, George

Knibbs thought, but that needed to be vastly extended by the new international agencies. What was needed was a "systematic survey of the whole position," in his mind, a stock-taking of humans, their needs for life, and the current resources of the planet. Such a world census of people and land might be modeled on the U.S. world soil surveys.[66] It would integrate vital statistics with an account of natural resources, and more: "Must it not take into account the migration and settlement possibilities of the earth, and the adjustment of the normal rights—if there are such rights—of races and nations? And will not such adjustments of mutual rights include the questions of the possibilities of food-supplies and conditions of mutual well-being of the peoples of the whole world?"[67] In a different tradition, but with similar scope, Berlin's geography professor Albrecht Penck had circumnavigated the world, pursuing an all-encompassing mapping system and gathering data on climate, soils, the "surface of the globe," and the possibility offered by the "different areas as a dwelling-place for human beings." This was a global audit, undertaken "from the point of view of feeding people." Penck the geographer and climatologist turned out to be more ambitious than East the plant geneticist in his estimates of the world's carrying capacity: there were resources for 8 billion, he thought.[68]

Optimum Density and Standards of Living

Edward East's calculations of 2.5 acres per person as the optimal density for the world drew immediate comment in Geneva about available land area, agricultural possibility, and global diet and consumption patterns.[69] Economists, agriculturalists, biologists, and emerging demographers and other social scientists, all sought to systematize and standardize the criteria by which optimum population might be recognized and subsequently realized. The very idea of "optimum density" already had generations of economic debate behind it, and scientists at the meeting thought that fresh biological perspectives on optimum population was one of the more important outcomes.[70]

Some thought optimal population might best be indexed by public health measures: morbidity, longevity, and in particular, infant mortality rates.[71] "The simplest principle is that of the highest physiological well being," commented Charles Vickery Drysdale, in one of his more succinct comments on the topic. Longevity was the key measurable index for him, the "average duration of life."[72] For others, the determination of optimum density assumed a more eugenic meaning of fitness, a greater proportion of fitter types in any given population. Geographer Sir Charles Close broadened the index to "the eugenic idea of the largest number of people who are perfectly fit, morally, mentally and physically."[73] This partly explains his insistence throughout the 1920s that Britain was overpopulated, that several millions, at least, should leave in various emigration schemes. Others again considered the idea in economic terms such as unemployment rates.

The corollaries of optimum population were over- and underpopulation. Some prewar Malthusians had argued crudely, "poverty is essentially overpopulation."[74] Reliably, Alexander Carr-Saunders corrected and refined such talk, indicating that Malthus, for one, never saw the possibility of overpopulation. "To him [Malthus] there could be no such thing," because population checks were always efficiently in operation.[75] The social and economic problem was the *means* by which population growth might be checked. Therein lay "misery" or potential improvement of welfare. For Carr-Saunders, this was the whole purpose of regulating any population. As he put it: "If the density fails to reach or exceeds the desirable density, then the community fails to attain that degree of economic welfare that is within its grasp."[76] Positively attracted by the population problem's imperative to fold together economics, biology, and politics, Carr-Saunders introduced the complex of "healthy social life" as an index of optimum density: "the increase in the density of population beyond a certain point will render healthy social life impossible."[77] The economic view would be that income per head would determine the best density. But from other perspectives—"a broader social outlook"—there certainly were limits, and it was all about standard of living.[78] For Carr-Saunders, a "reasonable standard of living" was the object of population policy, because it would materialize as greater health and welfare. Similarly, Henry Pratt Fairchild wanted to set aside the view that either density or population growth per se were problematic; they were only so in relation to economic well-being. Fairchild's paper "Optimum Population" moved the biological discussion on density that Pearl's papers raised well into the social sciences. The sociological and economic challenge was to balance the two Malthus-identified needs: "hunger and love."[79] Fairchild's encompassing schema identified four factors into which all social organization can and should be categorized: population, standard of living, land, and what he called "stage of the arts," by which he meant, the sum of knowledge, technology, devices and processes by which humans appropriate "the natural supplies of the land."[80]

If one limit of human population growth was the capacity of soil, literally the earth, to support more or fewer people in a given bounded area, this limit was obviously and necessarily affected by the "standard of living." Conversely, however, density itself could determine that standard. Corrado Gini asked an even more fundamental question: Whose optimum is to be considered? That of the individual or that of the state?[81] The immediate politics of Gini's question were evident to all, not least since he had just published "The Scientific Basis of Fascism."[82] Although the conference predated both the German fascist state and Primo de Rivera's Falange Española, fascist Italy was five years old, and the question of just how the individual related to an organic state was already very much alive in European politics. Notwithstanding the conference's bid for a scientific, opposed to a lobbyist, approach to population, the Geneva World Population Conference was hardly a politically neutral space.

An irreconcilable politics of population stemmed from fundamental disagreement about just what constituted "welfare" and a "reasonable standard," how that might be calibrated, and especially to what extent variation of that standard was tolerable within and between nations and regions of the globe. As George Knibbs put it, "Whether world-policy should be directed to the securing of larger numbers, living according to a more moderate standard than at present, or should aim at our being satisfied with a smaller world-population—not necessarily than the existing one—living according to a more highly elaborated standard." This, he added, was a moral question, not merely an economic one.[83] How, he asked, might creating this reasonable standard of living best be coordinated in an economically linked world? It was a question that directly prefaced, and led to, the politics of economic and demographic development after the next war.

Like George Knibbs, many participants in the world population discussion considered both the carrying capacity of the whole globe, and distributions of population densities within it. "What should be the normal standard of living?" he asked. And how does, but more problematically, *should*, this vary between particular countries and regions?[84] "Should" signaled the deep politics of the issue. French economic historian Adolphe Landry, another early demographic theorist of transition, provided one answer to Knibbs's question when it was discussed a decade later at the Peaceful Change conference in Paris: "A country is overpopulated in relation to another country when its standard of living is lower than that in the latter."[85] In this view, overpopulation was always and only determined relative to another country or region. Others, needless to say, proceeded on the presumption that the current differential standards of living would and should continue, apparently unproblematically. Geographer George Kimble wrote in his significantly titled book *The World's Open Spaces*, that population density needed always to be considered in relation to the "level of civilization"—in this context meaning standard of living—of the population concerned.[86] The lesser level of civilization in Asia meant that a different optimal density would be satisfactory. Articulating such a normalized global inequity might have been possible in such circles in the 1920s and 1930s, but after World War II, it was far less acceptable. Nonetheless, it was the idea of standard of living that linked an earlier idea of "civilization" with later "development."

Regional Densities: Problematizing Europe

Even though later twentieth-century critical discussion on "overpopulation" typically problematized south Asia, east Asia, and Central America, many population commentators in the 1920s and 1930s thought about Europe first. They would see no reason to disagree with Leipzig professor of political economy Karl Thalheim's blunt statement: "Europe is overpopulated."[87] Comparative—and differential—density was everything, and continentally speaking, Europe was quite plainly the densest continent on the planet.

Ravenstein had early calculated Europe to be so, with 101 people per square mile.[88] He registered the massive late nineteenth-century emigration of Europeans to North and South America, but even taking such demographically significant events into account, Europe was denser than other continents. It was still so when the postwar scholars turned in larger numbers to reckon differential density. George Knibbs informed his readers that Europe was the most closely packed continent, with 127.6 people per square mile, followed by Asia, with 65.3 people per square mile. The first substantive chapter of *Shadow of the World's Future* was, significantly because foundationally, "Distribution of the World's Population." When he broke this down to "national" groups for which there were reasonable figures, Knibbs showed that if India's density was 226 people per square mile, that of England and Wales was 671 people per square mile.[89] Carr-Saunders also took some time to explain that England and Wales was the most densely populated national census-unit in Europe, and globally could point to only two administrative units that were more so: Java, with 689 people per square mile, and Barbados, with 940 persons per square mile.[90] Geographer C. B. Fawcett said that in the habitable areas of the globe, densities ranged from 680 in Belgium to 2.2 in Australia.[91] Although the political implications of overpopulation in Europe, south Asia, or east Asia differed, it is important to note that in these years European density was conceptualized not just as *a* major problem, but was often ranked as the *first* major problem.

For this generation, the issue of the great modern demographic change after industrialization was at least as alive as the fairly recent and still inexplicable local declines in fertility rates. Early twentieth-century economists, demographers, statisticians, agriculturalists, and neo-Malthusians were all still coming to terms with the massive acceleration in the growth rate over the nineteenth century, which was unprecedented in world history. It had been so great, many commented, that even the residual population, after the fact of massive emigration, left a vastly overcrowded region. Returning to London from the Geneva conference, Julian Huxley began to publish articles announcing England to be overpopulated after industrialization, a situation that all industrializing nations would likely face, sooner or later.[92] He did not yet perceive the decline in fertility to be a delayed effect of that industrialization.

None of this precluded discussion of south and east Asian overpopopulation. It was common, however, for Chinese and Indian economists and demographers to challenge presumptions that Asian population trends were problematic by comparing European and Asian density. Mrs. Liu Chieh argued that since China's density was around 254 per square mile, and Italian density was 343.9, any undue focus on the former was quite misplaced.[93] Professor H. H. Chen also effectively criticized the tendency to problematize Asia at the expense of Europe. First, he said, the rate of increase of the Chinese population was insignificant compared with industrialized Western countries. Second, the average density was not high, comparatively. Third, he identified high densities in some areas of China as an internal distribution problem, not a national overpopulation problem, and one that could be addressed by internal migration policies. And finally, he argued that Chinese emigration to the rest of the world was "nothing" compared to the vast European movement beyond its borders that had taken place over the previous century.[94]

Differential density across the globe was coming to be the major comparative factor over which a number of arguments were based. Accordingly, density received considerable graphic and mapping attention. Although the world population problem has come to be most easily recognized through graphs that privilege a temporal axis—the great upswing over time—over the 1920s the problem was more likely to be represented cartographically, in world maps that privileged the spatial axis: comparative density.[95] Not growth over time, so much as distribution over space, created the world problem. When geographer Marcel Aurousseau advised Margaret Sanger on the 1927 World Population Conference, he suggested she set up a series of maps for the delegates' edification. This, he thought, would distill and explain the overlapping issues of territory, race, population, soil, and migration, each raised by the population question. One map should show the comparative natural increase in population, another the "habitable" parts of the world, and a third "world movement of wheat." The migration of Europeans and Asians over the last fifty years should also be represented cartographically, showing countries that welcomed, and those that excluded, immigration of east and south Asians. Although illuminating, Aurousseau suggested to Sanger, a few national delegates would be made quite uncomfortable. But before all of this, he said, there should be a foundational map, showing the density of population in various parts of the world.[96]

Ravenstein's 1890s hemispheric maps were already showing Europe, the Indian subcontinent, and east Asia as the three "Centres of Population." They were contrasted to "the cultivated, occupied and waste areas of Australia, British South Africa, and Canada" (figure 3.1).[97] This became a trend. H. L. Wilkinson's colored world map of density was the first plate of his book on the world population problem, foundational information for any reader (figure 3.2). Radhakamal Mukerjee opened his *Migrant Asia* (1936) with the characteristic map of global density, significantly based on Karl Mollweide's projection, which privileged representation of accurate area over shape of the continents (figure 3.3).[98] Maps from the U.S. Department of Agriculture's world surveys were commonly reproduced. Largely the work of Oliver Baker, they represented world wheat distribution, world population distribution, world soils, world agriculture, world forestry, and more. It was one of Baker's maps that Warren Thompson reproduced, the cartographic basis that explained his personal response to the Geneva meeting, the 1929 book *Danger Spots in World Population*.[99]

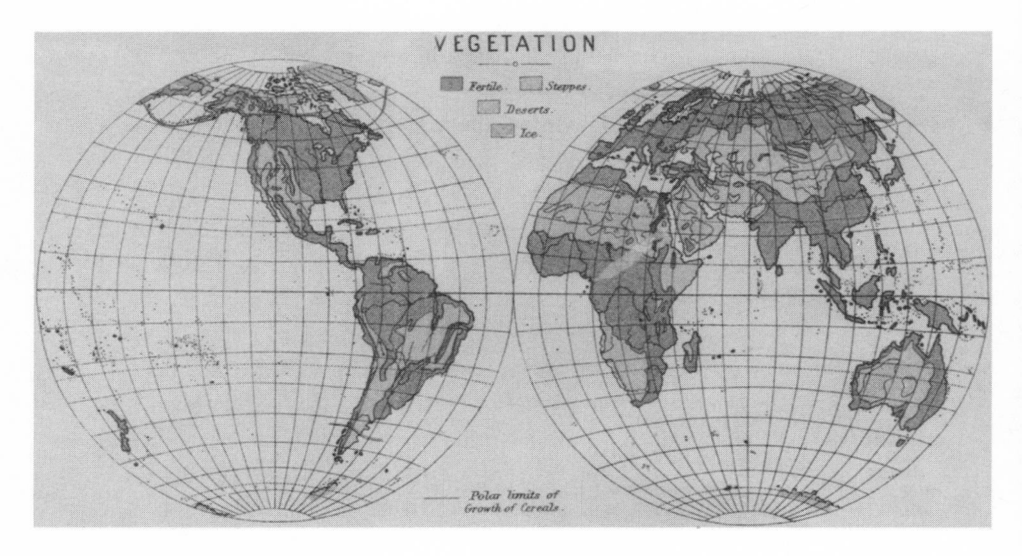

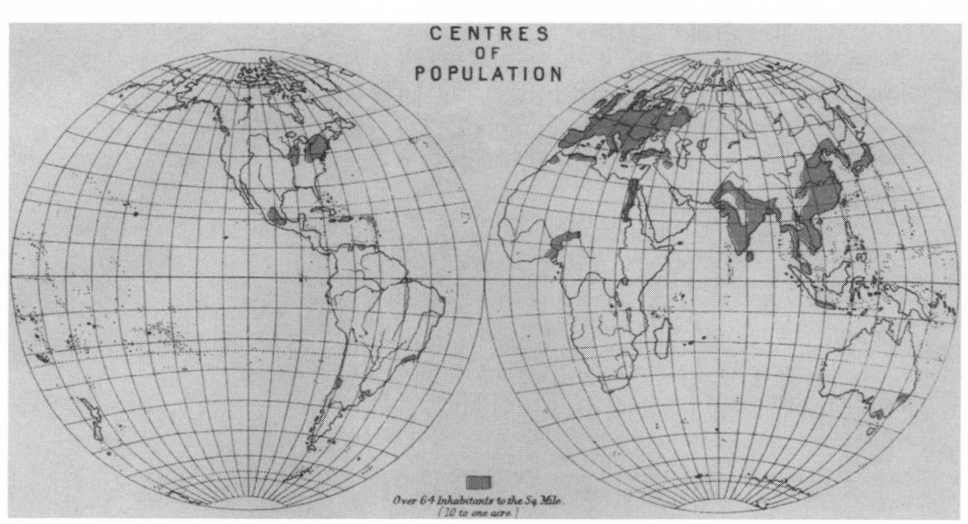

FIGURE 3.1 Available lands? Ernst Georg Ravenstein's 1891 maps showed the global "centers of population," in Europe, south Asia, and east Asia. (From E. G. Ravenstein, "Lands of the Globe Still Available for European Settlement," *Proceedings of the Royal Geographical Society*, 13, no. 1 [1891]: 64. Reproduced by kind permission, Wiley.)

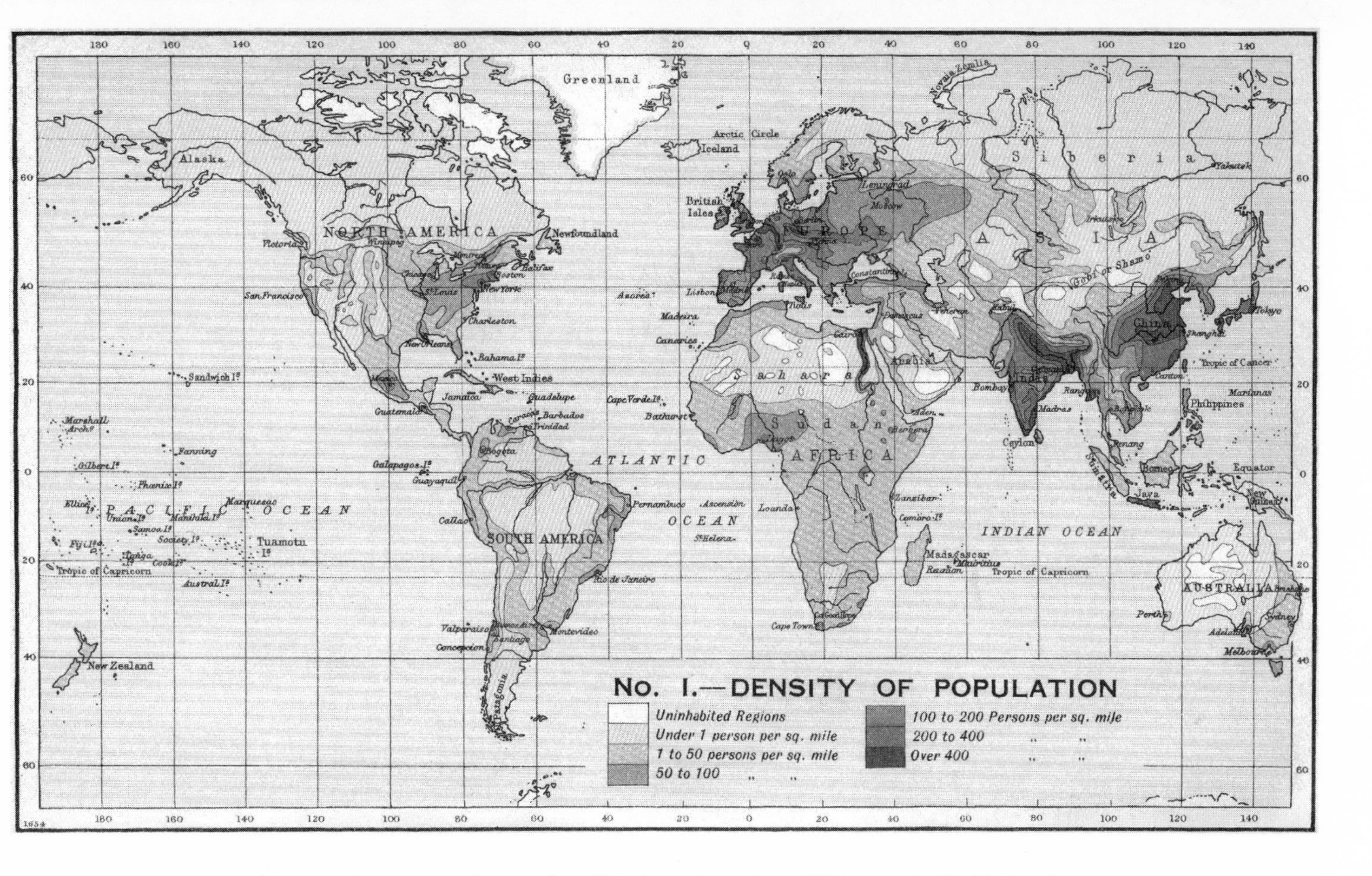

FIGURE 3.2 Comparative density of world regions. (From H. L. Wilkinson, *The World's Population Problems and a White Australia* [London: King, 1930], frontispiece.)

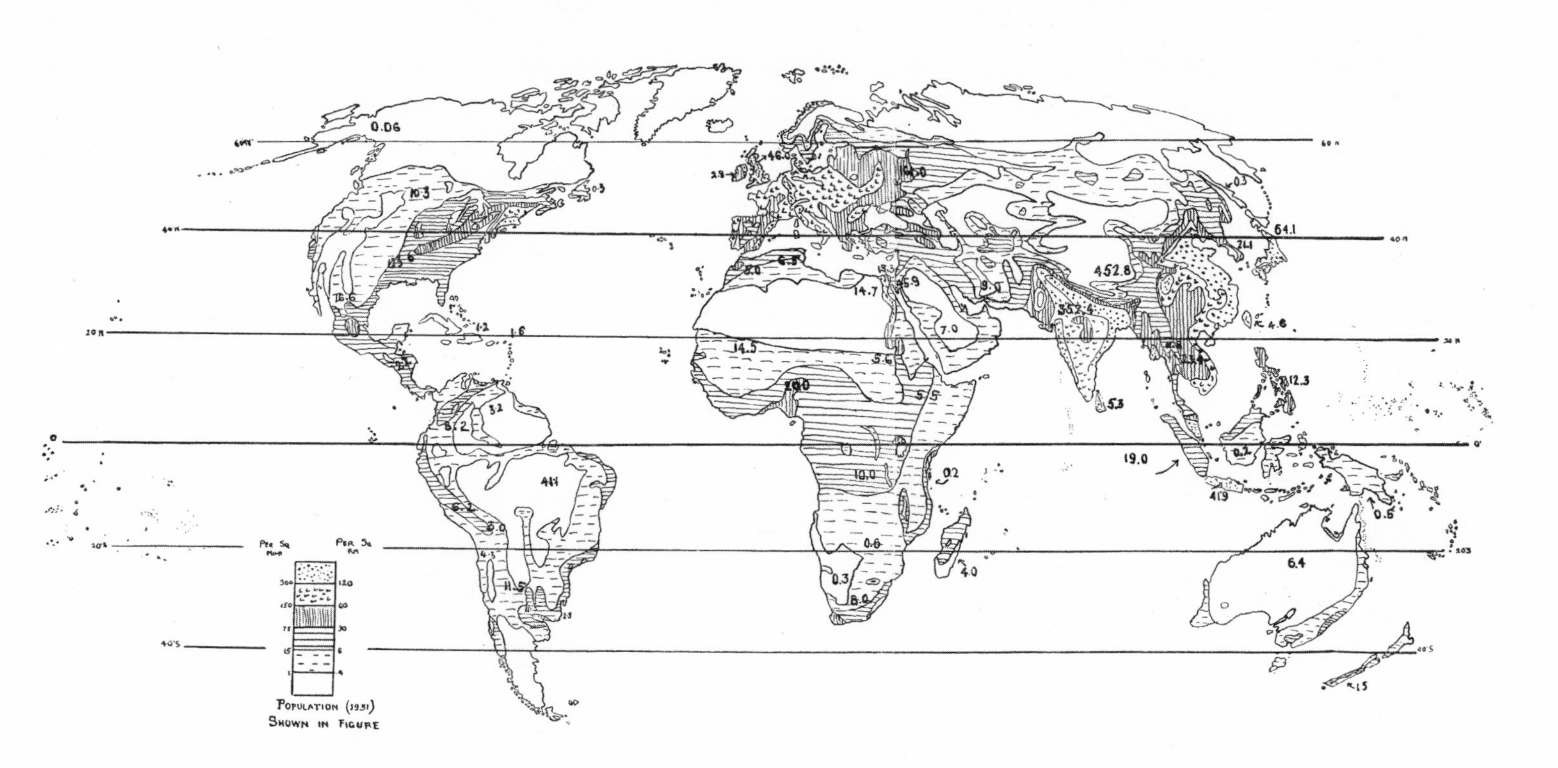

FIGURE 3.3 Distribution of World Population. World maps of population distribution introduced most books on global population in the 1920s and 1930s. (From Radhakamal Mukerjee, *Migrant Asia*, [Rome: Failli, 1936], 12.)

Warren Thompson's *Danger Spots in World Population*

This book opened, like so many, with a map of world population distribution (figure 3.4). In the corner, a pie chart explained percentages of world population by "nation," indicating the net dominance of China (18 percent) and India (18 percent). The real message, however, was neither total numbers nor projected growth, but comparative density between world regions. Thompson's primary agenda was to highlight overpopulated regions that were the political and economic danger spots: Europe, east Asia, south Asia. At a glance, imminent political tension could be identified, where white and black were adjacent: not white and black races (necessarily) but "sparse or empty" lands abutting "densely overpopulated" land.

Density and its relation to war and peace became the cornerstone of Thompson's work throughout the 1920s, when he began to observe the situation more internationally than trans-Atlantically. After receiving his Ph.D. at Columbia University, Thompson had become an instructor in sociology at the University of Michigan, not long after which the newspaper tycoon E. W. Scripps read his thesis on Malthusian economics, a lucky break for Thompson, who was invited to join Scripps on a long trip to "the Orient" in his yacht. On board and on shore, they talked about population and the world. Scripps clearly liked what he heard, since Thompson benefited from his patronage for the rest of his life. From 1922 until 1953, Thompson was director of the Scripps Foundation for Research in Population Problems, Miami University, Oxford, Ohio.[100] It was in that capacity that he participated in the Geneva conference.[101] Though small, the Scripps Institute grew steadily in influence, with Thompson and his deputy Pascal Whelpton producing some of the most cited scholarship on population in English. Thompson was also instrumental in introducing important French and German studies into the American discussion.[102] His 1930 text *Population Problems* became a standard in demography and sociology courses and remained so until the 1960s.[103] Making his first marks from and about a pre-1914 world, then, Thompson's career extended well into the Cold War era, when expert consideration of population, war, and peace, came to center forcefully on Japan and India.

Danger Spots was a study of political impact of differential density across the world and the measures likely to rectify this critical "unequal pressure," as he put it. It was a study in demographic international relations, premised on the "frictions" created over various kinds of acquisition and

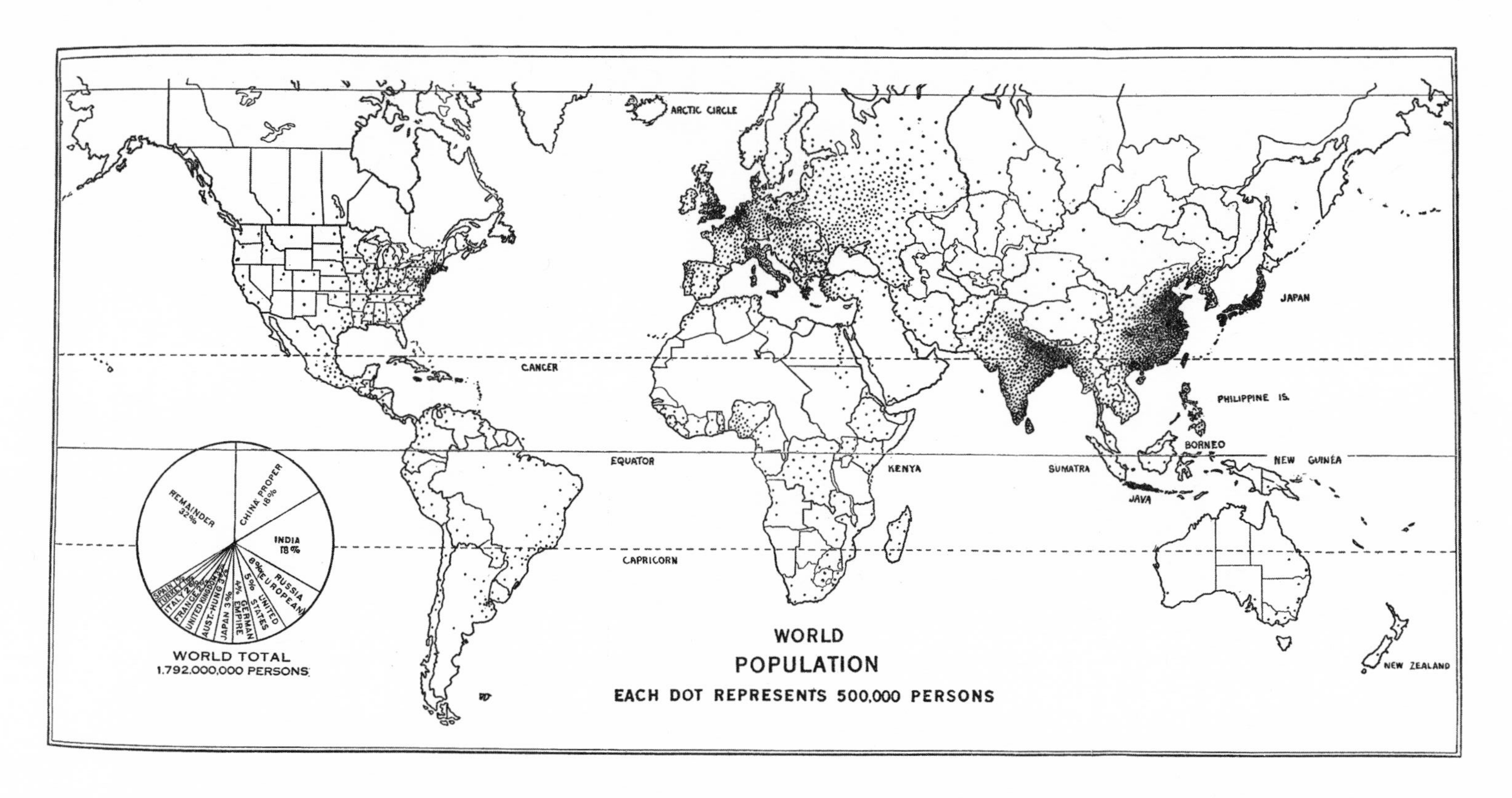

FIGURE 3.4 Danger spots. Adjacent sparsely and densely populated regions were Warren Thompson's key focus (From Warren S. Thompson, *Danger Spots in World Population* [New York: Knopf, 1929]. Thompson reproduced this map from V. C. Finch and O. E. Baker, *Geography of the World's Agriculture* [1917].)

maldistribution of land: "Will the efforts to equalize pressure result in war or will some other method of adjustment be found?"[104] The problem was at least as much people "holding low pressure areas unused," as it was people in "high pressure areas" seeking land. Indeed, for Thompson, responsibility for the creation and the solution of this global problem lay ultimately with the "low pressure" nations who kept others "pent up within their present boundaries indefinitely."[105] He problematized the geopolitical argument that justified territorial acquisition through war (what he called the force-system). An equalizing of densities was imperative, but an alternative to war needed to be agreed upon and implemented internationally. "The redistribution of the lands of the earth is the problem of problems that we must face in the world today as a consequence of the new population movements. . . . Can it be effected peaceably or must it be achieved by war?"[106] Land was both the problem and the solution.

Thompson focused on three key regions of the globe: the western Pacific, the Indian Ocean, and central Europe. In considering the first, he discussed Japan and China in relation to Australia in particular, but also the islands of the western Pacific—the Dutch East Indies, what he called British possessions (Borneo, New Guinea, Hong Kong, Singapore, and the Malay States), the Philippines, and French Indochina. "Can War in the Western Pacific Be Avoided?" he asked presciently. Thompson then turned to India, "Where Can the Indian Go?" One option, he thought, was to the Union of South Africa, noting its significantly light density of only around 15 people per square mile.[107] He then considered Europe: Italian density of 339 people per square mile was dealt with first, followed by central Europe, covering Germany, Poland, and the Balkans. The entire European situation had been made more tense, he noted, because of recent national quotas placed on immigration into the United States.

Thompson's demographic vision divided the world on axes that characteristically combined population growth and territorial limits: they were not separable. On the one hand, there were countries that had historically expanded their population and had simultaneously been able to expand territorially over the nineteenth century. On the other hand, in his contemporary world, there were countries beginning to expand demographically that were unable to expand in space. In his 1920s present, such "expanding peoples" were listed as the Japanese, Indians, Italians, and Slavs. As populations, they could not extend geographically, in part due to the immigration restriction acts, but in a larger sense because "the earth will not hold them."[108] A key figure due to his early version of the demographic transition theory that was later to drive so much development policy, Thompson always considered comparative birth and death rates in relation to closing global space. The time axis was only one element. What created urgency and pending crisis was the space axis, the differential densities across the world. This was the global geopolitics of demographic transition.

4

Migration

World Population and the Global Color Line

> The problems of emigration which today loom so large on the horizon of international politics, reflect man's effort to regulate his distribution on the land surface of the world. This is one of the means by which man is trying to solve the problem of his own multiplication.
>
> CARLOS PATON BLACKER, "THE MODERN CONCEPTION OF THE WORLD" (1928)

"The earth is filling up fast, and one of our questions is what to do about it," U.S. eugenics leader Charles Davenport announced.[1] The question was entirely disingenuous by the time he raised it with his population colleagues in Geneva. Some of his closest U.S. associates had not just designed but implemented their particular answer to the problem of an increasingly "filled up" world: the U.S. Immigration Act of 1924.[2] It regulated the entry of Europeans into the United States, compounding earlier restrictions of people from the so-called Asiatic Barred Zone.[3] Indeed the world's "neo-Britons" in Canada, Australia, and New Zealand were all busy racializing their claims to continental space in similar ways. Over these years, racial discrimination was internationally territorialized. A "color line" was dreamed and then implemented: the "Great Barrier," Edward Alsworth Ross called it.[4]

The geopolitics of immigration and emigration restriction was sometimes identified as the key factor that rendered the population problem one of international relations.[5] The connection was broadly threefold. First, economists had been concerned with "surplus population," and the migration and value of labor for more than a century, although as Warren Thompson commented, without "even a semblance of agreement about the effects of large numbers of immigrants on wages."[6] The vast nineteenth-century movements of Chinese, Indians, and Europeans were at core about labor, as, indeed, were most of the immigration acts. Remarkably similar legislation emerged in many jurisdictions, prompted by organized labor's drive to maintain wages; by venture capital's conflicting demands for temporary low-wage workers, especially for the large infrastructure projects, such as the North American railways; and by efforts to regulate indenture systems within the British Empire and Commonwealth.[7] The international movement of workers was the reason why the International Labor Organization (ILO) was so prominently involved in the discussions about world population, and why it put rather more effort than any other international body into gathering data on global movement and the new immigration and emigration laws.[8]

Second, the regulation of global movement that manifested as immigration restriction acts was, in one view, a response to differential rates of population growth. The connection between the well-known slowing of fertility rates in England, France, the United States, and Australasia from the 1880s and the apparent or relative increase in population growth rates in south and east Asia certainly concerned many. Anthropologist and race theorist Georges Vacher de Lapouge summarized: "It becomes clear that the number of men on earth (whites as well as yellow) has very nearly arrived at the possible maximum, that the density is badly distributed and that the redistribution of space can only be rectified by displacements, and finally that limitation of births in over-populated countries is the only means whereby the balance may be peaceably restored."[9] White, yellow, and black peoples and continents apparently had different demographic presents and futures that needed to be taken into account. The implications of "race suicide,"[10] as Theodore Roosevelt had early put it, were especially concerning when considered against "the yellow peril." It is a formulation of the world population problem that retained great currency, both for proponents and critics.[11]

However, creating "white men's countries" through race-based immigration restriction was not the only response at the time to the question of what to do about this newly "filled up" Earth. Rather more surprisingly, there were a significant number of demographers, economists, and geographers who were troubled by those very acts, the demographic rationale through which they were so often justified, and their implications for a divided world. This is the third and far less investigated reason why immigration restriction and population control were so often twinned. Not advocacy, but the strongest possible critique of immigration restriction was evident in one strand of Malthusian political and economic writing from the 1920s through to the middle of the twentieth century. Many protested that since differential density between nations was the source of so much "international tension" as it was often put, the mass movement of people was a solution that was literally being blocked. Far from restricting immigration and emigration, the intercontinental relocation of people should be encouraged: this required the dismantling of national regulations not their construction, or at the very least a closer alignment between national migration regulations and international needs and interests.

Population, "Race Suicide," and the Global Color Line

The idea of "race suicide" typically raised two population dynamics: white birthrate decline, and a relative nonwhite population increase that was sometimes comprehended as actual east and south Asian population growth. Before and during World War I this idea had settled as a demographically inspired Anglophone Nordicism. It referred to a global situation in which the United States and other "white" powers would be politically threatened by nonwhite populations whose fertility rates were apparently not declining. Colonial rule was at stake. Often enough, this seeming Asian population growth was construed as an effect of colonization itself: decreasing mortality rates due to imported health and sanitation measures. Roosevelt wrote, citing British–Australian historian Charles Pearson, and referring to Malthusian "checks," that "the very continuance of European rule, doing away with war and famine, produces an increase of population and a solidarity of the country, which will enable the people to overthrow that European rule."[12] This was the coming race war, many declared: fundamentally about population, fundamentally about territory.

Madison Grant's *The Passing of the Great Race* and Lothrop Stoddard's *Rising Tide of Color Against White World-Supremacy* are often analyzed in this context, manifestos for action against both "race suicide" and the "yellow peril."[13] For Harvard historian Stoddard, the decline in the white birthrate followed a prior phenomenon, equally important for commentators on population at the time, though less so for historians since: the great British, European, and American increase of the nineteenth century. As Stoddard put it, deploying enduring hydraulic metaphors, if industrialization had produced a global "white flood," materializing as white colonization, settlement, and rule, this was now "ebbing." The ebb in part created the apparent "deluge" in this formulation—the rising tide of color. No statistician, Stoddard calculated population crudely, in total global numbers first: 550 million whites and 1.15 billion "colored," he estimated in 1920. He linked this directly to rates of growth: "There can be no doubt that at present the colored races are increasing very much faster than the white." Stoddard continued the tradition of calculating doubling populations begun by Benjamin Franklin. White populations, he said, double in eighty years, yellow and brown populations in sixty years, black populations in forty years: "The whites are thus the lowest breeders." Stoddard claimed that "none of the colored races shows perceptible signs of declining birth-rate, all tending to breed up to the limits of available subsistence." But the "limits" were being removed by white man and his civilization: tribal war, epidemic disease, and famine were everywhere reduced as an effect of colonization, he worried.[14]

Madison Grant, who wrote the foreword to Stoddard's book, argued that the white race, the "Nordics," must fight to retain global control. "Fight it must, but let that fight be not a civil war against its own blood kindred but against the dangerous foreign races, whether they advance sword in hand or in the more insidious guise of beggars at our gates, pleading for admittance to share our prosperity." And in a reversed but derivative concern of Benjamin Franklin's early aspirations for a white North America, Grant noted: "If we continue to allow them to enter they will in time drive us out of our own land by mere force of breeding."[15]

All this, according to Stoddard was the world's greatest crisis, because it threatened to undo what was for him nothing less than a "Nordic *entente*—a Pan-Nordic syndication of power." He meant European colonization of the world: "The great explorers and empire-builders who spread white ascendancy to the ends of the earth felt that they were apostles of their race and civilization." And yet this dream was smashed by narrow-minded nationalist-imperialists: pan-Slavists and pan-Germanists put forth "literally boundless pretentions, planning the domination of the entire planet." They were thinking nationally, instead of racially, according to Stoddard, who understood World War I—the European war—to be "nothing short of a headlong plunge into white race-suicide."[16] It was a civil war, he thought, among the whites of the world, threatening their thin global hold on power. Stoddard and Grant's work was shot through, then, with a sense of imminent white decline: land and global power was literally, for them, a white birthright.

When historians draw such attention to the "yellow peril" idea with its crude demographic premise of population growth in Asia, what is overlooked is the distinct lack of clarity about the demographic situation at the time. Many statisticians queried, at the very least, Lothrop Stoddard's claims that "colored" populations were doubling their numbers more quickly than "whites." The trends were quite difficult to gauge in fact, and population experts knew it. Stoddard was actively ignoring much of the available work on Japan, the widely recognized uncertainty about China, and disputes over Indian birthrates.[17] In his 1915 study, for example, Warren Thompson indicated that "the birth rate is declining in nearly all countries," with the exception of Bulgaria, Romania, and Japan.[18] Later, British demographer Enid Charles stated that in the absence of good data, it was safe to say that the population of China was practically stationary and had been for a long time, while Indian net population was probably slowly increasing.[19] Carr-Saunders indicated that most authorities believed there had been no increase in the population of China during the previous eighty years, and importantly, he explained this in terms of spatial "saturation": "The population has not increased because the country can hold no more people as things are."[20] He thought that Japanese population growth was identifiable over the last sixty years, but was not particularly rapid.[21] Rajani Kanta Das, an economist at the ILO, indicated that if India's population had increased 20 percent in the previous fifty years, Europe's had increased 47 percent.[22] At the 1931 Rome International Congress on Population, Benoy Kumar Sarkar even argued that a decline of birthrates was already a fact of Indian demography, a position explained with reference to Pearl's law of population. In other words, he was already thinking through an Indian demographic transition. The rise and fall of "the birth curve" in India matched that of Europe, he suggested, but "at the chronological distance of a few decades."[23]

Japanese demographic trends were particularly difficult to pin down. Teijiro Uyeda, for one, projected a slowing growth rate.[24] Especially influential in considerations of the Japanese situation was the work of Ernest Francis Penrose of the Food Research Institute at Stanford University. Penrose had lived for equal periods in the United States, Great Britain, and Japan, offering, he claimed, "an impartial and cosmopolitan attitude, free from the influence of tribal sentiment." His aim was to dispute the recent proliferation of interest in the "population problems of the Orient."[25] Penrose's book *Population Theories and Their Application with Special Reference to Japan* emerged from the 1927 Honolulu Conference of the Institute of Pacific Relations, where discussions between the Chinese and Japanese were as tense as discussions between the Australians and everyone else, with Australia's immigration act very much on the table. Japanese birthrates were in fact falling, Penrose claimed, and so, although there was certainly a population problem, it was a medium-term one that would resolve over several generations. Like others, he thought Japan's birthrate was comparable to, not wildly higher than, European birthrates, and therefore not especially troublesome. In 1931, the Japanese birthrate was 32.16 per 1,000.[26] The peak English birthrate had been 34 per 1,000 in 1880, and the German birthrate was about this in 1904.

Stoddard was obsessing about fertility rates in a manner inherited from the Roosevelt era. Yet even then there had been significant disquiet about capitulating to a "yellow peril" anxiety. Keynes had grappled with this aspect of the global problem, explaining to his Cambridge students the various "national prejudices and policies," as he perceived them just before World War I: "Alien acts, military strength, imperialism, yellow peril, South African, Australian, and American feeling against coloured immigration. In these forms the primitive instinct for the preservation of one's own race, whatever it may be, now shows itself." But some leaders had freed themselves from that primitive instinct, he wanted his students to know: "They feel sympathy with the aspirations of the other and very alien races; they are pleased by the existence of varied civilisations and would assist weaker nationalities; they are less convinced than the former that their own race contains within itself all that is in the world most desirable; and they are occupied by the task of improving the moral and material conditions of their own and other races."[27] Stoddard, though, was having none of this fence-sitting. To substantiate the "rising tide of color," he had to actively ignore the multiple factors that constituted population trends, most importantly mortality rates. Fertility growth or decline was not the same as net population growth or decline, although then as now, one could fairly easily be sold as the other to receptive and inexpert audiences. The lowering rate of infant mortality, led by New Zealand but evident also in the United States, the United Kingdom, Australia, and Canada, was why population growth rates were in fact higher in "white man's countries" than anywhere else. Carr-Saunders saw clearly, for example, that even if the Japanese population had recently increased, it was "little more than half that shown by the United States and the British Dominions during the same period."[28] His own maps of population increase showed the settler colonies, alongside Japan and eastern Europe, as "black," that is, as having the highest rate of natural increase (figure 4.1).

Perhaps more important, various experts with a stake in world population discussion critiqued Stoddard pointedly, because he represented old-school, "race suicide" anxieties that seemed outdated after the war. Edward East's *Mankind at the Crossroads* was an attempt to introduce good quantitative data to temper the ongoing scares about "racial danger" represented

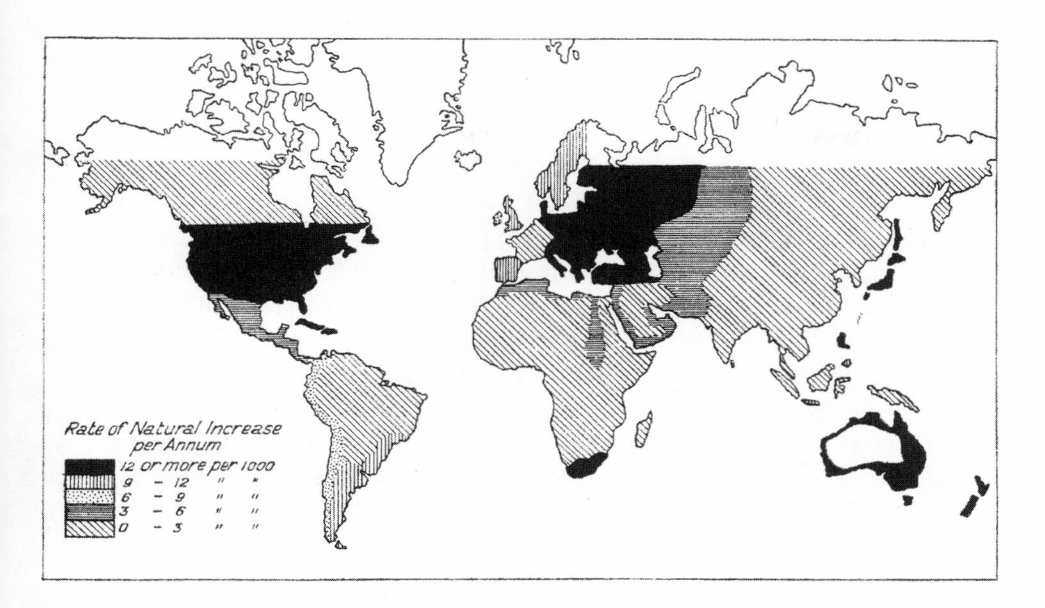

FIGURE 4.1 Comparative natural increase. Global maps of natural increase sometimes countered the presumptions of Asian population growth. (From A. M. Carr-Saunders, *Population* [London: Oxford University Press, 1925], 75. Reproduced by kind permission, Oxford University Press.)

in and by Stoddard's book. "No such dangers are impending," the geneticist wrote. East annihilated Stoddard's figures, beginning with the world population data and its constituents of "whites," "yellows," "browns," and "blacks."[29] Using George Knibbs's and Warren Thompson's data for projected increases in the white race, East put it far higher than Stoddard. Calculating white increase at the minimum for which there was decent evidence, and colored increase at the maximum, he concluded that "the white race is even today numerically superior to any other. Its growth is remarkable. . . . Before 1950 . . . the white race will have a true majority instead of a plurality. This will be true notwithstanding the fact that white birth-rates are falling in many countries, because death-rates have also receded." Importantly, his explanation for this massive and ongoing white demographic dominance was geopolitical. Land, characteristically, was everything: "The white race is increasing rapidly. . . . Simply because it has political control of nine-tenths of the habitable globe, and because it has the ability to utilize the space it holds." And even more starkly: "Due to the fact this race holds all the remaining under-populated territory, it will soon outnumber all other peoples combined. Thus, there is no necessity of the white race increasing its birth-rate in a vain competition for race survival. It will survive, simply because the other races have no room to expand."[30] At one and the same time, East dismissed Stoddard and offered an alternative argument for immigration restriction.

Though critical of the Grant–Stoddard school, Edward East was certainly no anticolonialist or antiracist, and he fully endorsed the recent immigration restrictions that his federal Congress had passed. And so, while East might have disposed of Stoddard's statistics (there was no "rising tide of color"), he certainly did not undermine Stoddard's position politically: "white world supremacy" was appropriately intact. East, for one, held the line, and wanted to see the "color line" hold, both with regard to race-based exclusions and with regard to "intercrossing among the primary races," which he assessed as a geneticist as a likely future problem.[31]

Critiques of Immigration Restriction

Warren Thompson was happy to read Edward East dismissing his Harvard colleague Lothrop Stoddard's work as "inaccurate vaporings."[32] Yet he was to dispose of the *Rising Tide* nonsense not just statistically but politically as well. If Stoddard and East both presumed that modern European population increase and the concomitant territorial expansion had been a global and civilizational good, Thompson thought otherwise. "From constituting one-fifth . . . of the human race in 1800," he wrote, "Europeans had become about one-third in a little over a century."[33] In this period, the great European increase created a situation in which Europeans moved outward to the rest of the globe, as emigrants and as forced labor, in the great colonizing moves of the nineteenth century. As we see in the next chapter, anticolonialist Thompson thought that the very right of white control over land was up for question. It was from this position that he spearheaded demographers' critique of the immigration acts. He asked why the current "expanding peoples" of the globe should "sit quietly by and starve" while the previously expanding peoples enjoyed the lion's share of global wealth and land, acquired by virtue of the fact that their population grew at a time when the globe was open rather than closed. Thompson considered acquiescence on the part of densely populated nations unjust as well as unlikely, and a situation aggravated rather than solved by the immigration acts. He was a fierce opponent.

The earliest Chinese restriction laws had been enacted in the self-governing colony of Victoria and in California in the 1850s, and by the 1920s restrictions were far more widely and insistently applied in both Australia and the United States. Because of this original legislative activity, and because of the great expanse of those continents (so much of the earth was at stake), it was the United States case and the Australian case that garnered particular critical attention. The Australian Immigration Restriction Act (1901) was part of a broader "white Australia" policy that had been internationally scrutinized in Paris in 1919. There, the right of sovereign nations to exclude people on the basis of race was formally questioned by the Japanese delegation.[34] The "right" was upheld, however, and Australians at various international fora over the 1920s stood firmly, even stridently, by this international decision as well as their own long-standing domestic policy. When the Institute of Pacific Relations addressed the topic of population and migration at its 1929 meeting, for example, the Australian delegation insisted that in the light of the Paris decision, its national population policies would not be up for discussion: "The English-speaking countries of the Pacific will not willingly permit their population policies to become a matter of international politics, or even of international discussion. In pursuance of those policies legislative measures have been taken with the object generally of preserving the economic standards as well as the desired racial character of the population."[35] It was a weak assertion, because the Australian insistence on reserving such a massive part of the earth's surface for such a small white population was already a matter of international politics.

The consolidation of U.S. immigration restriction in 1924 also came under scrutiny. In this context, the law's global effects, as opposed to national effects, were examined.[36] Many criticized American law and policy as limiting exit from European regions widely constructed as dangerously overpopulated. British eugenics leader Carlos Paton Blacker put it presciently: "Now that restrictions have been imposed on the entry of Italians into America, it is conceivable that Italy may at some time discover an Imperialist mission on the shores of the Mediterranean."[37] Differential population densities were exacerbated and even created by the immigration restriction acts, thought Thompson, similarly isolating the Italian case. While 25 percent of Italy's "natural increase" had migrated to the United States in the prewar years, now such movement was largely barred, acutely compounding Italy's overcrowding. The U.S. Immigration Act itself, and especially the "expert" debate surrounding it (which Thompson himself put in quotation marks) reflect "but little credit on either our intelligence or our honesty. But we may rest assured that it will not be forgotten by the Italians." Alternative outlets must be found. While he thought Italian colonies in north and east Africa "of little worth agriculturally," South American destinations were to be encouraged ("there is no land available there for annexation as colonies, but there is large opportunity for migration"). He thought the League of Nations, which had mandated Syria to France and Mesopotamia to Britain, should consider encouraging those national governments to take up a policy of Italian movement to "Asia Minor," especially since the climate was similar. Indeed, the control of Syria might be usefully transferred from France to Italy, he thought, as a way to redress Italian population density.[38]

There were several grounds on which the immigration acts were problematized. Many thought, like Thompson, that they were likely to compound the possibility of future war.[39] Likewise, French political economist Étienne Dennery was convinced that the acts would heighten and even create, rather than solve the problems of a racially divided world. The people of Asia have been "shut within their own territories." The ramifications of the immigration acts would likely have worse consequences than emigration and immigration itself, he wrote, since "the prejudice against colour denies him [the Indian] and the Japanese alike access to nearly all the lands where they would like to earn their daily bread." This produced anticolonial sentiment and political activity, especially in India, with far-reaching consequences. "A solidarity of the coloured races is aroused, far more intense than mere nationalist feeling."[40] On this view, not just anticolonial nationalism, but pan-Asian political solidarity was an effect of the exclusionary immigration acts.

In offering such an argument, Dennery drew directly from the 1920s literature on world population: the Geneva conference proceedings, Warren Thompson's work, the Institute of Pacific Relations' proceedings, Rajani Kanta Das's *Hindustani Workers on the Pacific Coast* (1923), and Harold Cox's *Problem of Population* (1922).[41] Dennery also cited Sun Yat-sen on *The International Development of China* (1922), interested in the Chinese leader's counter-argument to Stoddard and his critique of colonization: far from whites being overrun, Sun Yat-sen thought the history of colonization and diminution of indigenous numbers more relevant to the modern global picture, and conceivably the "end of the Chinese people," who might be extinguished like the "Red Indians." In citing the Chinese nationalist leader, economist Dennery noted that in the mind of the "Asiatic," the struggle with the white man was beginning to be understood as "a struggle for life itself."[42] And Dennery referenced another period altogether: the German geopolitical tradition on Chinese and Japanese migrations, including Friedrich Ratzel's *Die Chinesische Auswanderung* (1876) and Ernst Grunfeld's *Die Japanische Auswanderung* (1913).[43] Dennery's deep knowledge of the global situation recommended him as one of the experts appointed to gather the vast set of information at the Peaceful Change meetings held in 1937.[44]

Armed with considerable experience of east Asia (in the light of which he was later appointed French ambassador to Japan), Dennery published *Foules d'Asie: surpopulation japanaise, expansion chinoise, émigration indienne* in 1930, and a year later it was translated and published by Jonathan Cape as *Asia's Teeming Millions and Its Problems for the West*. The title of this book would seem to place it within the Grant–Stoddard tradition: certainly modern scholars often assume this to be one of many books to focus on Asian crowding, the latter-day "yellow peril" discourse that gave rise to the immigration acts.[45] In fact, Dennery's main message was critical of the global proliferation of laws that limited Asian entry and movement. He called up the idea in order to dispel it. "Pent within the narrow limits of their homes, will these masses ever be able to break down the barriers that surround them and pour forth into other continents? Truth to tell, this yellow peril, as viewed by the people at large, is not an actual danger."[46] In this instance, density in south and east Asia was not justification for, but argument against the "dikes," as Lothrop Stoddard proposed them.

Likewise, Taraknath Das was insistently opposed to Stoddard in his 1925 address to the New York Neo-Malthusian and Birth Control Conference. The menace was not a rising tide of color, but a "white peril," he pronounced, referring to eighteenth- and nineteenth-century European population growth and territorial expansion. He dismissed the so-called rising tide of color as so much mischief, with no foundation in fact. Das refused to accept that India had reached "the so-called saturation point of population," considering simply incorrect any presumption that India was overpopulated. India's rate of growth was comparatively slow, he pointed out.[47] Problematizing an imagined Indian overpopulation and not an actual English overpopulation was both nonsense and politically sinister when contextualized by the history of British expansion of territory. This argument was deeply connected to Das's anticolonial nationalism, as Matthew Connelly has shown.[48] For Das, the long history to be drawn was not just eighteenth- and nineteenth-century British-directed deindustrialization of India (by substituting its own spinning and weaving industries with Lancashire cotton),[49] but also an insistence on the territorial implications of British overpopulation. Since

1848, he argued, European nations had acquired 13 million square miles of territory from non-European peoples, and were presently tending "to fence a large part of the world in North America, South America, Africa, Australia and certain portions of Asia against the people of Asia, including India."[50] Parts of the world had been fenced in very effectively indeed, as he spoke in 1925, and Das knew this better than most. He had spent many years on the west coast of North America—both in Canada and the United States—informally representing the interests of migrating Indians, and for seven years he had been employed as a translator in the Canadian Department of Immigration in Vancouver.[51] His journalism and activities became increasingly anti-British, and in 1925, at the time of the Neo-Malthusian Conference, he had recently been released from jail, having been tried and convicted at the 1917 Hindu-German Conspiracy Trial. Das had far more than book learning when it came to migration questions and their connections to colonial and world politics.

It is unsurprising that so many south Asian economists and demographers gravitated to this question. The long-standing strength and status of Indian economic scholarship and the British investment in census technologies had produced generations of scholars on India and population.[52] Perhaps the most complex was Lucknow economist, ecologist, and sociologist Radhakamal Mukerjee. Like Das, Mukerjee turned Stoddard's arguments around entirely: "It is the aggressive policy of America and Canada, and particularly of Australia, against the Asian migration which is responsible for the rising tide of colour."[53] And with reference to similar work by geographer J. W. Gregory, Mukerjee wrote: "The menace of colour is as much a scaremonger's phantasy as the common notion of the thoughtless reproduction in China, India and Japan is fallacious."[54] Mukerjee pressed arguments against immigration acts into a systematic and strong anticolonial critique, squarely in the context of world population growth.

Born in western Bengal in 1889, son of a barrister, Mukerjee was educated and later taught at the University of Calcutta and University of the Punjab in economics, sociology, and political philosophy, eventually becoming professor and vice chancellor of the University of Lucknow. His young life, by his own elegant account, was shaped by the visit of mystic saints to his household and by an education in Sanskrit literature and English history, literature, and philosophy. Mukerjee's expertise was presented to various late-colonial economic enquiries—the 1924 Todhunter Taxation Enquiry and the 1927 Royal Commission on Agriculture—and he rose within the Indian National Congress to chair its important committees on population policy, influencing Nehru both before and after independence.[55] Well known in international academic circles, all of his earliest books were published by prestigious London houses, and his articles circulated through Indian, U.S., and European journals, in several languages. He spent time in Rome and London, engaging with experts connected to the International Union for the Scientific Investigation of Population Problems (IUSIP), and was invited professor on lecture tours of the United States, visiting Columbia University, Chicago, Michigan, and Wisconsin. It was his contributions to the Rome meeting of the IUSIP in 1931 that alerted its chair, Corrado Gini, to this radical set of Asian ideas, and it was Gini's committee that published just one of Mukerjee's forty books, *Migrant Asia,* in English and in Italian.

Like so many studies, Mukerjee opened *Migrant Asia* with a map of density (see figure 3.3). Such a map announced that "Asiatics," who represented half the human race, were artificially and problematically confined to 4 percent of the globe's surface. This immediately foregrounded his proposition that population density was in large part the effect of the barriers, not the reason for them. Immigration restriction artificially stopped a natural dynamic of population movement from more to less dense areas. Exchanging hydraulics for something like a Boyle's law of population, he explained: "As air flows to regions of low density, the excessive pressure of the Asiatic population seeks release through its overflow into all the thinly inhabited regions of the Indo-Pacific area."[56] More than other commentators, Mukerjee presented positive arguments for Asian emigration and immigration, as much as he critiqued the restriction acts. For most countries, he noted, an immigrant who can labor is an asset, and especially so where land is not yet developed. Unlike the nations that tended to send industrially trained emigrants—*more* likely to cause social unrest, he thought—Asia could send compliant but skilled agriculturalists and land settlers.[57] For Mukerjee, the immigration restriction laws represented a "lack of normal adjustment in the field of the migrations of labour" and systematically hindered world economic development, as well as the development of each particular excluding nation. In the same way that Indian labor had produced so much British imperial wealth in the late nineteenth century, it could produce global wealth in the twentieth century too. This tradition of wealth production should be extended and promoted through open global movement, not limited through nonsensical race-based exclusion acts: "The policy of exclusion fanned by race prejudice cannot last in the face of the gradually increasing economic interdependence between India, China and Japan, on the one hand, and the American and Australian on the other."[58] In Mukerjee's mid-1930s, Depression-inflected view, allowing Indian laborers to migrate, putting them hard to work, would mean nothing less than a recovery of the world economy.[59]

It is important to recognize that Mukerjee's arguments against the immigration acts, and Das's similar position, did not spring only from the anticolonial nationalism that they certainly shared. The same arguments were clearly discernable in Anglophone population scholarship more broadly. Warren Thompson was unequivocal: "I do not take any stock in the idea that there is now a conspiracy of the coloured races against the white race. But if the white race continues to exploit the coloured races, and if it denies them a fair share in the resources of the world, then it would be but natural to find an alliance growing up against the present masters."[60] Harold Wright asked in 1923: "Can we tell them that they must limit their numbers while Europe continues to increase and spread its children over the world?"[61] And it was Harold Cox who wrote the foreword to the English edition of Dennery's book, summarizing the way in which the immigration acts created the problems they ostensibly sought to solve.[62]

Over the 1920s and 1930s, the question of density, pressure, and over- and underpopulation, also came to be read within the terms of early twentieth-century mass or crowd psychology. With increased globalization, people in south and east Asian societies would become newly aware that their standard of living was lower and their local world more dense than adjacent regions. "Over population is essentially a state of mind—a condition of '*felt* population pressure,'" thought Frederick Sherwood Dunn in 1937, at which point the issue was far from abstract.[63] This was argued over Japan in particular, whose consciousness of population pressure was acute, Anglophone scholars told each other: the Japanese were unwilling and unlikely to remain quiescent.[64] The conclusion was that this would engender both local and eventually international tensions. Apprehensive about the effects of overpopulation, "expansion" was cast even as "a psychological remedy for their fears."[65]

Other arguments deployed against the immigration acts concerned principles of freedom of exit and entry. Many at the 1927 World Population Conference sought a return to a nineteenth-century model of unrestricted movement, framed as a golden age when people were unrestrained by national legislation—the migrant was "free."[66] Paris-based Russian professor of Slavic studies, Dr. A. Koulisher, pronounced the nineteenth century "the only epoch in world history when international migration was free. There existed then a kind of universal citizenship right."[67] By the early twentieth century, this narrative continued, multiple nation-states had erected barriers around their territories, "the closing up of these immigration countries." It was a chronology framed far more around restrictions on Europeans than restrictions on Chinese and Indian movement. This line of argument was pursued at the 1933 International Birth Control Conference in London, the meeting dedicated to "Birth Control in Asia." A Japanese delegate complained: "Man-Made expedients, such as political frontiers and economic barriers, are obstructing the free movement of people and wealth, though it is a commonplace that the world is large enough to accommodate in comfort many more people than are in it today."[68] At the 1937 assembly of the League of Nations, a Polish delegate spoke similarly on the need to return to "pre-1914 freedom of movement" as a way to equalize problematic differential population densities between nations.[69]

There were aspects of this issue that took a form specific to the British Empire and Commonwealth. While most of the "Dominions"—the white settler colonies—had excluded Indians one way or another, this sat uncomfortably in an imperial context, not least for Queen Victoria, empress of India, and her descendants. There was weighty argument that Indian subjects should be free to travel and reside anywhere within the empire, and Indians were rightly resentful that they could not. Not a few British scholars and commentators supported the Indian position—over settler colonials—and saw this as grist for the Indian anticolonial mill. "It is certain," proclaimed Harold Cox, "that the bitterness thus engendered has helped to stimulate the spirit of revolt now raging throughout India against British rule."[70] Indians "want the right of freedom of movement within the empire as free men," Das announced in New York in 1925, citing Cox's work in support of this position.[71]

A final argument against immigration restriction that emerged most commonly in interwar world population discussion, and that retained real purchase through the post–World War II period, concerned the right to land according to need, where that need was food. This was a version of lebensraum, a reiteration of one of the core geopolitical ideas from Ratzel through Haushofer. Indeed Haushofer himself was deeply critical of U.S. and Australian immigration restrictions, both in defense of the Japanese and because immigration restriction cut across the natural law of the expansion of vital peoples. One of his maps in *Geopolitik des Pazifischen Ozeans* stamped Australia and North America "*Einwanderungs Verbot*" (migration forbidden) in solid caps and uppercase (figure 4.2). It signaled, for him, a problematic Pacific entente between the United States and Australia that cut across a Japanese right to move freely, indeed to dominate the region.

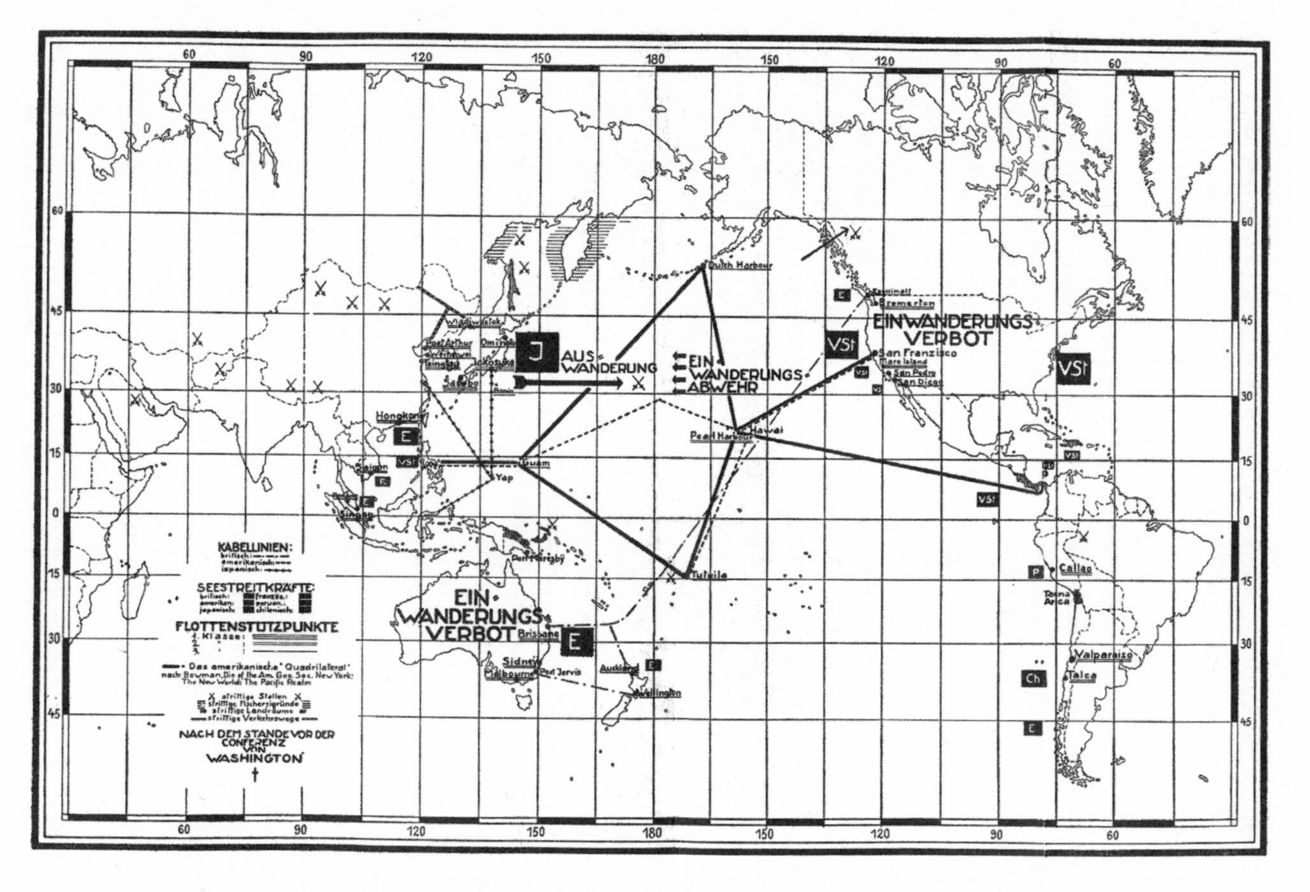

Figure 4.2 Karl Haushofer opposed the restrictions on Japanese movement in the Pacific "living space": "Migration Forbidden" in white Australia and America. ("Wehrgeopolitische Skizze des Pazifik nach dem Stande vor der Konferenz von Washington," in Karl Haushofer, *Geopolitik des Pazifischen Ozeans* [Heidelberg-Berlin: Vowinckel, 1938], 245.)

Emigration, Immigration Restriction, and Birth Control

In the "race suicide" tradition, the global color line was understood to be a response to an apparent increase in Asian population growth in relation to the U.S., British, and Australasian decline in fertility. The solution to a global problem lay in holding on to continents as "white" and in increasing white fertility rates through various pronatalist programs. But in the British imperial context, there was also an argument for white settlement and sovereignty that stemmed from concern about white population *growth*, not "suicidal" decline. This was a territorial argument not about the reproduction of humans but about production of food to sustain them, in particular a British anxiety about its reliance on wheat importation. Just when Roosevelt was inventing and inflaming his concern about race suicide,[72] British chemist William Crookes declared as catastrophic the massively increasing population of the "bread-eaters of the world." That 75 percent of British wheat came from elsewhere, he cautioned in 1898, spelled looming crisis when placed against diminishing wheat-growing lands.[73] Concern about food security manifested as argument *for*, not against, ongoing regulation of fertility rates in the West, and as a rationale for the securing of the white man's land across the globe.

Many interwar Anglophone population commentators advocated not Roosevelt-style pronatalism, then, but birth control: Western fertility decline was the model that the East should, and in some versions would, follow. The U.S. Department of Agriculture's Oliver Baker, for example, argued strongly in this tradition: control of fertility in the West would result in both food and land security.[74] In some versions, differential population growth between East and West, orient and occident as the regions were still frequently nominated, justified immigration restriction in the medium term. Edward Alsworth Ross placed this most squarely within a demographic argument. Less than one-sixth of the Earth's people "have their reproduction under control" and the rest of the world should follow that trend. Until then, it was likely that the "population surpluses" of the "congested peoples" will emigrate, and "the advanced peoples will be forced in sheer self-defense to bar out mass immigration."[75] In other words, the demographic argument for immigration restriction was not necessarily accompanied by pronatalism in the West (Roosevelt's version); rather, it sometimes accompanied birth control advocacy.

Ross saw birth control and migration as so connected that he structured his long book accordingly. The first half of *Standing Room Only?* was titled

"Population," and dealt with population pressure, war, and what he called "adaptive fertility," meaning birth control. The second half was titled, significantly, "International Migration," with two final chapters on the acts titled "The Closing Gates" and "The Coming Great Barrier." Like the Great Antarctic Ice Barrier, Ross wrote in a not entirely successful simile, somewhere between 35 and 40 nations have erected laws restricting immigration from Asia and Africa. "There will emerge a fairly uniform dike against pressure from the chief reservoirs of human beings." The exclusion, Ross insisted thinly, was not based on race prejudice even if that argument had been deployed historically. Rather, the policy of exclusion was based on the principle of homogeneity, including standard of living, and what he called "the laws of human increase:" "Until the Asiatic thinkers wake up to the population tendencies in the world today and perceive the folly of curing famine and pestilence without at the same time delaying marriage or restricting births, they must expect that in sheer self-defense the Western peoples will meet the menace of a deluge of Asiatics by throwing up barriers."[76] Asia must "learn the lessons of voluntary parenthood as the West is doing it," he argued, a lesson that would be delayed by opening the doors of the Australian and North American continents.

The lesson was constructed as one of modernity. In Ross's volume, too, lay an early expression of a demographic transition and its links to theories and politics of modernization and westernization. "Civilization" had a demographic meaning. The "revolution [of adaptive fertility] is destined to spread over the earth in time, relieve population pressure, and pave the way for a universal sharing of the blessings of civilization." Ross told his readers that the Great Barrier was in no sense to be imagined as a permanent global state. Rather, it would be lifted once modernity, or "civilization" was achieved.[77] This was precisely how Kingsley Davis was to frame world demographic transition in 1945.

If Ross wrote of a "League of Congested Peoples," which, if provoked, could divide the world into two great camps, others wrote of the corollary, a much more desirable League of Low Birth Rate Nations, which should lead the global way, defending and maintaining their admirable demographic trend. "League of Low Birth-rate Nations" was Harold Cox's paper at the 1925 Neo-Malthusian Conference in New York.[78] But unlike Roosevelt's "race suicide," and even Stoddard's extension of this tradition, in this Malthusian argument, Western fertility decline was a praiseworthy development that the East would do well to follow. Charles Drysdale, that mainstay of the British Malthusian League, argued often for a federation of "low birth rate nations" that would gradually encompass the world, demographically and politically, in a cosmopolitan and globalized federation of world states: "Remove all barriers between the low birth rate nations, and aim finally at a universal federation where nations will appear as states of a Union and frontiers will cease to be barriers or to have any special significance."[79] Harold Cox agreed, adding that the League of Low Birth Rate Nations would usefully be able to focus on the quality of their populations, having mastered the quantity issue.[80] Eugenics, in this model, was phase two of world population control.

There was a fairly common argument that for all nations, emigration of "surplus people" was a second-order solution to problematic national density; nations should first aim to restrict births domestically, whether through birth control methods and devices or through later marriage. Emigration would only result in relief from population pressure if that country or region could first retain a stationary population. Henry Pratt Fairchild was clear about this: "No overpopulated country can claim the right of emigration until it has demonstrated its ability to maintain a stationary population without emigration. Then there is reasonable hope that it could utilize emigration to reduce population."[81] Along the same lines, Ross argued that Asian countries should be made to experience the consequences of their own overpopulation and growing birthrates. The immigration restriction acts would give notice to Asian peoples that "enlightened humanity is not willing to cramp itself in order that these peoples may continue to indulge in thoughtless reproduction." But as we have seen, many others were deeply troubled by the immigration acts and the differential densities they would compound. They argued that birth control measures were inappropriately forced on some countries precisely because of the restrictions on global movement; in a problematic way, birth control thus became their only alternative, the only option open to Indians, Chinese, and Japanese.

At the Indian Population Conference in Lucknow in 1936, much of the discussion was shaped by the absence of a migration option "practically closed to the people of India."[82] This was Radhakamal Mukejee's position. He thought that birth control "checks" should be encouraged in India, China, and especially Japan, but that "this must not conceal or minimise the importance of finding fresh and convenient outlets for Asiatic emigration."[83] Margaret Sanger argued the other way around: "Emigration ceased to provide a space for the problem of overpopulation." This left only two alternatives, "death—nature's method of birth control; and birth control proper."[84] Retrospectively, however, Sanger constructed a critique of immigration restriction aligned with some of her interwar contemporaries. Immediately after World War II, she ventriloquized Warren Thompson: "Everyone knew that Japan would have to have war if she kept on increasing her population. . . . All the doors of the world were closed against the Japanese, and they had no place to go except South America. Australia, Canada, and the United States would not have them." She recalled her invitation to Japan to speak at a roundtable on population and peace. She was to join Einstein on relativity, H. G. Wells on reconstruction, and Bertrand Russell on "something or other that I have forgotten," a comment more acerbic than amnesic. Famously, Sanger was herself refused a visa to enter Japan.[85]

Some proponents of birth control recognized that the controversial status of contraceptive knowledge and practice made its promotion difficult and unrealistic, leaving population movement or emigration the only likely short-term solution. George Knibbs wrote rather more generously than Ross, that emigration measures needed to be pursued by various countries precisely because birth control advocacy was politically difficult and sometimes impossible, and reminding us that the international discussion was not solely about east and south Asia.[86] He explained that "over-peopled" nations declaring against birth control are thereby "compelled to study all existing opportunities for expansion of migration."[87] Warren Thompson, with a twist to this argument, thought that the international movement of "crowded peoples" was a highly desirable short-term (perhaps three-generation) solution to an immediate global geopolitical crisis, but that ultimately birth control was the remedy for overcrowding. Moreover, emigration itself—the reduction of pressure and the increase in standard of living, even if only momentary—"produces the conditions in which birth control thrives."[88] This was a position later largely shared by the Princeton Office of Population Research as they developed theories of demographic transition and economic status: emigration was a palliative but not a permanent solution.[89] Ultimately, it was to prove no solution at all.

It is important to note that emigration was set up as a policy alternative to birth control not just by those like Knibbs and Thompson, who were firm advocates, but also by those who were actively opposed to birth control. Along these lines the Roman Catholic Church promoted the spatial rather than sexual management of people as its response to the problem of world population growth. This was the case increasingly over time, as birth control became more acceptable to states and intergovernmental organizations, and in the face of growing interest in birth control and population control exhibited by the United Nations and United States after World War II. At the 1954 International Catholic Migration Congress, for example, "European over-flow" still shaped the church's discussion.[90]

World Population Policy: Managing World Movement

For some in these years, "world population policy" meant a unified front on birth control and the need to diminish birthrates globally. This was the "really pan-human question of population growth and birth control," as a Czech participant at the Geneva meeting put it.[91] Oliver Baker also thought that the time might be approaching when world plans might begin to replace national plans and policies.[92] But Baker was talking about the spatial dimension of the population problem; the immigration acts, not contraception. Because maldistribution of people across the globe was a key problem, a "world population policy" was needed to govern migration and movement from an international perspective. The nations of the world, however reluctantly, would be compelled to combine international consultation and action, it was announced in an early issue of the journal *Population*.[93]

The national immigration acts were extending into the interwar period the kind of belligerent nationalism that liberal internationalists had long sought to overcome. According to Jean Bourdon, director of International Studies at the Sorbonne, if national interests were set aside for wider points of view—"say the point of view of the League of Nations"—people from Japan, and possibly overcrowded Russia in the future, would be allotted and moved to Canada, Australia, New Zealand, the temperate zone of South America, and the south of Siberia.[94] Some scholars argued for a radical deregulation of international movement, allowing for a natural flow of population like a free exchange of goods to reassert itself. This was a liberal internationalism in the population sphere that would imitate the liberal internationalism that favored free trade in the economic sphere. In this manner, Norwegian economist Wilhelm Keilhau thought that far from developing supranational mechanisms, people should not be directed at all. He argued that all humans have the right to move freely "based on the fact that we are all compatriots of the same planet."[95] Into the 1930s, though, many thought it as futile to talk of returning to free migration of the nineteenth century as it was to imagine restoring that era's economic liberalism and free trade.[96] Not less regulation but different regulation of movement was needed.

In this context, the ILO and Albert Thomas could imagine and propose a world court, a fully supranational body with authority to direct the movement of people around the globe. For Thomas, such plans were not only worthwhile but also realistic. In his view, of all the variables that affected population and human organization, migration was the most manageable, "the most susceptible to direct intervention and control." International migration might thus be regulated in global interests by "a supreme supernational authority . . . a Higher Migration Council with the power of deciding the right of overpopulated countries to populate other territory."[97] In making such a statement, Thomas spoke, as he saw it, on behalf of the "overpopulated" countries, signaled chiefly by laborers without work or without enough work. The ILO advocated an international migration board for many years, driven largely by Thomas's socialist and cosmopolitan politics and his long-standing term of office from 1919 to his death in 1932. He thought that migration had problematically assumed an increasingly national character, and that this posed real and direct problems for a world population policy based on the redistribution of people. Such ideas might be regarded as utopian or at least premature, he conceded, but they were beginning to be discussed by some nations and were implicit in much of the discussion at the 1927 International Economic Conference that the League organized in Geneva.[98] The ILO's own Permanent Migration Committee constantly discussed international movement in terms of population, resources, and labor, recommending a Permanent International Committee on Migration for Settlement.[99]

The ILO's work and claims were substantiated in the data gathered by Imre Ferenczi, its population and migration expert from 1920 to 1939 and author of the authoritative *International Migrations*.[100] Even before World War I—at a Zurich Conference for international associations for labor organization, social insurance, and unemployment—Ferenczi had put forward suggestions for international agreements on the migration of workers.[101] And indeed it was the ILO's standard-setting projects over many decades that came closest to a program for the international regulation of movement. Its cumulative documents, conventions, migration statistics, and collations of migration statutes were certainly relied upon in the post–World War II period, when international control of migration was again proposed.[102] It worked with the new agencies and economic frameworks to try to ensure planned and regulated migration of workers. It recommended to the new International Bank for Reconstruction and Development, for example, consideration of "the solution of migration problems" when proposing projects of economic development.[103] The ILO was sometimes put forward as a base and even a model from which world-level policy of all kinds might be developed and even implemented.[104]

Ferenczi and Thomas's visions for an internationally regulated labor market were not realized, but they do give a strong sense of the extent to which the world population problem was conceptualized economically as one of surplus or deficit labor. Ferenczi did not comprehend "international population policy" in terms of health and fertility at all. The League of Nations avoided population problems, he said, because "migration touches sensitive political issues, such as sovereignty." He conceded that there was some work being done in terms of protective measures and international migration data collection, but the League sidestepped what he saw as a real international population policy that would entail international regulation of the labor market, international employment services, "curbing of national sovereignty in matters concerning migration," and international aid to settle migrants overseas.[105]

Warren Thompson was also in favor of such authority being delegated to the League. It might consider "the claims of the nations needing more territory."[106] Likewise, Radhakamal Mukerjee thought that the problems that the immigration acts were causing, economically and in terms of raised international tensions, could only be solved by some kind of international machinery. Migration was a domain that should not be "left to drift or [be] solved by sovereign national powers," but one that called "for international adjustment."[107] For him, future world cooperation required "a revision of the tenaciously held doctrine of territorial sovereignty and the restrictionist immigration policy flowing from it, so that the overcrowded and dissatisfied nations in both Europe and Asia might obtain facilities of colonization and settlements in the wide open spaces of the earth." Open frontiers plus equal opportunity in the treatment of immigrants were the key solutions to world economic development, for "world amity" and for a lasting peace.[108]

At the 1937 International Institute of Intellectual Co-operation meeting, the repeal of restriction on colonial immigration was again presented as one possibility for "peaceful change."[109] Professor William Oualid proposed an international organization that would adjudicate between exaggerated ideas of "national conservation" on the one hand, and the inalienable rights of individual liberty and movement on the other. When the League of Nations finally appointed a Committee of Experts on Demographic Problems in 1939, it was charged with assessing the possibilities of internationally facilitating emigration and immigration "to reduce the effects of demographic tension."[110] As chair, Carr-Saunders was certainly interested in both freer movement of populations across national frontiers, as well as greater "freedom of movement of capital and of international trade . . . [to] relieve the pressure of population."[111] Yet in practice, the whole trajectory was toward more, not less, regulation of movement, and eventually toward biological control of reproduction, not spatial redistribution of people.

The rationale of relieving population pressure did drive some population transfers and resettlements implemented throughout the 1920s and 1930s. One-and-a-half million Greeks were successfully moved after the war, claimed Albert Thomas, and "put to work on a soil which had been drained and improved."[112] Some bilateral agreements concerning temporary and occasionally permanent migration had a "relief of population pressure" rationale behind them. Carr-Saunders noted a treaty between overpopulated Poland and underpopulated France whereby both nations agreed on the number and "class" of persons to be selected, sent, and received as a "collective migration."[113] This kind of controlled large-scale management of migration, "not the individual unselected migration of the past," was one model whereby overpopulated states could be assisted, though a difficult one, according to Yale Institute of International Studies economist Frederick Dunn.[114] In a different context, policy and practice within the British Empire and Commonwealth was explicitly implemented to relieve English overpopulation through the controlled promotion of migration to the settler colonies, but in a way that would keep people, resources, and wealth within the limits of the empire. The emigration promoted by the Empire Settlement Act (1922) was an interwar version of late nineteenth-century Greater Britain.[115]

Sir Arthur Salter, head of the Economic and Financial Section of the League's secretariat thought that world overpopulation was not inevitable, but was preventable through labor and migration schemes. Local surpluses of population were the pressing problem, in his view. "It should be possible, by national and international arrangements, to prevent these local surpluses driving the world to war. But if explosions are to be prevented, the peoples of the world must work positively and co-operatively at finding a safety-valve for the explosive forces, and not merely defensively at national policies designed to direct the explosion elsewhere."[116] Another solution that might be considered, he thought, was an international administration of territory. Land in Africa might be set aside for development and administered as an international protectorate, under the authority of an international body. The suggestion in 1884/1885 that the Congo Basin be neutralized and "held in trust for the benefit of all peoples" was cited as a precursor to this idea.[117]

The intergovernmental management of refugees represented one instance, and for some a model, of how people might be reallocated on a world plan that incorporated population, resource, and economic imbalances. The League's High Commission for Refugees, which became the Nansen International Office, managed refugees from Russia, Armenia, Assyria, and Turkey. For some time to come, "resettlement" was rationalized by the idea that refugees constituted a "surplus" population.[118] More directly connected, in principle and in action, were the long-standing plans for resettlement of European Jews, and in the United States, it was Isaiah Bowman who was presidentially charged to recommend worldwide solutions. Just when the geographer was collating *Limits of Land Settlement*, Franklin D. Roosevelt commissioned him to look for "uninhabited or sparsely inhabited good agricultural lands to which Jewish colonies might be sent."[119] This became the so-called M Project or Migration Project, driven by the founding idea that "population pressures had often caused wars throughout history."[120] The M Project reworked much of the expertise on land and population that geographers and demographers had covered over the prior twenty years. As Bowman put it, this was a body of work that considered various regions in terms of their "structure and growth, their standards of living, their freedom (or lack of it) to expand territorially, the possibility of increasing the development of undeveloped land."[121] Demographer Joseph Schechtman's *European Population Transfers, 1939–1945* and his *Population Transfers in Asia*, analyzed wartime and postwar versions of the territorial and geopolitical solution to population problems deriving from this discussion of the 1920s and 1930s.[122] It all prefaced "a territorialization of postwar planning," the most significant manifestation of which was the Jewish state in Palestine.[123] The linked demography and geography of population and land was deeply integrated into the science as well as the politics of Jewish resettlement.[124]

When Isaiah Bowman wrote in 1937 that "the population of the world has become stabilized," he was not referring to population growth rates over time but to the peaking of global movement, the end of "an epoch of migration," during which global lands had been discovered, occupied, and exploited. World War I and the consolidation of immigration regulations in the period immediately after "emphatically closed an era of western expansion that had opened three hundred years before."[125] That certain populations needed "outlets" in the form of migration or resettlement schemes was a modern global imperative. The world population problem was almost automatically the world migration problem, too. For many, this was precisely what constituted the geopolitics of population, caught between high nationalism and its twin, interwar internationalism, a problem for anticolonial nationalists, liberal internationalists, and lebensraum theorists alike.

5

Waste Lands

Sovereignty and the Anticolonial History of World Population

> No people has any moral right to hold lands out of use which are needed by other peoples.
>
> WARREN THOMPSON, *DANGER SPOTS IN WORLD POPULATION* (1929)

The claim that British population increase—that island's modern "overcrowding"—was a key factor in Atlantic and Pacific expansion, settlement, and colonization was common enough.[1] John Stuart Mill had said as much in the mid-nineteenth century, advocating "the removal of population from the overcrowded to the unoccupied parts of the earth's surface."[2] Time and again, population writers acknowledged that New World land was the necessary condition of and for nineteenth-century world population growth. This was the literal ground from which populations doubled and economies thrived—in South America, the Pacific, and especially on the famous frontier of the North American continent.[3] In the age of Malthus, "there was a superfluity of empty spaces on the globe," wrote Georges de Lapouge, as if New World land was there for the taking. The early twentieth-century moderns considered themselves confronted with a different problem: "Now these empty spaces are becoming rare, even in America."[4] Eastern and western, northern and southern hemispheres, New and Old Worlds, had been either claimed or actively settled with people, or both. As Henry Pratt Fairchild put it, "Whatever technical advances may still be in store, there will never be added to them another hemisphere."[5]

It was in this context that global empty spaces—"waste lands"—became both critical and politically fraught. In general, it was not the question of prior indigenous occupation that created political discomfort for population commentators, however. It was the fact that in a geopolitically closed world, all "fresh" land was part of a modern political system. "As the known earth has grown at the cost of the unknown," Ratzel had written, "political territories have continually increased in size and number, one unknown region after the other taking on political value."[6] If sparsely populated or unpeopled land used to be claimable through a commonly understood, if disputed, system of colonial occupation, now it had to be negotiated with political entities recognized within the law of nations. The history of world population growth had everything to do with the history of colonialism; indeed, this was a commonplace observation over the 1920s and 1930s. There was also a clear, if surprising, strand of anticolonial politics that ran through Malthusian thought, though not one centered on indigenous land rights. It derived, paradoxically, from Lockean justification of possession on the basis of land use. Some argued that "overpopulated" societies without enough food had a particular and special claim to wasted global lands. In an overpopulated and underfed world, what right, then, did "settlers" have to reserve that land for themselves? This raised nothing less than settler-colonials' right of occupation. World population growth was not just about colonialism, it was about sovereignty.

Exploring the Limits

In geographer Ravenstein's terms, what lands of the globe were still available for European settlement? For any human settlement? Some thought, to be sure, that the limit was not even on the horizon. William Beveridge, directing the London School of Economics in the 1920s, considered that there was plenty of land. Only the surface of the earth had been scratched, he said: a mere 8 percent of Canada's territory was utilized, 18 percent of Siberia, 6 percent of Australia, and 3 percent in South Africa.[7] This list of "waste lands" was itself becoming a canon: Canada, Argentina, Siberia, Australia, and parts of east and southern Africa were constantly referred to, each dramatically "underpopulated" and even cast as "empty," albeit clearly peopled and claimed.[8]

Increasingly, extreme zones came under expert geographical, economic, and demographic consideration: sub-Arctic and Antarctic islands, the polar continents, the vast deserts, and the jungle tropics. How far might the zones already considered "frontier" be pushed farther outward and upward, how could they accommodate more people, and grow more food? An entire generation of geographers was busy mapping global limits and new human–natural frontiers: world wheat lines, rice lines, and maize lines; lines of rainfall; latitudes and altitudes where average temperatures fell beyond the possibility of cultivation or even habitation. How far had humans *actually* cultivated land, and how far *could* they? This was the global "shore-line of settlement," according to agricultural economist Oliver Baker.[9] To Isaiah Bowman it was "the limits of land settlement," his great collaborative project that documented the climatic and environmental determinants of human settlement in Canada, Soviet Siberia, Australia, Africa, and South America. This "pioneer fringe" was the global zone that occupied geographers' mental space for a generation, now that the North American frontier was closed, indeed now that the world was closed.[10]

This is why two explorers circulated among the biologists and economists of population in the Geneva summer of 1927. John Walter Gregory and Vilhjalmur Stefansson were both taking a break from the rigors of extreme expeditions. Gregory came to exploration (and population) through geological and geographical training in London. He was a professor of geology at the University of Melbourne between 1900 and 1904, and subsequently at the University of Glasgow for twenty-five years. Stefansson, based in Canada and the United States, came to the population question through anthropological and ethnographic training. Both were fascinated by sparsely populated lands, and were commissioned to explore these spaces with new migrant populations and "resettlements" in mind. They were surely the last generation to self-identify as "explorer," but they were certainly qualified to do so, by any measure. Gregory had spent much of 1892 as a naturalist and geologist on an East African expedition. Shortly afterward, he traveled across central Australia and then Angola in 1908 and 1912, and he walked through Burma, China, and Tibet in 1922.[11] Five years after the Geneva conference he was drowned after his canoe overturned in the Urubamba River in Peru.

Recognized as a field expert on the global "people and land" question, he was also a key actor in the related project of Jewish resettlement. Gregory's expeditions to Libya and Angola were commissioned by the British-based Jewish Territorial Organisation to advise on possible colonization.[12] Gregory reported on suitable localities in terms of soil, climate, water supply, and mineral wealth, as well as likely political availability. He discussed possible Jewish settlement with the Bedouin locals in terms of a "concession" of territory.[13] Initially, it had been reported that the land was more or less empty: "Almost all the soil seems well adapted for cultivation, although only a very small portion is ploughed by the Arabs."[14] It was conceded eventually, however, that the current inhabitants "could not be expected to welcome alien colonists." In the end, it was not a positive result for the Jewish Territorial Organisation, because the region was plainly occupied, or as Gregory put it, "not so unoccupied as we had expected."[15] One outcome, though, was the early twentieth-century British offer of other land to the Jewish Territorial Organisation, a new settlement in an "East African Zion."[16] Gregory was later involved with the British-based Palestine Exploration Fund, signaling the important continuity between settler-colonization, migration, and resettlement of European Jews. Gregory's exploration and territorial resettlement work was for him of a piece with his major contributions to the international migration and population discussions.

Idiosyncratic anthropologist Vilhjalmur Stefansson went to equally unlikely places. His lifelong mission lay in revealing to policy makers, and to those anxious about population growth, the great possibilities for the future of the human species on the planet. Empty and sparsely populated spaces were vastly underutilized, he argued, and could be brought into production without too much difficulty. There were great tracts of unused land all over the globe, from the Arctic to the interior of Australia, that could and should be peopled. Stefansson was an extreme man, however, who felt comfortable in extreme places. His main expertise lay in the Arctic, where he had worked from 1904 in a series of notorious expeditions, many of which came undone in questionable circumstances. He led the Canadian Arctic Expedition of 1913–1916, leaving the ice-bound ship, he said, to hunt for food; others said he deliberately abandoned his crew. It was not the only expedition in which he ended up traveling alone or with Inuit guides. In 1921, he sought to "colonize" Wrangel Island for Canada, even though it was recognized as Russian; all the men he recruited for that expedition died on the island. Despite such disasters, Stefansson managed to publish (apparently without irony) *The Friendly Arctic* in 1922 and a swathe of journalistic pieces that many of his scientific colleagues in Geneva would have read.[17] Stefansson saw major potential for future populations in the northern zones of the Arctic circle—Greenland, Spitsbergen, Siberia, Alaska. The potential lay not just in the opportunity for a denser population but also in the use of this global zone for the cultivation of food for the world.[18]

Stefansson and Gregory were explorers in a period when continental and island land masses were known at their edges, but their interiors were not necessarily mapped. Some of these territories still had an unclear "political value," to use Ratzel's phrase. The 1920s and 1930s was a time when polar expeditions were becoming increasingly regular, when the Antarctic continent became nationally sectioned, and when Norwegian, American, Australian, British, and Russian expeditions by ship, sled, motor vehicle, and plane, accelerated in pace as well as technology. Scientific agendas thinly veiled occupation agendas.[19] New claims and old were challenged in international law that was itself, at this point, undergoing revisions with regard to the process and the right by which any nation could claim new territory, either peopled or unpeopled. Governments knew that cases would be strengthened by the demonstration of "effective occupation." Accordingly, the tradition of wintering on Arctic and Antarctic territory began to become standard.

Important principles of international law were sorted out over the disputed Arctic island of Spitsbergen in what became the Norwegian Svalbard Archipelago.[20] Before the 1920 treaty that settled Spitsbergen as part of Norway, it had been a no-man's-land, a site where the legal concept of *terra nullius* was contested. One geographer declared in 1919 that the island held "an almost unique position today in being a *terra nullius,* one of the last remaining territories on the face of the globe to be unclaimed by any state."[21] *No Man's Land* was quite rightly the title of explorer Sir Martin Conway's history of Spitsbergen. He had walked across it during 1896 and 1897.[22]

Gregory and Stefansson had Spitsbergen in common. Gregory, as a member of Conway's team, had detailed its geology and co-authored the account of this island's "first-crossing."[23] Stefansson's interest was raised when Spitsbergen became part of the territorial readjustments at the Peace Conference in 1919. He jumped at the opportunity to include Spitsbergen in his public relations bid to re-create the northern latitudes as the "Friendly Arctic." With journalistic flair that typically overreached itself, he told the *New York Times* that the island might well become "the new Pittsburgh of the Far North."[24] Meanwhile, George Knibbs registered Spitsbergen as the least densely populated region on the planet, apart from territory that was actually uninhabited. His 1927 calculation of "the distribution of human beings over the earth's surface" ranged from 0.05 person per square mile in Spitsbergen to 18,718 persons per square mile in Macao.[25] For Stefansson, this made Spitsbergen precisely the kind of wasteland that was underutilized, waiting to absorb more of the earth's population, much as Engels had argued of the Mississippi valley back in 1844.

Closer to the Spitsbergen than the Macao end of the all-important density scale lay Australia. This was another island—bigger, hotter, drier—that Gregory and Stefansson had in common. Gregory's account of his expedition to the inland was titled *The Dead Heart of Australia*.[26] Undertaking this exploration only a few years after his Spitsbergen trip, Gregory was faced with dry desert, not ice and snow. And critically, unlike Spitsbergen, inland Australia was well and truly occupied. Spitsbergen really was *terra nullius*, while Australia was just pronounced to be so. It was the mythic logic of Australian emptiness, the very idea of its "dead heart," notwithstanding Aboriginal presence, that underwrote the Australian government's commissioning of Vilhjalmur Stefansson to tour, research, and report on its desert regions in 1924. Desperate to show to the world that the whole continent could be occupied, settled, and rendered productive, Stefansson was their man. He made no secret of his driving personal policy: "I am thinking of making a life study of those parts of the earth that are supposed to be worthless either because they are too cold or too hot (humid and disease-ridden) or too dry." Anyone who could declare icy Spitsbergen an industrialized Pittsburgh of the north, who saw the possibilities of human settlement most anywhere on Earth, would also likely declare desert Australia ready to absorb some of the world's growing population.[27] On a selective basis.

Who Owns the Land? Sovereignty and World Population

It was John Walter Gregory—explorer of Spitsbergen and Australia—who entered an uncomfortable principle at the 1927 World Population Conference: "No nation has made the land it occupies, or has the right to prevent its adequate use."[28] The sharp end of the debate among those who sought a redistribution of the planet's growing population came up against disputed principles of sovereignty. Gregory himself applied this to question U.S. and Australian restrictions on European immigration.[29] Others thought universally applicable the principle that "no nation makes the land it occupies" and therefore had no right to prevent its (agricultural) use. In fact, it was not at all uncommon for participants in interwar discussion of the world population problem to argue that the need of some nations/people for more land trumped other nations' claims to that land when it was demonstrably uncultivated or undercultivated. It was an open question, Albert Thomas thought, whether a nation should or even could claim sovereignty over territory "which it does not exploit and from which it is incapable of extracting the maximum yield."[30]

The population question thus became the sovereignty question. For George Knibbs, there was certainly a legal principle to be thought through alongside a moral principle. If lands were not cultivated and used, what precisely was the "right of the occupants to hold the territory as against all comers, and equitably so from an international or world standpoint?"[31] Even Alexander Carr-Saunders wondered about the viability of claims to land that remained unused. In the first issue of *Population*, he summed up the significance of a world history that connected European demographic and geographic expansion. Much of the world had become subject to Europe as a result of this expansion: "But if permanent ownership of land is only to be had by cultivating it, and there is much to be said for that view, it must be doubtful how much of the territories falling to Europeans in the last three hundred years will turn out to afford new permanent homes for them."[32] He cautioned that "it would clearly be hard to defend the retention of regions by Europeans which they could not use." This was repeated and closely discussed at the League's 1937 Peaceful Change meeting on population and resources.[33]

These were significant twentieth-century expressions of natural law arguments that defended European settlement outside Europe on the basis of the right to free and peaceful access to all parts of the globe and on the principle that proper occupation required the cultivation of land.[34] This places the insistent discussion about soil, cultivation, and population not just within a Malthusian political economy tradition but also within a long tradition of international legal defense and critique of colonization. Perception of the inadequate use of land by indigenous people constituted one English and Spanish defense of colonization of the Americas in the early modern period and the English rationalization of land availability, and therefore displacement of Aboriginal people in Australia, in the late eighteenth and nineteenth centuries. By the late nineteenth century, when competing claims in Africa were to the fore, legal argument was being developed in terms of "effective occupation"—in relation to European presence in the Congo in particular—although just what "effective" meant was unclear.[35] Legal scholar Benedict Kingsbury has explained that "settlement (under the juridical name 'occupation') regularly provided a basis for claiming title to vastly larger areas that were not 'settled' by Europeans, or in many cases by anyone." A key irony of the Lockean system in practice, he continues, "was that sparse European settlement might underpin vast claims to *imperium* and *dominium* over lands the Europeans did not use."[36] It was precisely this irony that many demographers pointed to in their discussion of empty spaces on the globe. Such wastelands might not just be better utilized, some argued, they might even be ceded to other peoples.

This argument was occasionally, though rarely, put with respect to indigenous people and their land. The real turnaround in the twentieth-century world population version of this natural law, however, was its application to white-settler colonies—even white nations. Their failures to successfully cultivate land after they had claimed it raised their right of occupation. Further, considered climatically out of place, it was argued that white settler-colonists had been markedly unsuccessful and were even unable to cultivate the land in a way that demonstrated effective occupancy and, therefore, sovereignty.

When things became specific, Australia was not the only case in point, but it was often the first cited. Harold Cox compared the 77 million people in Japan with the 5.5 million in the nearby Commonwealth of Australia: "If a Japanese Sir Thomas More were to draw the plans of a new Utopia he would justifiably declare that in Australia a vast area of ground was being held 'voyde and vacaunt to no good nor profitable use,' and that therefore Utopians from Japan would be morally justified in invading it."[37] Radhakamal Mukerjee did not bother filtering the argument through an east Asian Thomas More: "It is well known that Australia was first explored and mapped by the Dutch, but was later annexed by the English on the theory of non-use and non-settlement. The same argument may be used for the settlement by Indians, Chinese and Japanese of various islands of the Pacific, including Papua and Australia."[38] It was the simultaneous sparseness and largeness of the Australian continent that drew such attention, combined with the international notoriety of the white Australia policy. It was the great "empty" land adjacent to densely populated east Asia that had never been fully occupied by settler-colonials; its "frontier" had never closed, and vast tracts of land had never been disturbed by a plow.

Australia was one of the least cultivated and least densely populated continents on Earth, "apart from Antarctica," as it was typically put, a troubling comparison for successive Australian governments. In the maps of density that constantly marked population scholarship, Australia looked graphically and statistically precarious. Only slightly more promising than Spitsbergen, its average density was usually calculated at 2.2 people per square mile. This would seem to mark it as ready for both more people and more cultivation.[39] From one angle, the Australian government agreed, and officially so: its policy to encourage migrants from Britain, accompanied by a long-standing set of pronatalist policies, laws, and practices, was explicitly based on the problem of filling, and therefore claiming, a very large continent indeed. This is why the Australian government invited Stefansson in particular to assess the settlement suitability of the desert regions. From another angle, the regular citation of Australian emptiness in world population talk was decidedly worrying for Australian governments, because it was often accompanied by calls to increase immigration from dense places on the planet that were rather closer to it than the British Isles. Malthusian Harold Wright summed up the problem: "The 'White Australia' policy, by which a population considerably smaller than that of London claims the whole continent and excludes Asiatics not only from the districts now inhabited, but also from the tropical north where European settlement has not yet been successful, is a typical, if extreme, instance of attitude which the white man has adopted."[40] Such views increasingly brought into question the white Australia policy, because Australia was daily implementing one of the longest-standing and most strident and public of the race-based exclusionary acts. Everyone from biologists like Jack Haldane to sociologists like Warren Thompson, from geneticist Frank Crew to ecologist Radhakamal Mukerjee, cast Australia in this way.[41] In short, it was the world's best-known wasteland.

This particular "danger spot" to use Thompson's phrase was compared most often to Japan. In *The World's Population Problem and a White Australia*, Japan's birth and death rates were graphed, vis-à-vis Australia's rates, projected to the year 2000. These data were to be read against the map of "Density of Population," which included important "uninhabited regions," as well as those with less than 1 person per square mile (most of the Australian continent).[42] One Japanese delegate to the 1933 Birth Control in Asia conference agreed with this Lockean argument as applied to the West. He argued that land, not birth control, was the fundamental solution for his overly dense islands: "We all know that the doors of all the empty lands, which should be developed and cultivated are closed against Asiatic immigrants."[43] In the light of such "claims for national expansion," the "distribution of population" was an Australian security concern. Thus population policy was the business of defense policy, not implicitly but explicitly. The country's defense minister's speech to an imperial conference in 1937 used just these phrases, betraying an acute awareness of international criticism, as well as a resentment that similarly placed nations seemed to be out of the spotlight: "South America is presumably passed over because of [the] Monroe Doctrine." By May 1937, it was all about Japan, as far as the Australian government was concerned. And yet the minister presented his national population policy as apparently lofty principle: "Population is of vital importance . . . the consideration of security is doubly important where there are national ideals to be preserved, such as racial purity and minimum economic and social standards." These ideals, the defense minister continued, were "as dear as life itself, for they are, in fact, inherent in our conception of a nation." A mere decade later, such statements would be impossible to parade, certainly not in an international public sphere. Even in these earlier years, governments were already acutely aware that maintaining strict immigration restriction at the same time as defending claims to a vast continent, with a population of only 6.8 million people as this minister spoke, was almost untenable.[44]

The whole issue of occupancy, density, and land use took on an explicit anticolonial agenda as well. Warren Thompson was unequivocal: "I believe that wars of expansion cannot be avoided in the near future unless certain traditional nationalistic and imperialistic modes of conduct are much modified by the great powers holding lands not in use."[45] For Thompson, the land-based problem of population required a territorial solution, but one that would also be peacefully implemented. Failing the repeal of immigration restrictions, he argued that underutilized land should be peaceably ceded from underpopulated sovereign nation-states to overpopulated nation-states on the basis of need and in the interests of international security. Failing the redistribution of people in migration projects, there should be a political redistribution of land, he thought. This was an international "reallocation of territory" to those who needed it because of overpopulation and scarce resources.[46] As we have seen, Thompson's early demographic transition work was geopolitical as much as biopolitical. The real global problem was that the peoples first (in time) to expand demographically had also been able to expand territorially. They consequently held vast amounts of global land, which "they cannot settle," but which subsequently they would not allow anyone else to settle or occupy either.[47] Warren Thompson's strong position was that finding Japan "new colonies" was the best alternative to war and should be pursued and endorsed by an international polity until domestic birth control became a more permanent solution. Various western Pacific islands, including the Philippines, might be ceded by the United States to Japan. Australia in particular, Thompson thought, "should be prepared to make large concessions." With an "enlarged Japan," thus established by the voluntary cession of territory, a war in the Pacific might be averted. Failing that, it was almost inevitable, he argued presciently, that the acquisition of new territory would happen by force, unless world population pressure was "equalized" by systematic and voluntary cession of unused land.[48]

Geographer Isaiah Bowman was alive to such a plan and called it a "modern-day Louisiana Purchase." In his view, however, any such major redistribution of territory was unthinkable. "It would shake the world," he warned.[49] Yet in fact, partly at Bowman's own hand, any number of transfers of sovereignty had been advised and implemented after World War I.[50] In the "competing visions of world order" and the "Wilsonian moment,"[51] territorial redistribution was constantly discussed and put into operation, first within Europe itself. The mandate system was also entirely concerned with the reallocation of territory and people. The Versailles Treaty had divided and distributed the German African colonies to Britain, France, Belgium, Portugal, and South Africa. The Pacific colonies north of the equator were granted to Japan, with German Samoa ceded to New Zealand and German New Guinea, the Bismarck Archipelago, and Nauru to Australia. The Mandated Territories, previously held by Germany, were thus deemed "available," and the idea that these islands could be reused by altogether different populations under international agreement held both official and popular currency. There were some aspirational claims for their reuse, as the whole idea of global reallocations came to seem viable and reasonable. In 1924, for example, the League of Nations Secretariat received a petition from the International Convention of Negro People of the World that former German colonies might be developed as independent Negro nations.[52] Likewise, Zionist ambitions were attempts to reuse various lands. The Zionist rejection of East Africa for Jewish settlement re-emerged in the 1917 Balfour Declaration, affirming British commitment to a national Jewish homeland in Palestine, the League's 1922 British Mandate for Palestine, and ultimately the state of Israel. Elsewhere, the 1935 Franco-Italian agreement on East African colonies reallocated to Italy parts of French Somaliland and ensured the Italian citizenship of Italians living in Tunisia, in apparent exchange for future Italian support against Germany. In all of these ways, adjusting and rethinking global lands and peoples shaped international relations in the decades after World War I. This was a context in which territorial and population redistribution, far from being "unthinkable," as Bowman disingenuously put it, was in fact in operation.

This explains in part why so many population writers could imagine territorial redistributions as a solution to density and could do so with seeming ease. An overpopulated nation-state was theoretically justified in "re-establishing its population equilibrium on the basis of a larger circle," as Henry Pratt Fairchild approvingly put it.[53] Adolf Hitler liked this idea too. And it is no coincidence that by the late 1930s, the Anglophone community of population thinkers collectively posed the question of sovereignty, population density, and land with direct reference to the National Socialists, referring back to *Mein Kampf*. At the 1937 Peaceful Change meeting, it was put thus: "Unless the expansion (proposed by Herr Hitler) is to take place on land now uninhabited . . . his programme would seem to imply that the native population of any land claimed by Germany would have to be removed or exterminated. For, how else would it be possible to establish the natural and healthy proportion between the numbers and the increase of the nation and the size and quality of the land in which they dwell which he has contemplated?"[54] Well might this League-endorsed meeting seek to criticize Hitler's expansion in 1937. But it was Hitler himself who drew rather more honestly the parallel between his own Greater Germany and eighteenth- and nineteenth-century British, French, and U.S. colonial rule. Each was based on population redistribution rationales, policies, and practices that entailed a negotiation of some kind with people already there. Greater Britain, *Outre-mer*, and the manifestly destined United States on the one hand, and *Großdeutsche*, *Italia Imperiale*, and *Dai Nippon Teikoku* on the other, may have been imminent opponents in war, but they were each colonizing entities with interests and histories in the population-absorbing capacity of lands, especially those that could be deemed empty.

Indigenous Land

Where did indigenous people fit in such plans to redistribute territory, and to reallocate people to empty places not through migration and citizenship infrastructures, but through changes of sovereignty? Early twentieth-century demographers occasionally perceived that moving Europeans into other spaces displaced indigenous people. T. N. Carver, political economist at Harvard University, said in 1927 that the expansion of population resulted in "robbing lower races of their land."[55] Mabel Buer, lecturer in economics at the University of Reading, linked European population increase with colonialism and a problematic domination of "large parts of the world."[56] For most, sparse occupation by indigenous owners seemed to be a minor problem. J. W. Gregory, for example, was quite confident: "As there are large areas in the world which are uninhabited or sparsely occupied, nations with a large surplus of population should do their share of breaking-up new land."[57] In such statements, the descriptors "uninhabited" and "sparsely occupied" were rendered problematically equivalent. Vast areas known to be populated could be pronounced empty,[58] because emptiness was comprehended less as absence of humans—pure nature—and more in economic terms as lack of cultivation. This represented an agrarian missed opportunity to render nature properly productive. It was a kind of waste.

Warren Thompson's critique of colonialism was more damning than most.[59] His view of British rule was almost postcolonial: "It appears to the man from the East that the Westerner has never hesitated an instant to dispossess the natives of any land he has wanted for his own use. If he has not wholly driven out the natives or exterminated them, he has enslaved them as far as was possible and made them serve him." One of his examples was the clearing of Kenyans in the highlands of East Africa by and for the occupation of whites. Yet those whites were increasingly seen by locals purely as exploiters of resources and labor. Anticolonial sensibilities were on the rise, he warned, and with good reason.[60] Yet this did not necessarily mean giving land back to the Kenyans. It meant redistributing land on the basis of need. For Thompson, in other words, "need" did not necessarily translate to any principle of indigenous or native title to territory.

This was not just an act of intellectual omission, it was his active conclusion. Thompson was deeply critical of British rule in East Africa but still considered Kenyan land legitimately available for south Asians on the basis of the land's underutilization. The idea that the land was the "birthright of the Negro" was invalid, he argued, because no people had a moral right to land that was not cultivated when it was needed by other peoples. "This would apply to the Negroes as well as the whites, and, indeed, the whites in going into thinly settled lands and assuming control and undertaking settlement have generally assumed the soundness of this position."[61] Like many of the early moderns, anticolonial Thompson applied this principle to natives as well as newcomers.

In itself, this absence of argument about indigenous African claim to land was not unusual for the period.[62] Yet it sat strangely with Thompson's strident critique of British colonial exploitation.[63] Far from recognizing any indigenous sovereignty in national or international law, he actively rejected it. Thompson literally swept aside indigenous presence when it fitted what was to his mind the higher purpose of global population redistribution. In arguing for the cession of New Guinea to the Japanese, for example, he wrote: "The population probably does not exceed 1 million and could readily be placed on reserves not exceeding a few thousand square miles if race amalgamation did not prove feasible."[64] Aboriginal people in Australia were already in precisely this kind of reserve. Official projects for "race amalgamation" were also under way; Thompson's framework indicates just how tightly modern "assimilation"—race amalgamation—was tied to the politics of land. And yet, even though Australia was one of the key "empty zones" at stake, Aboriginal people rarely figured in this international discussion. Their claim to land also went unrecognized after World War II, even by anticolonial nationalists with a strong interest in the country. Demographer Sripati Chandrasekhar for example, stridently opposed the white Australia policy, but failed to appreciate the colonial implications of the "emptiness" discourse that he inherited from the previous generation of scholars. Arguing for the need for Indians to migrate away from their dense region—just like Thompson—he constantly pointed to the large Australian area that was "unused Crown Land," "entirely unoccupied," and "economically valueless."[65] In a later era, it was precisely this Crown Land that became open to native title claims.

Chandrasekhar's elder countryman Radhakamal Mukerjee was more aware of "the rights of the native peoples." He suggested that new land systems in Africa should recognize pre-existing "imperfectly developed [but] quite precise" legal conceptions of property. He pointed to some successful examples in French West Africa and Tanganyika, peasant proprietorship in Java, cooperative tribal enterprise among the Maori in New Zealand, and a leasing system in Tonga. But even for Mukerjee, the League's "sacred trust" for the welfare of the "junior races" was justifiably in operation. International codes, begun in Berlin and Brussels, should be carried forward, not only in the territories dealt with under the Paris Peace Settlement but in all territories "where native populations are held in subordination or tutelage."[66] Mukerjee's internationalism, in this instance, trumped his anticolonialism.

The Lockean irony was also relentlessly brought to bear on European colonials themselves in the 1920s. If prior occupation was not valid as an argument for indigenous people's claim to land, neither was it valid for colonizers and so-called settlers. In a strange twist, this principle was ultimately deployed to delegitimize white control of territory. Thompson continued: "The same argument for the better, more complete use of lands would certainly justify Oriental peoples in taking the land which the white man is not using as well as they could."[67] Thompson, Mukerjee, and Chandrasekhar were each outspokenly anticolonial, but the overriding principle was need determined by relative population density, not indigineity: the global was privileged over the local.

Incapacity: Climate, Physiology, and the Limits to White Man's Claims

There was a further aspect to this international interwar discussion about the right to claim long-unused land. This concerned not whether land *had been* cultivated or rendered productive but whether it *could be* cultivated;

that is "effectively" occupied by those who claimed it. This was a question not of the land per se, but of its relation to people via climatic environments. The geography of racial physiology was at issue, a twentieth-century manifestation of long-standing sciences of the relations between climate and human constitutions, capacity, and difference.[68] Could white people work and reproduce normally and effectively in the tropics? This question was as old as European travel to the Indies. And it had entered legal doctrine as early as Montesquieu, who distinguished between laws in general and laws in particular that bore a relation to "the climate of each country, the quality of its soil, to its situation and extent."[69] Just how different humans fared in the torrid zones was an enduring question at both expert and lay levels, and one that was pursued well into the twentieth century. If anything, the integrated sciences of human–climate interaction became increasingly technical over the first half of the twentieth century, and its politics heightened. In the 1930s, the Medicine and Hygiene Section of the International Union for the Scientific Study of Population still integrated the "possibilities of expansion of white races in glacial and tropical climates" into its core business.[70]

The bioscience here was very specific, involving the physiology of respiration, excretion, and metabolism. This was an object of inquiry researched and calculated along axes of environmental difference, human difference, and comparative adaptability, with a focus on "white" physiology. Anglo self-assessment was not looking good. In a 1923 issue of *The Lancet*, tropical medicine specialist Andrew Balfour summarized the widely accepted, although not undisputed, position: "So far as the [British] race is concerned, I am persuaded that the hot and humid tropics are not suited to white colonisation and never will be, with our present knowledge, even if they are rendered as free from disease as England."[71] Versions of this conclusion entered adjacent disciplines as well. If statistician George Knibbs thought that the major global question for mankind was "how best to distribute him over the earth's surface,"[72] one of the "Great Barriers" to such distribution was the zone between the tropics of Capricorn and Cancer, the environment that posed difficulties for permanent white settlement. It should come as no surprise that one of his earlier publications was his *Mathematical Analysis of Some Experiments in Climatological Physiology* (1912).[73]

In the context of an accelerating population growth, and the pressing planetary limit that so many acknowledged, the tropical question was at once physiological, geographical, and deeply political. It was no longer just an imperial issue, but an international one. None of this was fringe discussion. Sir Arthur Salter, head of the Economic and Finance Section of the League of Nations and soon to be professor of political theory at All Souls, Oxford, asked: "Should the admitted and natural right of a country to limit immigration at least enough to retain its own racial integrity be regarded as subject to qualification if it cannot, with its existing population, develop its own territory, or can develop it only very slowly: and especially if some of its territory is for climatic reasons uninhabitable by its own nationals while suitable for those of another race?"[74] Warren Thompson, for one, had a straightforward answer: "The whole of the white man's tenure in the tropics has no vestige of right." He insisted that there was simply no evidence or likelihood that tropical land had been or would be brought into proper productive use by "Anglo-Saxon labour."[75]

The old problem of the "white man in the tropics" complicated the thesis and agenda of redistributing the world's population, much as it had long complicated white peoples' claim to legitimately occupy tropical zones.[76] The physiological base of this problem meant that the right to hold land in the tropics was and would continue to be contested, Thompson predicted. Such a contest needed to be anticipated, so it could be avoided. Hence, in Thompson's extensive discussion of the Australian situation, he considered the granting of the tropical sections of the continent to the Japanese and the Chinese a sensible foreign policy. He also argued, though, that this should be done in full awareness that they "are a southern people rather than a northern people." They could not settle in Siberia, for example, even if free to do so. By contrast, they could agriculturally settle the tropics of the Australian continent more effectively than white Australians, and therefore more legitimately.[77]

Little wonder that Australian tropical medicine research boomed in these years. The government brief to the new Australian Institute of Tropical Medicine was unequivocal: "Is White Australia Possible?" This was a research question pursued systematically by pathologists and physiologists, who sampled and analyzed young, white laborers' sweat, urine, bile, and blood for evidence of resilience, decline, or adaptation. Long interpreted by historians within a national and occasionally imperial framework,[78] this physiological inquiry had in fact a much larger global and international import. Can white men live and labor in the tropics? Many Australian scientists tended to answer affirmatively; given what was at stake, this is perhaps unsurprising. Viable tropical physiological and reproductive functioning was shown to be quite possible, Australian scientists concluded, a little too insistently.[79]

In the apparently empty continent at the heart of it all, Australian policy makers and politicians watched and listened closely. Unlike the United States, which could and did claim that its frontiers were closed, and thus it had reached its optimum population, Australia was not doubling its population quickly enough; it was nothing like the United States in this respect. On the eve of World War II, its total population was minute: only 7 million people. Governments responded to Stefansson's confirmation that the desert areas could be populated with multiple infrastructure programs, assisted emigration schemes (from Britain), and close settlement policies. Yet at the same time, Australian governments had to deal with the ongoing claims from men of science far more qualified than Stefansson. One national response was a counter-physiology of racial adaptability. Another was to overturn the idea that the continent was held "out of use," in Warren Thompson's phrase. "Populate or perish," became official Australian policy. Nationalist anxieties trumped world problems, but vainly so in this instance; as late as 1965, the continent was still announced as "Asia's Safety Valve."[80] And it was held to account: the white Australia policy was to slowly unravel after World War II, a process influenced by the world population argument of anticolonial demographers.[81]

Environmental and racial determinism reached well into the mid-twentieth century, retaining remarkable currency and constantly impacting the terms of world population debate. Many physiologists and tropical specialists retained, unqualified, the long-standing theory that tropical conditions were antithetical to white constitutions. For some a convenient truth expediently deployed, for others open to question, and for still more a physiological given, the apparent incapacity of whites to properly cultivate and therefore potentially even claim land out of their global place raised an uncomfortable corollary. Who, then, had that capacity?

Climatic Capacity and Anticolonial Determinism

For Radhakamal Mukerjee, the entire discussion was an opportunity to argue for Asian physiological advantage, and over many decades, he twisted racialized environmental determinism in unexpected directions. Any ambition for world-level optimum population depended "in large measure on the migration of surplus populations to empty but similar climatic regions."[82] This explains the climatic maps that accompanied his density maps. "Climatic Regions of the Continents," for example, divided the world into tropical, temperate, and cold climates, with finer distinctions ranging from equatorial, tropical monsoon, and tropical semi-arid climates, to cold temperate, humid continental climates with long summers, and temperate semi-arid, subpolar and polar climates. It all mattered deeply, because he argued that there was a correspondence between Indians and many of the undercultivated tropical and subtropical areas of the earth. This gave Indians a huge physiological advantage. Similarly, he proposed a physiological fit between the northern Chinese and the environments of Canada and Siberia. Insisting on calling potential Asian migrants "colonists"—once again the signal that humans could labor themselves, not deploy slave or indentured agricultural laborers—Mukerjee wanted to see the color bar dropped so the potential of the Asian migrant could be realized globally in a kind of physiological free market. Far from racial specifics functioning as an automatic bar to movement, a scientific world population policy would consider racial physiology on its merits: differences should sensibly be taken into consideration in immigration selection. This would recommend rather than discount most Asians "for the agricultural transformation of vast untenanted areas of the globe."[83]

Mukerjee went on to document this advantage in close physiological detail, though not quite at the cellular level of the tropical medicine studies of white men in Australia conducted in the same period. The basal metabolism of the Asiatic peoples is 10 to 15 percent below the English and American standard, he said, describing a physiology of skin: "more sweat glands, with smaller rate and volume of respiration, with smaller blood fat and secretion of bile and urinary excretion." They have smaller body surface and weight, needing less protein. This was an advantage over the European settler, he claimed, whose diet shows "an excess" of 50 percent calories and 40 percent protein "beyond what is physiologically indispensable in the tropical environment."[84] This physiology presented an enormous untapped potential in pursuing economic reconstruction. Indian agricultural laborers were a ready-made package for any nation willing to overcome its baseless prejudice: they could bring not just energy-efficient bodies, but also climate-appropriate crops and agricultural techniques.[85] Recognizing this physiological advantage, putting it literally to work rather than artificially limiting it through race-based exclusion acts, would mean the turnaround of global wastelands.

In Mukerjee's opinion, different races had different capacities for acclimatization. Drawing on the work of Ratzelian Ellen Churchill Semple, he argued that European colonization had been limited by latitude, but that altitude in part counteracted the problem. Indian emigration, equally, had been largely determined by latitude, and "a similar lack of adaptability" was evident in the Japanese, even though they were scattered in some small numbers from Alaska to the tropical plains of Brazil and the Dutch East Indies. But they have settled best in Hawaii, he said, where they made up more than 40 percent of the population. The failure of the Japanese as farmers in Hokkaido and Manchuria and in subtropical Formosa, he wrote, was partly about climate and lack of adjustment. This failure was the combined result of the limit of a particular human physiology and a climatic limit of rice agriculture. The Chinese, by contrast, show greater capacity for acclimatization. From the 35,000 in Siberia to over 1 million in the Dutch East Indies, he claimed, they flourish over a range of at least 95° of latitude.[86]

Ultimately, then, Mukerjee's global vision, like Thompson's, matched people to place: the Chinese and Japanese to tropical Australia, to Canada and the Unites States west of the Rockies, the temperate zone of South America, and the south of Siberia; Indians to the Straits Settlements, Dutch East Indies, New Guinea, Syria, Mesopotamia, Madagascar, East and South Africa, tropical South America, and Australia. Indeed such natural capacity for acclimatization became a kind of right to land: "If there be an honest attempt to distribute the surplus population of the world from an international rather than from a racial or national standpoint it is evident that these regions will be allotted to the Asiatic peoples." This would be a far improved regional system of "non-exploitative development of the tropics by coloured settlers."[87] The whites who cannot labor or cultivate the land for themselves, cannot function as the master class for long: the system is too unstable, Mukerjee thought. As a senior figure in the Indian National Congress, Mukerjee presented racial physiology as entirely compatible with anticolonial nationalism. It was, as we shall see, part of his global political ecology.

An idea almost always attached to the colonial project, racial and environmental determinism occasionally re-emerged as argument against colonialism when filtered through the problem of population growth and limited land. It was a problem that received its fair share of loose scholarship, but those at the sharp end of the intellectual community were also thinking hard on it. In this high moment of determinist thinking, and of explicitly political bioscience, distinguished geneticist Frank Crew went so far as to claim that "biological fitness is the only valid title to territory." One of the most eminent scientists to pursue this biology of human distribution, Crew was an animal geneticist at the University of Edinburgh, and from 1927, he held the

Rockefeller-funded chair of genetics. Alongside his punishing administrative load and his genetic research on the inheritance of plumage color in the turkey, the duck, and the Old English game bantam, Crew found time to think about humans, to attend the Geneva conference, and to contribute a biologist's synthesis on the topic of distribution of humans and their relation to climate. For Crew, the biological aspects of migration were to be heeded, and he spelled out the political implications starkly: "It is doubtful that pure bred British stocks can ever claim to be in biological harmony with habitats outside the temperate zone. No doubt they can exist in most places, but to live fully, to reproduce freely, to develop and to function smoothly, they must be where they belong, in a habitat that biologically belongs to them."[88] This was an extraordinary statement that seemed to seal the fate of an empire already in decline. He went on to re-present the argument that biology could undermine sovereignty, typically with respect to Australia. "If men of Anglo-Saxon stock cannot colonize them whilst others can claim a biological harmony with them, then it is to be expected that sooner or later the type that can only exploit will give place to the type that can colonize." That "type" might be the Italian, Japanese, or Chinese, who could cultivate the land through their own labor—"colonize in a more harmonious way," is how Crew put it. There was a kind of natural law, even a law of nature, being stated: "The biologist, in his capacity as biologist, must agree that the latter have the prior right to these regions." For Crew, such arguments might apply to Australia, East Africa, or South Africa. By contrast, the claim of the northern European to North America and New Zealand was biologically sound. They were at least as fitted to the climate as the populations already there.[89]

Crew considered migration a biological issue, then, because humans, like other animals, have an instinct for movement. But it was also an ecological question: it was a relationship with a particular environment that was necessary to examine, because humans as natural organisms were driven to establish an essential harmony between number and conditions of habitat. There were sustainable and unsustainable relations between new humans in new environments, he claimed, and any form of imperialism was to be considered in this light. Western Europeans, through migration, had come to politically possess "almost every part of the world that was sparsely populated and that could be colonized or exploited. The distinction between colonization and exploitation is really a biological one." If animal populations migrate and select habitats that suit them, humans selected habitats often entirely unlike their own, chosen for fatness not fitness, as Crew put it. What characterized much exploitive imperialism was a total lack of fit: "Migrants from temperate zones, biological types fashioned by a peculiar habitat . . . have settled in tropical and subtropical regions which demand an entirely different type of man." Humans can modify their environment, and might even be modified by it: humans adapt, he argued, like Carr-Saunders, but there were definite limits to this adaptation. "All individuals, and all types of mankind are not equipped with equal powers of adaptation to marked and varied differences in environmental circumstance."[90]

For many of this generation of Malthusian internationalists, both those who were troubled by nationalism and those troubled by colonialism, there was an ethical question about unused land in an overpopulated world in which famine existed. Not a few, alongside Mukerjee, were exercised by the principles of occupation in the light of a limited food supply and its equitable distribution: the dominant principle of land occupation was one of need. Even for George Knibbs, the esteemed Commonwealth civil servant, sometime advocate of white Australia, the "need" principle in his world Malthusianism trumped all. The very nature of the right of occupation was called into question, he thought, when assessed through "the point of view" of people with "insufficient territory."[91] This was not a right based on present occupation or even prior occupation (native title), but a duty to recognize others' right to land based on need for food, what even came to be called a right to life. What was deemed ineffective occupation signaled a moral failure in an economy increasingly understood as global. The politics of earth manifested as the politics of life.

Part III

The Politics of Life, 1920s and 1930s

6

Life on Earth

Ecology and the Cosmopolitics of Population

> From a world point of view . . . human beings are relatively but the merest specks on the earth's surface.
>
> GEORGE KNIBBS, *THE SHADOW OF THE WORLD'S FUTURE* (1928)

After the Malthusian League's fiftieth anniversary dinner in late July 1927, John Maynard Keynes jotted a postcard to another of the Malthusian diners, Julian Huxley: "The impressions and excitements of this world are the instruments with which the Supreme Being forms matter into mind." Although cryptic to many, Huxley hardly needed Keynes's scribbled prompt: "1st edn of An essay on the p. of p."[1] They both knew their Malthus and recognized that they were his intellectual descendants: Keynes directly so, as reigning Cambridge economist; Huxley indirectly, as neo-Darwinian biologist and grandson of Charles Darwin's great defender, Thomas Henry Huxley. That Malthus's "p. of p." had been an inspiration for Darwin's theory of evolution by natural selection was a fact not lost on either of these already eminent men. In attending the Malthusian League dinner, indeed in being neo-Malthusians of any description, they were living embodiments of the "life" and "earth" of population, its linked political economy and natural history.

By Keynes and Huxley's generation this connection had a new name. "Ecology is biological economics,"[2] Huxley wrote. This was the new scientific approach to assessing the interdependence of life on Earth, at once technical and precise in its methodology and global in its application. Ecology was named in 1866, systematized from the 1890s, and haltingly institutionalized in the United States, Germany, India, and Britain in the early twentieth century. By the 1920s, when Huxley picked up the idea, ecology was in its second generation of serious scholarship, still faltering, but now professionally grounded with a few dedicated scientific journals.[3] At the most basic level, ecologists counted organisms in defined spaces. They watched how those organisms reproduced and died in interaction with one another and their environments. As historian Sharon Kingsland has characterized early ecologists, they studied the process by which organisms lived together in apparently stable communities. As a new breed of natural historian, therefore, the ecologist was "part taxonomist, part census taker."[4] It was all about the life and death of populations of interdependent organisms in environmental spaces, in limited universes of various scales.

Those who first called themselves ecologists worked outside the human and social sciences altogether, in plant and animal biology. Their ideas crossed over, however, and sometimes ecologists themselves crossed over into human population studies. Indeed, it became characteristic of scholarship on human population to integrate ecological concepts on the relationships among death, sex, fecundity, and environment, and increasingly to assert that studies of human population were, necessarily, part of the new ecological inquiry. Everyone was taking note of the new "embracing science." It is no wonder that when economist A. B. Wolfe reviewed the run of world population books over the 1920s, he identified the trend to be addressing at core "an ecological problem."[5]

At a conceptual level, the human population problem was so complex, with so many variables and feedbacks that it begged this kind of synthetic approach. Methodologically, ecology was both appropriate and useful. But there was more to this connection. Ecology was conceptualized as the economy of nature, not just analogically or metaphorically but substantively as well; populations with high fertility and high mortality were "systems" that wasted energy in the biological economics that was ecology. Indeed, this was the precise formulation that Kingsley Davis deployed later (but not much later) in his theory of world demographic transition, as we shall see. Thus, without an understanding of ecology, our understanding of the theory of world demographic transition as first articulated is incomplete. This complex of ideas needs careful unraveling.

Ecology's curious link to a popular "environmentalism" was several generations away, but it was nonetheless already being fashioned politically in the 1920s and 1930s. One manifestation of political ecology emerged from its connection to cosmopolitanism. They shared a conceptual holism. For the science of ecology, this holism reflected a particular intellectual moment in the history of science primed to think in integrative ways about any given object of inquiry, including complex human societies, at both local and global scales.[6] Similarly, cosmopolitanism was built from a kind of holism. An old political idea that had a fresh significance after World War I, cosmopolitanism manifested as ideas about world citizenship and federations of world states.[7] In such "federalism," the whole was greater than the sum of the parts, and while the parts might or might not be national units, some system of integration of governance at a world level was its point. Immediately after the earth-shattering events of World War II, ideas about world federation hit a high point. Julian Huxley, by then at the top of the new United Nations as director-general of the United Nations Educational, Scientific and Cultural Organization (UNESCO), presented a manifesto for his organization and for a world population policy. There, he re-presented the long tradition of work by ecologically inclined scientists who integrated population within cosmo-inclined politics. This is the tradition in which Huxley's controversial UNESCO manifesto is most appropriately understood.[8]

Population and Ecology: Biological Economics

The moments at which T. R. Malthus's political economy and Charles Darwin and Alfred Russel Wallace's theory of evolution by natural selection met and fused is one of the more famous interactions between the social sciences and life sciences. Yet there was a later moment when a derivative exchange took place. On the icy shores of Spitsbergen in 1921, Alexander Carr-Saunders was deep in thought about population. He was still lecturing in zoology at Oxford, and this was the first of several Oxford-based expeditions to the Arctic, not long after the archipelago's status in international law had been resolved. For such expeditioners, islands like Spitsbergen were delightful, enabling population surveys of multiple species, whose interactions could be relatively comprehensive and simple to count. Carr-Saunders was technically there to observe aquatic life,[9] but he was spending most of his time thinking about humans. He had already begun to cross over with his prewar work at Karl Pearson's biometric laboratory at University College, London. After the war, and despite his zoology lectureship, he had a new project and problem to the fore of his active and organized mind: the evolution of human populations. His first book on the quantity and quality of humans, *The Population Problem: A Study in Human Evolution* (1922), was taking shape on the Spitsbergen trip. It rehearsed the Malthus–Darwin/Wallace synthesis as foundational. And it sensed the ecological mood; the study, he explained, was not a contribution to any single strand of analysis, rather it aimed to sort out just how multiple individual factors related to the population problem as a whole.[10]

One of the greatest ecologists of the early twentieth century was also on the Spitsbergen expedition. The young Charles Elton—still a student—exchanged ideas about population and density with his teacher, as he intricately mapped the ecological zones around their frigid camp.[11] The Oxford scientists were thinking through what human population problems meant for populations of all organisms, in an intellectual repetition of human scientist Malthus's impact on natural scientist Darwin, but in a vastly different time and place. One outcome was Elton's theory about the influence of climatic cycles on fluctuating animal populations and his later influential work on animal population cycles in the Canadian Arctic. Another was the idea of a nitrogen cycle.[12] Elton's "biological economics" of animal life—his "production biology"—counted the number of organisms in an area, rates of reproduction, consumption and growth, excretion rates, metabolic rates, and rate of death. With these data, writes Frank Golley in his history of the concept of "ecosystem," a "detailed cost accounting for the flow of energy or a chemical element into and out of a population" could be determined.[13]

Julian Huxley was also on the Spitsbergen expedition in 1921; he had organized it.[14] Huxley was thinking mainly about birds, as always, but like his friend "Alec," was becoming intrigued with the new genetics and its relation to human quantity and quality, and with the idea of ecology in relation to human populations.[15] Returning to Oxford, Huxley read most of the draft chapters of Carr-Saunders's *Population Problem*, and like him was keen to synthesize everything as widely as his grandfather had done, but with quite new tools at hand. Carr-Saunders's book "made me think hard," he remembered. "It was another case of growth of man's numbers relative to his resources." Unlimited multiplication would lead to deforestation, shortage of food, overcrowding, and although Malthus had come to the same conclusion, he later said, there were now better means by which numbers could be kept down.[16]

If commentators on Raymond Pearl's population law (itself part of the institutionalization of ecology in the United States) had been troubled by the application of biological studies to human–environmental systems, Carr-Saunders and Julian Huxley were not just intrigued by the complexity of adding human organization into the problem of organisms in space, they were positively drawn to the challenge. It was Huxley who injected the quickly growing body of ecological ideas and data into the discussion of human population at the Geneva conference. Responding to Pearl's paper on density, he alerted colleagues across the disciplinary spectrum to animal ecology, in particular to Elton's recent work on cycles of growth and decline in rabbits, lemmings, and field mice. "The population increases to a maximum, and then, through an epidemic is brought rapidly down to a minimum."[17] He detailed H. Eliot Howard's work on a "territory system" in bird life, and the biological mechanism by which population density was reduced. By 1920, Howard and Huxley had already been discussing with each other "the mighty big inter-related whole."[18] Drawing from animal ecology, then, students of human populations were registering the connections between density, territory, food supply, and aggregation and their effect on reproduction and mortality.

The Science of Life and Human Population Control

Julian Huxley met H. G. Wells in 1926, just after the famous author, always larger than life, had finished another large book, *The Outline of History.*[19] Wells was already planning his next massive synthesis, which became *The Science of Life*, coauthored with his son, G. P. Wells, and the talented Julian Huxley. It applied to the natural world Wells's tried-and-true method of joining "part to part," of addressing life in large-scale space and time, and of placing humans "in relation to the whole scheme of things."[20] Scandalously, but in the end true to its ecological spirit, Wells wrote of the three authors as a trinity, "three in one, constitutes the author of this work." The whole, he wrote, represented a summary of the "vaster Genesis" that had little by little been produced by scientific investigation of life on Earth.[21] Wells's signature provocations were everywhere in the introduction; no reader was left wondering which of the three was god the father. And while the actual son of this authorial trinity might have been G. P. Wells, the real intellectual son was Huxley, about to embark on a lifetime of popular communication of scientific ideas. Huxley gave up his chair at King's College London, promised by Wells that they would secure £10,000 each. In the writing process, Wells relentlessly pressed Huxley for copy and testily warned him against distractions—"sideshows like Geneva."[22] Huxley later complained that Wells had largely forgotten his biology, which as it happened had been taught by Huxley's own grandfather, Thomas Henry Huxley. It was old-fashioned, the twentieth-century Huxley dismissed, "pre-Mendelian" in fact, implying that it might as well have been prehistoric.[23] And so it was just as well that Huxley was distracted, listening to and reading the work of his biology colleagues in Geneva. The "sideshow" was putting human population into *The Science of Life*, just as Huxley was putting ecology into the World Population Conference.

Beginning with an explanation of "life" in their new Genesis—"metabolism and movement are the primary characteristics of living things"—the trinity set out their project as a great exercise in ecological thought that incorporated the inorganic, organic, and human worlds. And what was ecology for them? "A fresh way of regarding life," that considered "the balances and mutual pressures of species living in the same habitat." From the Greek for house, they explained, ecology has a kindred word in economics, a science one hundred years older than ecology. But whereas economy was limited to the social world of humans, ecology carried this inquiry into every province of life: "Economics, therefore, is merely Human Ecology."[24] Indeed political economy might have been a better and brighter science, they noted, had it begun biologically.

Their foundational message for the whole work was to consider "the limitation of life in space." And in their long explanation of ecology—nominated also as "the chemical wheel of life"—they suggested that complex communities of organisms that preyed upon but also depended on each other should be considered as one thing, a "super-organism:" "Life-communities develop and evolve as wholes. They might be called super-units of life."[25] Huxley contributed examples from contemporary ecological studies of the Great Lakes in North America and from Spitsbergen. Jotting an outline for a chapter on "the organism and its environment," he noted the need to explain ecology, regulation of environment, and geographical distribution.[26] From Elton's work in particular, this materialized as a section titled "Storm Breeding and Death," in which they explained that the idea of a balance, albeit a swaying balance, was always in operation. The population of any species was dependent upon the rate of reproduction and growth on the one hand and upon death rates on the other. And, straight from Malthus, "were it not for these two opposing forces at work, multiplicative and destructive, life's power of increase would be overwhelming." In animals, multiplication was mainly controlled by predators and less often by disease and starvation. They instanced sudden "stormings," rapid rises in populations—in rabbits, in locusts, in mice, in lemmings—that inevitably then fell as a result of food shortage, disease, or migration. Further, violent epidemics of disease seemed relative to density of populations, they noted, referring to Sir Ronald Ross's work on malaria, which showed its correlation to population densities of either (or both) mosquitoes and men. Cycles of abundance and scarcity were observable, they explained, drawing on Elton's *Animal Ecology*, in which climate appeared to be a key determinant.[27] Rehearsing some of the work he presented in Geneva, Julian Huxley added Eliot Howard's territory system as another kind of natural population control: "When population-pressure seeks relief in migration, opportunity is given for the colonization of new areas." For birds or locusts or lemmings, attempts at breeding in new areas were often unsuccessful. This, they explained, was part of the real nature of the "ordinary Struggle for Life." Readers, however, should realise that "struggle" was not a military one for victory or extinction of another species; that would only lead to overpopulation of one species and end in its own demise. "To multiply and replenish the earth unchecked may be only the prelude to decay." And finally, as if added as a last-minute coda, the chapter on the new science of ecology ended with humans. "Unrestrained breeding, for man and animals alike, whether they are mice, lemmings, locusts, Italians, Hindoos, or Chinamen, is biologically a thoroughly evil thing."[28]

"Thoroughly evil" was a peculiar way to wrap up their exposition of a natural, amoral, and secular process, their scientific "new Genesis." But it laid the ground for explaining how population control in humans should be seen as a "good." Their chapter on ecology, with its inflammatory ending, was followed immediately by "Life under Control." This detailed the way in which a change in one species of the "life-community" may transform the whole. This was the context in which human introduction of pests and subsequent efforts to biologically control them had produced as many problems as solutions. Still, such efforts represented an emerging applied ecology that was humankind's particular role in and for the "super-organism," a role that was sometimes advantageous. Biological invention and application could bring efficiencies to the chemical wheel of life: discoveries such as Justus Liebig's artificial manures, Abbé Spallanzani's method of artificial insemination, or, they added, "the invention of safe and simple methods of preventing conception," that had momentous consequences. Population control was part of a human-managed total system that could bring "life under control" in a way that recognized the impact of one type of organism on another.[29]

The authors' passing comment about Italians, Indians, and Chinese certainly points to their prejudices, of which more below. But focusing on that alone can obscure the more technical crossover between ecology and human population control, which rested less on fertility than on rates of infant mortality. Population changes in nonhuman organisms, they explained, involve high death rates, an "appalling wastefulness" in nature: "It is, humanly speaking, stupid that each year three-quarters of all the young that singing birds produce must come to nothing." For humans themselves, there should ideally be no such waste of energy in breeding offspring that have little chance of survival. From the standpoint of ecology, then, contraception needed to be considered within a larger system of waste and use: "to make the vital circulation of matter and energy as swift, efficient, and wasteless as it can be made." The recent "breeding storm" of *Homo sapiens* had been enabled by human overuse of energy: "In the last couple of centuries he has accelerated the circulation of matter—from raw materials to food and tools and luxuries and back to raw matter again to an unprecedented speed. Humans have over-killed other species, drawn too extensively on energy supplies, stripped lands of trees, and soil." Humans needed to plan total food and energy circulation and population "as one community, on a world-wide basis; and as a species." This ecological assessment of human population growth as wasted energy within a total system entered demographic transition theories directly. To survive the current acceleration of population growth which Huxley and the Wells well knew had been led in modern times by their own countrymen, *Homo sapiens* needed to take control not only of "his" own biological destiny, "but of the whole of life." Life under control was the final message of the encyclopedic *Science of Life*. The species needed a definite reproductive policy, one that would reflect the ultimate unification of "a collective human organism" that consciously controls "the destinies of all life upon this planet."[30] Humans might be animals, dependent on other species, but in this ecology they were undisputed masters of the planet who had misused their power. They had to get life under control, not least through control of the birth and death of their own species. Human ecology was population control, and for these authors, population control was human ecology.

Human Ecology

Of all the world events unfolding in 1915, the Calcutta town-planning exhibition did not register for many. But it proved critical and deeply interesting to the young Radhakamal Mukerjee, afforded an opportunity to assist Patrick Geddes, the Bombay-based sociologist and urban planner. In the process, Mukerjee learned a good deal about this new integrative ecology. Yet another student of Thomas Henry Huxley, Geddes had settled in Bombay after a period in Edinburgh as a zoologist and then in Dundee as a botanist. Geddes's work struck Mukerjee as deeply relevant to his own tasks, and he was watching and listening closely. Just how did this ecological approach relate to humans in urban environments and in Indian environments specifically? And how exactly did it fit with the economics in which Mukerjee was originally trained? While working for Geddes, Mukerjee was writing his first book, *The Foundations of Indian Economics* (1916). An enthusiastic mentor, Geddes wrote a grand and sweeping introduction that sent Mukerjee on his intellectual and professional way.[31]

If Julian Huxley, H. G. Wells, and Alexander Carr-Saunders considered human populations ecologically from their foundations in natural history, Radhakamal Mukerjee did so from a base in economics. He had perceived the connections from his earliest student years, when he read "a great deal of Haeckel, the founder of bionomics or ecology."[32] From the beginning, then, Mukerjee was steeped in the polymath world of ecology, seeking to work out both general principles and specific applications to the peculiar and troubling context that was colonial India. He prefaced his first book with "India's Message to the West": "India stands for living Humanity as against inert matter; for more equitable distribution of wealth; for less luxury and more brotherhood."[33] It was plain just how much was at stake in this political economy that was also an economy of nature.

Over several decades, Mukerjee put his observations from Geddes's biologically inspired urban ecology to work, increasing in scale and ambition from the local to the global. In his *Regional Sociology* (1926), Mukerjee set out the scope of human ecology applied to local communities. In his *Regional Balance of Man* (1938), he worked through the then-governing ecological concepts of symbiosis, balance, optimal population, dominance, and succession.[34] And in 1939, Mukerjee published *Man and His Habitation*, in which ecology was even more comprehensively applied. It "traces the effects of the harmonious vegetable and animal aggregations on the nature and kind of human food supply and the character of man's occupation and social and economic interaction."[35] The idea of regional balance was inspired, Mukerjee said, not by his work on the crowded Ganges plain but by the American Midwest and prairies. Invited to deliver a course on ecological methods in sociology at Columbia University, he visited also universities in Chicago, Michigan, Wisconsin, and Minnesota, the intellectual heartlands of American animal ecology, regional ecology, and urban ecological thought.[36]

Through all this, Mukerjee was deeply influenced by the animal ecology then bursting onto the intellectual scene. Warder Clyde Allee's work on animal aggregations was important to him,[37] and like Huxley, he saw a special relevance for Elton's studies with the Malthusian idea at their core.[38] He rejected an economic definition of optimum population as problematically restricted; in its place, he developed an expansive "Ecologic Optimum," a spatial optimum that was regionally specific, determined in part by climate.[39] The economic measure of standard of living had to be qualified by and compatible with, ecological balance. There would be two ultimate measures of such an ecologic optimum: the maximum duration of (human) life and the maximum stability and balance of the region.[40]

What Charles Elton later systematized as invasion ecology was one of the more important concepts that Mukerjee applied to humans. There was also the question of unsuccessful invasion; inappropriate immigration and land policy wherein "stock and clime" did not correspond.[41] What and where humans could "be" was determined in part environmentally.[42] This, too, related to animal ecology; climate was key for Charles Elton's explanations of population booms and busts. He had applied the geography of Yale's Ellsworth Huntington, who argued deterministically that the differential complexity and evolution of human populations were fashioned environmentally, not least by temperature and humidity.[43] The term "habitat" (just one of Huntington's many studies was *The Human Habitat*) was deployed increasingly regularly with respect to humans.[44]

The tenacity of the scientific connection between biological types and environmental place in the 1930s is curious at one level, because biological "race" was being strongly unraveled by not a few scientists in the interwar period, often by those centrally involved in formulating world population as a problem—notably Julian Huxley.[45] But on another level, the ongoing conflation of physiology and geography was being strongly reorganized within the new logic of ecology. Indeed, one effect of ecology was to highlight rather than diminish the deep connection between people and place, to naturalize humans (arguably to renaturalize them), and conceptually to tighten humans' bodily relation to, and even differential determination by, "climate" and "environment." We have seen how antique ideas about climate and human constitutions still shaped debates and possibilities about where people should and could be on the globe; whence the tenaciously racialized overcrowded peoples might disperse as a solution to the war-producing problem of differential density. This was increasingly underscored with ecological reference. The barring of Indians from places like the Australian tropics, Mukerjee insisted, was not just political but ecological nonsense, because the white Australians themselves—"Northern peoples from the West" as he put it—"are, and must remain exotics." It was "the policy of Oriental exclusion" evident over much of the globe, that prevented a balanced world distribution of people to appropriate lands, the kind of informed plan, he wrote "which a scientific human ecology envisages."[46] A "symbiosis" was required: "the co-operation of the different parts of Animate and Inanimate Nature for the uplift of the entire regional complex."[47] All of this was beginning to add up to an important intellectual shift: the centuries-old determinist discourse on people-in-place, on physiological capacity and incapacity, became part of and was filtered through an emerging human ecology. One outcome was that ecology entered demographic transition theory.

Ecology, Demography, and the London School of Economics

Throughout the 1930s, Carr-Saunders also framed the population problem squarely as ecological, concerning humans' capacity to adapt to the limits that any given habitat set. But he typically argued for human capacity to change: it was not about inability to survive outside a given climatic zone, but about relative adaptability, higher or lower mortality or fertility rates.[48] Considering the two cases of the movement of African Americans from the South into the colder northern parts of the American continent and the health and fertility of whites in tropical Australia, Carr-Saunders was inclined to argue for possibilities of adaptation, socially and physiologically.[49] Even he had to put the issue tentatively: "While there may be as yet no proof that white men can live and work in the tropics and remain as healthy and vigorous as in a temperate climate, there was no evidence that they cannot do so."[50] His study *World Population* was a key document for consideration at the 1937 Peaceful Change meeting of the International Institute of Intellectual Co-operation (IIIC), and those gathered in Paris to discuss population and resources were also asked to consider different races and distribution of humans in ecological terms: like the different types of flora and fauna, different humans could thrive and multiply only within the limited climatic regions to which they were especially adapted. If that were the case, it had an important bearing on the problem of population and peaceful change. Lebensraum, high on the meeting's agenda, also envisaged an ecological match between particular humans and particular lands.

Carr-Saunders was soon to become the director of the London School of Economics, and it was there that ecological studies of human populations were particularly influential and successful within the British context. Political economy and political ecology were entwined in this institutional home. The idea of social biology had been strongly supported under William Beveridge's leadership, and zoologist Lancelot Hogben held the chair in social biology from 1930. Hogben's wife, Enid Charles, branched into statistics and demography in the early 1930s,[51] working alongside Robert René Kuczynski after his flight from Germany in 1933. Both H. G. Wells and G. P. Wells were part of Hogben and Charles's circle. And this was also the context in which the young sociologist David Glass was trained. In this milieu, left-wing politics mixed with left-wing science. In the 1930s, when Hogben was recommended by Julian Huxley to the left-wing John Boyd Orr, the director of the Rowett Institute was attracted to his politics: "He will find the whole of my staff at the Institute here with the same rather 'forward' ideas as he has."[52] It was the British left far more than conservative circles that constituted the training ground for population science and population politics in these years.

Population studies and demography took off at the Fabian-created London School of Economics, and it was there that Huxley and Wells's formulation of economics as "merely human ecology" was seriously thought through. While the London School of Economics and Cambridge economists were undertaking their own battles, the demographers and biological scientists charged to think through human population were doing so increasingly through ecology. Their politics broadly endorsed planned economies. One outcome of the London School of Economics' close ties to demography and its particular line of biological and political economy emerged as the Political and Economic Planning Committee. Established in 1931, this was a Depression-induced response to ecologist Max Nicholson's work, pushed forward strongly by Julian Huxley. An influential think tank, its population policy group devised recommendations for British and colonial policy, as well as world policy on population growth and transfers. The inclination toward left-wing planned economies eventually sat neatly with a philosophy of family planning at both national and international levels.

Socialist Enid Charles fitted the London School of Economic's biological and political culture perfectly. She was intellectually taken with ecology from an early point, and her foundational education and research in zoology, as well as her Ph.D. in physiology, is significant. Charles took considerable pains in much of her writing to explain the new method as the basis for understanding human population problems.[53] She was interested in Raymond Pearl's density studies, but wondered why the logistic curve did not admit the possibility of a decreasing population, merely a stabilized one, when it was known that human populations had died out in the past, something she thought possible, even likely, for the British, and already in process with the diminishing Melanesian populations of Oceania. There was a particular problem with his density concept, which she thought had little biological significance, because it failed methodologically to take into consideration interaction of organisms or "the materials of the space in which they live." Pearl's claims about fertility and mortality were not, she thought, "universal laws of ecology which apply to all population organisms, including man himself."[54]

The biological optimum was not static but was constantly changing, Charles offered, precisely because of the extent to which humans could interfere with their environment. The dynamism of this system is what Malthus also missed, she thought: humanity interferes with the selective power of nature on a scale that no other animal has the power to do. And so, while humans might be part of nature, the parts were not necessarily equally placed. As a socially organized whole, she thought that the human species had the power to seal its own doom and that this made it unique among all species. These were signature Huxley arguments too. Thinking of the human species in relation to other species and the environment, population was really an ecological problem that had implications for economic and social systems. But that was not the end of it. Socialist Charles constantly folded her ecology back into her left-wing views on political organization. Although "industrialism" had reduced mortality considerably, she thought it a biological failure, because the most industrialized society—her own—was losing the capacity to reproduce itself. "If the human race is to continue as a dominant species, man must now devise new economic arrangements more adapted to the requirements of his own ecology." Charles herself sought a socialist state far stronger than the kind of welfare state that her director, William Beveridge, was just then envisaging. But it was perhaps her political ecology, rather more than her socialism, that was most enduring. For Enid Charles, it was necessary to analyze the history and future of human populations in terms of interdependence with all other organisms. This entailed conceptualizing the human population as part of a "planetary life community."[55]

Population and Planetary History

Ecological research between the wars was undertaken on vastly different scales. Field ecologists like Charles Elton were studying the interactions of organism and environment in local, sometimes very small spaces. At a larger scale, the ecological "region" was actively developed in both human ecology and animal ecology. Mukerjee was key here. And a further tradition of ecological thought concerned the planet as a whole. Indeed, the scales of ecology ranged from the atom to the universe, as Alfred George Tansley noted in the 1935 paper in which the term "ecosystem" was first used.[56]

A long line of natural historians, geographers, meteorologists, chemists, and polymath astronomers were interested in the proliferation of life and its relation to the planet. Alexander von Humboldt's *Kosmos* (1845–1862) was a massive project that ultimately sought a system to the patterns of global distribution.[57] Even earlier, eighteenth-century Benjamin Franklin held a planetary vision to such an extent that he could "see" the globe through the eyes of the inhabitants of Mars and Venus. Writing about population growth in North America, he penned an extraordinarily powerful image: "We are . . . *scouring our planet, by clearing America of Woods*, and so making this Side of our Globe reflect a brighter Light to the Eyes of Inhabitants in *Mars and Venus*."[58] A similar visualization of the surface of Earth from space was later detailed by Russian geochemist Vladimir Vernadsky in *La Biosphère* (1926), translated from Russian to French in 1929. He saw an organic surface, a zone of life: "The face of the Earth viewed from celestial space presents a unique appearance, different from all other heavenly bodies. The surface that separates the planet from the cosmic medium is the biosphere."[59] All of this, of course, prefigured the ideas of a later planetary thinker, the chemist James Lovelock and his Gaia hypothesis.

While some in this period were reinscribing temperate and torrid zones as appropriate or inappropriate habitats for different kinds of humans, others were considering the planet as the habitat for the species. In "The Human Habitat," British geographer Halford Mackinder wrote of the skin of the planet, the "fluid envelopes" of water and air that "by their circulations, their physical and chemical reactions, and their relation to life, impart to the earth's surface an activity almost akin to life itself."[60] For Mackinder, the hydrosphere was itself a closed system that interconnected humanity within it.[61] In studying such a system, he considered the geographer to take over from and incorporate the work of the astronomer, physicist, chemist, geologist, biologist, historian, economist, and strategist, building a vision of the dynamic system "of the world whole."[62] Others were reversing the telescope, contemplating not life on Earth, but life on Mars. "Is Mars habitable?" elderly Alfred Russel Wallace wondered. In his 1903 study, *Man's Place in the Universe*, he had decided not, a claim later extended in his *Is Mars Habitable?* (1907).[63] This was a response to an American astronomer's suggestion that "canals" and therefore life could be discerned on the planet. Wallace countered that the idea failed to consider how enough food could possibly be produced to support a large population on Mars with enough time to be so inventive, to rise from the "low condition of savages to one of civilisation, and ultimately to scientific knowledge?"[64] Even as an old man, Wallace was still putting Malthus's population theory to work.

All this was the workaday world of H. G. Wells. He was thinking about the earth in space too, the globe as a single organism and as part of an integrated cosmos. Both Wells's science and his fiction brought together visions of the singular planet, ideas of space travel, interconnected life and death, and the promise of human perfection (or else extermination). The expansion of human living space and the narrative and biological tension produced by limits to that space drove forward not a few of his famous fictional works. Much of this was compatible with, even derived from, his Malthusianism. All of it was compatible with his eugenics.

Identifying the limits of the planet was an important basis of *The Science of Life*. The summits of the highest mountains and the depths of the sea are effectively lifeless, the authors explained, as is the inner matter of the planet. This meant that "life" was confined to a few miles above and below the earth's surface. There is no life beyond earth's atmosphere, they continued, nothing that "feels and moves of itself and reproduces its kind." Moving well beyond Isaiah Bowman's world, the limits not just of settlement but also of human physiological existence intrigued them. In this instance, they thought of the global biosphere through an altitudinal axis, not the standard latitudinal axis of torrid and temperate zones. The issue was not the relative adaptability to the tropics of Cancer and Capricorn, but the extreme climates that pressed on the limits of any human life. They meant this literally, too, considering atmospheric pressure. How high had George Mallory and Andrew Irvine climbed on Mount Everest in 1924, before expiring? What altitude had balloons reached before expeditioners were rendered unconscious? They knew that in November 1927, a U.S. Army Aviation Service expedition had reached 42,470 feet, the captain losing his life through oxygen depletion on descent. Just over 8 miles high is one limit to life on Earth, they concluded, while the "barriers set to air-breathing life" are soon reached in oceans as well. Divers might reach 300 feet, while other life continues beyond, perhaps to 7 miles. The limiting factor, at altitude or depth, is the increasing solubility with pressure of the atmospheric gases in the blood. Their point was consistently to emphasize not just the smallness of the planet in the vastness of space, but also the even smaller zone of life for humans. "Life, therefore, so far as we know, is confined to a layer of air and a layer of water, having a total thickness of less than fourteen miles, on this comparatively small planet Earth." And is there extraterrestrial life beyond Earth's biosphere? For the authors—Wells in the main—outside the double film that enveloped the planet, there was the smallest possibility that this was so.[65]

It was partly in this tradition of thinking about the planet that George Knibbs conceived his population problem. The astronomy-trained statistician explained that the earth-globe, to the amateur eye a sphere, was really a sphere with a bulge around the equator. This ellipsoid's mass, surface area, situation in space, and internal energies all impacted on the human population question for George Knibbs. He saw the earth from space in a similar way to Franklin. But whereas Franklin's perspective was close enough to the earth—as it were—to still see divisions between kinds of humans (this was his point), Knibbs's vision of the globe and the population problem was both more distant and more catastrophic. Indeed, Knibbs's perspective was so distant, so literally universal, that not just human difference, but humans themselves effectively disappeared. Trumping even Malthus's reputation for gloom, he thought that from the widest point of view, "the obliteration of the whole solar system is insignificant." This was not just a world or global perspective, but a planetary one, should Earth even survive. At issue here were not only nations and empires, or even a supranational comprehension of social and biological geography: it was about the bounded Earth, a singular planet in space. At times, Knibbs would write out the human capacity to perceive the planet altogether: the issue, conceivably, was not even about humans. "Man's view of the world is frankly anthropocentric. . . . Apparently aeons passed in earth's life-story before even the crudest progenitors of the human race appeared."[66] George Knibbs was a post-Darwin Stoic.

Knibbs's own planetary vision—and his neo-Malthusianism—began neither with the sciences of life, nor with the human science of economics, but with the sciences of Earth. He was trained originally in surveying, and later lectured in astronomy, physics, and geodesy (from the Greek: "division of the Earth").[67] This was the discipline devoted to the analysis of the planet's shape and position in space, in essence the application of the surveyor's technology of coordinates and triangulation on a planetary scale. Geodesy dovetailed with geologists' inquiries on the geodynamics of the planet's crust, of tides, and of the poles; with geographers' and cartographers' mapping projects; and with physicists' work on the planet's gravitational field. Steeped in this discipline, then, Knibbs was both by training and by inclination equipped to bring a planetary vision to the population question. He introduced to world population thought "developments in astronomy and astrophysics, leading to the great surveys now being made of the solar system and of the stellar universe." He added a stellar scale to the climatological dimensions of population expertise.[68]

It was the question of the age of the planet that directly linked the traditions of thought about life on the one hand and Earth on the other with the world population problem. The debates regarding the age of the earth helped establish the age of humanity, and vice versa, through geological and paleontological records. For all kinds of commentators who speculated on actual world population growth over many generations, knowing both the age of the earth and the length of time that humans had been reproducing on it was important. Knibbs explained: "A study of the early traces of Man, and of general geology in connection therewith, shows that he has been a denizen of this earth for at least hundreds of thousands of years, possibly even for millions of years. Human history, however, goes back at the most only something like ten thousand years." Estimates of the length of time humans had been on the planet was the factor from which rates of increase could be determined or hypothesized over very long periods of time. What could be shown, to great effect, was that the rate of increase was "extraordinarily slow" given the thousands of years that humans had had the opportunity to reproduce, and fill the earth, like Franklin's fennel. Accelerating nineteenth- and twentieth-century rates of growth signaled unquestionably a new era.[69] It was unprecedented and potentially catastrophic, and entirely overshadowed the more recent localized decline in birthrates.

Knibbs wrote often about humans as denizens of the earth. The ultimate aim of all statistical inquiry should be "the study of man's destiny" as the "denizen of a world of limitations."[70] Implying an alien or a resident conditionally admitted, denizen also had an ecological usage, signaling organic life from outside, naturalized or acclimatized perhaps to a particular region, but retaining some alien status. "As a physically insignificant denizen of the earth," he wrote, "Man is dependent upon the energy, resources, and the vicissitudes of the system of which his earth forms part. He is dependent upon the energies radiating from the sun, perhaps more generally those arriving from space." But humankind was also reliant upon energy contained within the earth, radioactivity, or internal heat.[71] The population problem was ultimately about energy and its transformation.

George Knibbs's work comprehended the world population problem—and solution—as folding together an ecology of life and Earth, but it was a political ecology as well. His integrative holism connected elements of the population question across the politics of life, Earth, and human organization: "(a) the efficiency of human organisation, (b) the appropriate localisation of human beings upon the earth, (c) the standard-of-living adopted, and (d) the degree of freedom of migration attained, for the purpose of permitting any degree of concentration reached in any region to be adjusted to the local population-capacity of other regions."[72] The world population problem drew in the relationship between nature and the whole social field, leading potentially toward not just internationalism but a cosmopolitan solidarity of humanity.[73] It is because of this, he concluded, that the population problem was not "merely a mathematico-physical one," but one that "involves our whole conception of life." The chemicophysical factors he thought relatively simple compared to sociopolitical complexities, which included the need to eliminate "unscrupulous egoism in the life of nations and in the relations of races."[74] In Knibbs's particular neo-Malthusian thought-life, the biopolitics of vital statistics met the geopolitics of a singular planet. He summarized all this in a *Science* article (1923) as a new human recognition "of the significance of environmental changes on his future earth-life."[75] This was an early articulation of what some climate change scholars have latterly called the Anthropocene, irreversible human impact on the planet's biosphere.[76]

Another idiosyncractic and influential polymath, Alfred Lotka (1880–1949), was the U.S. counterpart to George Knibbs, a holistic scientist initially based at Johns Hopkins University who went on to become a critical player in the methods and institutions of emerging demography, in national and international population associations, and in the adjacent applied field of life insurance. Like Knibbs, Lotka was a mathematician, but if Knibbs channeled his inclinations for political arithmetic into the work of the state, Lotka's statistical skills were channeled into the Metropolitan Life Insurance Company, where he worked from 1924 until 1947 alongside another giant of U.S. demography, Louis Dublin. The project of vital statistics was core business for insurance companies, which were arguably even more concerned with demographic patterns and projections for morbidity and mortality than states themselves. Lotka became president of the Population Association of America—he was present at its founding meeting[77]—and was deeply involved for many years in the International Union for the Scientific Investigation of Population Problems (IUSIP) that his Johns Hopkins mentor, Raymond Pearl, had established. Lotka was spotted early by Raymond Pearl, less for his biostatistics than his peculiar crossover work on energy in the early 1920s. It was Lotka's holism that predisposed Pearl to him, and for which he was brought into the Johns Hopkins fold. There, at the same time that Pearl was writing up his new law of population, Lotka produced the far more expansive ecology classic on integrated energy systems, or the energetics of evolution, as Lotka termed it. *Elements of Physical Biology* was published in 1925.[78] This work also viewed Earth as a single system and as Sharon Kingsland notes, Lotka's outlook was openly Stoical.[79] The relevant point for both Lotka and Knibbs is that Stoicism was a version of political cosmopolitanism: humans were not just organisms linked on one planet, but social beings of one world.[80]

The Cosmopolitics of Population

This generation was trying to work out how to bring it all together. Julian Huxley quipped in "A Journey in Relativity" that while everyone had heard the definition of life as "one damn thing after another," this might more accurately be put as "one damn relatedness after another."[81] G. H. L. F. Pitt-Rivers, too, strove for conceptual models to bring economy and ecology together, in his instance with anthropology. Honorary general secretary of IUSIP, he suggested combinations on endlessly expanding axes of analysis. The synthetic field of "human econology," was one attempt.[82] And elsewhere, the all-encompassing nature of "race-population-culture in change"—the science of human ecology—he proposed to call "ethnogenics."[83] In doing so, he was pointing to the integrated nature of the factors involved. Amid all this, one solution was found simply in the idea of "holism." The term itself was offered by a political ecologist who doubled as an internationalist. It was South African prime minister Jan Smuts's innovation in *Holism and Evolution* (1926).[84] Synthesizing, integrative, and interdependent, the whole—whatever its nature and however large or small—was greater than its parts. "Holism is an attempt to solve a scientific and a philosophic problem," Smuts explained to Julian Huxley in 1926 "the problem of synthesis which meets us everywhere in nature."[85]

The population problem came to be deeply shaped by the conceptual holism of this emerging global political ecology and by another holistic enterprise, political cosmopolitanism, a universal human community and the formation of federated or even supranational world government.[86] Thinking about human populations constantly pressed forward, as we have seen throughout, the question of relations between nations. In his 1914 lecture, Keynes squarely identified the world population problem as cosmopolitan in this sense: "The problem . . . is made much worse and far harder of solution by having become, since Malthus's time, cosmopolitan. It is no longer possible to have a *national* policy for the population question."[87] For George Knibbs, global economic integration in a limited space pointed politically toward integration: "mankind" has yet to regard such issues "from the standpoint of the good of the whole."[88] Whether leading toward disintegration and war or toward integration and peace, the prospect of overpopulation prompted major discussion about just how world citizenship and world government might play out in the twentieth century.[89]

This conflation of population, ecology, and cosmopolitanism was further underscored by the pacifism that drove many proponents. War and peace might be the business of politicians and diplomats, but it was also the business of biologists and ecologists. Darwin's theories and their Malthusian derivation were constantly used to explain or justify struggle and war but also, as Gregg Mitman has shown, to argue for peace.[90] This all came together over the substance of the problem of world population growth, both politically (because overpopulation caused war) and scientifically (because ecology derived from the combination of Malthus and Darwin). We might say, then, that in the cosmopolitan response to the world population problem in the 1920s, the Malthusian global "struggle for room and food" met the descendants of Immanuel Kant's *Perpetual Peace*, penned in the same years as Malthus's population essay. Kant's pacifist cosmopolitanism was being applied to a new problem in a new time, in a world continually assessed as smaller, faster, and more connected, and in which world peace was the urgent project after world war.[91] A cosmopolitics of world governance of population matched and sometimes conflated with an integrative ecology, producing a powerful new global political ecology.

The politics of those trying to integrate life and earth in such ways was not predetermined or uniform or predictable: Huxley and Knibbs's new liberalism was one facet, and Wells and Enid Charles's socialism was another. A third was Pitt-Rivers tendency toward fascism (he was to be interned by Winston Churchill during World War II). But as we have seen, the lebensraum idea within fascist thought also invited holistic approaches, whereby blood and soil, people, and environment and spirit were deeply connected.

It is no coincidence, for example, that what attracted Mukerjee to Patrick Geddes's work was its ambition to incorporate the interrelation between "Place, Work and Folk, or Environment, Function and Organism."[92] Nor is it coincidental that lebensraum was sometimes translated as "habitat."

"Holism" and "cosmopolitanism"—a bit like "ecology"—sound politically wholesome. Plainly, however, these were ideas that could be attached to a range of political philosophies. They shared a logic of the whole being greater than the parts, but this did not mean that the parts were necessarily perceived to be the same or equivalent in value. Indeed, if anything, proponents of "holism," "cosmopolitanism," and "ecology" tended to value "parts" differently. The political ecology of Jan Smuts is both important and instructive in this respect, not least because he was a key player at the Paris Peace Treaty and the early UN, famously authoring the preamble to its charter. He was, on the one hand, a constant advocate of the pacifism that was so often understood to be both driver and result of internationalism and, on the other hand, he was architect of South African spatial and racial segregations. Smuts's internationalism, pacifism, and defense of the League of Nations were, in his view, perfectly compatible with his insistence on racial difference and segregation in a united South Africa and his authorship of the preamble.[93] Humans were different from one another, evolving in varying biotic communities, even as they should and would also form part of an integrated system. That whole might be a united South Africa, a federation of world states, or a regional ecosystem. Smuts's holism was quite consistent, on its own terms, with hierarchical difference of its constitutive parts. It was a version of unity in diversity, but one in which hierarchical difference was both presumed and acted upon, something that seasoned anticolonial commentator Taraknath Das among many others, did not fail to point out immediately.[94] Just how to understand and assess individual and racial sameness and difference within the fast-reproducing species called *Homo sapiens* puzzled and bothered neo-Malthusians and eugenicists as they linked the "quality and quantity" dimensions of the world population problem. The idea of human equality was being trumpeted anew in UN circles, in the light of the war and the Holocaust. But even biologists committed to racial equality were not necessarily committed to the idea that humans were biologically of equivalent value. Accordingly, democracy was a problem for quite a few of this generation of cosmopolitan ecologists, not least Huxley and Wells.

H. G. Wells was not just an author of science fiction and fact, he was also one of the most famous political cosmopolitans of the period. Just before he published *The Science of Life*, he had written *The Commonsense of World Peace* (1929), and just after it, *The Way to World Peace* (1930), *The Fate of* Homo sapiens (1939), and *The New World Order* (1940).[95] Even before World War I (and the peace) Wells had forthrightly advocated a world state of various kinds. Global unity was on his mind as early as 1901, when he wrote of the "establishment of one world-state at peace within itself."[96] Anticipating but possibly even shaping the dreams of the International Labour Office, he imagined in *A Modern Utopia* (1905) demographic control in terms of world-level interventions into labor shortages and excesses that would involve relocation or resettlement.[97] During World War I, he precociously publicized the idea of a League of Free Nations, "a Peace League that will control the globe" with powers to "re-draw every frontier."[98] The league that was actually formed in 1919 disappointed him greatly, having powers lesser, not greater, than its constituent nations. And yet, as his political biographer John Partington has shown in *Building Cosmopolis*, Wells continued to press the idea of a cosmopolitan world governance through the 1920s and 1930s, indeed past the end of World War II, until his death in 1946. He was, in short, one of the period's most significant popularizers of the links between cosmopolitanism, ecology, and populations in space. An H. G. Wells Society established in the 1930s called itself "Cosmopolis."[99]

Wells is often assessed as a lone cosmopolitan voice in the early twentieth-century years, articulating a "unique strand in twentieth century internationalism," even "the only outstanding prophet of world order in the first third of the century."[100] Wells's particular vision had no internationalist predecessors, claims Partington. What is missing in this analysis of Wells's political thought is the international, pacifist, and cosmopolitan politics of population: his place in the Malthusian intellectual line. As we have seen, he was strongly involved in the early twentieth-century British Malthusian League, supportive of the prospects and politics of birth control, permitting his influential name to be used by the National Birth Control Association, which became the British Family Planning Association.[101] As early as 1903, he was one of the bolder and more public proponents defending neo-Malthusianism vis-à-vis Rooseveltian pronatalism: "So far from encouraging any increase of the kind I would encourage a decrease. That I strongly hold. I have no belief whatever in the value of mere numbers, and have no sympathy at all with the Rooseveltian views on the question."[102] By the 1920s, such sentiments were shaped rather more by pro-Sanger than anti-Roosevelt inclinations. Wells was sexually and romantically, as well as politically and intellectually, involved with Margaret Sanger and wrote the introduction to her 1922 *The Pivot of Civilization,* in which birth control and internationalism were linked, almost as a matter of course.[103] "There is no other subject of such importance as Birth Control," Wells wrote, gushing. "Knowledge of it marks a new and happier phase in the history of civilization."[104]

The Malthusian catastrophic tradition, which had long functioned on a planetary scale, is one to which Wells firmly belonged. Setting out seven points for a "blue print for a world revolution," Wells's point four was "recognition of the necessity for world biological controls, for example, of population and disease."[105] And in *The Open Conspiracy* (1928), he reiterated ideas that many of his Malthusian colleagues had long held close:

> We have to make an end to war, and to make an end to war we must be cosmopolitan in our politics. It is impossible for any clear-headed person to suppose that the ever more destructive stupidities of war can be eliminated from human affairs until some common political control dominates the earth, and unless certain pressures due to the growth of population, due to the enlarging scope of economic operations or due to conflicting standards and traditions of life, are disposed of.[106]

This was his core idea in a chapter significantly titled "Broad Characteristics of a Scientific World Commonweal." The aim of human ecology, for Wells, was efficient global management.[107] So often the prophet, Wells seemed to see population control of the 1960s as a vision: "Suffice it for us here that the world community of our desires, the organized world community conducting and ensuring its own progress, requires a deliberate collective control of population as a primary condition."[108] Just as Charles Elton integrated the elements of earth and life into a nitrogen cycle—later an "ecosystem"—for Wells, synthesis was everything. His plans for the education of new world citizens built from the basis of biology, history, and human ecology, with history itself conceived biologically, as a world story beginning at the subhuman level and proceeding to "the social organisation now growing to planetary dimensions, of the human species."[109] It is clear why Edith Charles would point to "Mr. Wells" in particular as "recognizing that the human population problem is first and foremost a department of ecology."[110]

Wells died in 1946, witnessing the new UN, with its cosmopolitan charter, and witnessing also Julian Huxley's place in the brave new world, as first director-general of UNESCO. This was the postwar incarnation of the IIIC, which had a decade earlier considered world population problems at its Peaceful Change meetings. Huxley wrote a controversial manifesto for the new UNESCO, without so much as checking in with the new and delicately placed UN Secretariat, let alone his colleagues. It was ecology that framed the UNESCO manifesto. Malthusian and neo-Darwinian evolutionary theory together provided Huxley's conceptual framework for the future of humanity: "An evolutionary approach provides the link between natural science and human history. . . . [I]t not only shows us the origin and biological roots of our human values, but gives us some basis and external standards for them among the apparently neutral mass of natural phenomena." In general, he wrote, UNESCO must constantly be testing its policies against the touchstone of evolutionary progress.[111] If humans thought of themselves ecologically, that is, in relation to other living organisms, a more peaceful and productive international community would result. Humans were the "trustees," as he put it, of life on Earth, and political organization should reflect that responsibility. UNESCO policy should thus include the application of medical science; studies on agricultural productivity, including soil erosion; and studies on social welfare. It would also have to include the provision of birth control facilities. "The recognition of the idea of an optimum population-size (of course relative to technological and social conditions) is an indispensable first step toward that planned control of populations which is necessary if man's blind reproductive urges are not to wreck his ideals and his plans for material and spiritual betterment." Huxley saw before him a vision of efficient human-controlled and planned progress.[112] The League had failed to bring about peace, but there were now new prospects for world federation and cooperation. He wrote of UNESCO's potential not just to promote world unity but the possibility of a "transfer of full sovereignty from separate nations to a world organisation."[113] These were terms that would not only be recognized by but were inherited from the liberal internationalists and (it turns out) ecologists who had imagined a new world in 1919.

Political ecology was the new field from which Huxley, for one, could imagine and press for a world population policy that minimized waste, as well as a future planned world in which quality would trump quantity. This was even a world in which "truly scientific eugenics," as he put it, would be realized. It was a deeply personal statement. It was also, in so many ways, *The Science of Life* rewritten on Wells's death and for the next new world. Between one world war and another, the problem of world population had emerged at the intellectual and political crossroads of ecology and cosmopolitanism.

7

Soil and Food

Agriculture and the Fertility of the Earth

The prosperity of the human race depends, in the last analysis, upon the soil.

EDWARD EAST, "FOOD AND POPULATION" (1927)

Throughout the 1920s, and certainly through the economically and militarily tumultuous 1930s, the nature of soil and conspicuous demonstration of its cultivation were politically significant signs of belonging and of claim to territory. The restrictions on global movement of people and ongoing anxiety about global population growth, not to mention famine and food shortages, all brought to the fore local, national, colonial, and even world imperatives for the intensification of agriculture. Commentators such as William Beveridge thought Europeans overseas should cultivate more intensively and perhaps differently the soil they claimed,[1] while Warren Thompson thought those not cultivating the soil they claimed should give it up. Adolf Hitler was also thinking about soil, food, and belonging over the 1920s, with regard to Germans at home and as a rationale for Greater Germany: "Never forget that the most sacred right on this earth is a man's right to have earth to till with his own hands, and the most sacred sacrifice the blood that a man shed for this earth."[2] "Blood and soil" may have become a specifically fascist slogan, but in other respects it was a sentiment true to the times more generally.

Symbolically, politically, economically, and literally, soil was the substrata of the population problem. It makes sense, then, that agricultural science was the expertise base for many population commentators. In both the United States and Britain, they were likely to emerge from the agricultural experiment institutes that were springing up on several continents. Edward East held early positions in experimental agriculture and horticulture, including at Harvard's Bussey Institute. Raymond Pearl worked in experimental genetics at the Maine Agricultural Experiment Station.[3] Warren Thompson was a midwestern crop farmer as well as a demographer, as fascinated with yields per acre as he was with births and deaths per year. His colleague Pascal Whelpton also came to population via agricultural economy.[4] Oliver Edwin Baker did research at the University of Wisconsin–Madison Agricultural Experiment Station before transferring to the U.S. Department of Agriculture (USDA), where he undertook his world soil surveys.

Within different bureaucratic and university traditions, the agricultural pattern was the same in Britain. When Julian Huxley suggested an experimental project to Alexander Carr-Saunders in 1922, the latter wrote back excitedly from Oxford: "re Pigs: the project you mention is extraordinarily attractive. It would combine the two occupations I am really keen on—farming and biology."[5] Carr-Saunders was deeply engaged in the implementation of alternative agricultural economies. In nearby Rothamsted, directors of the oldest and longest-running agricultural experimental station in the world, Daniel Hall and E. J. Russell, both took up the population question in the interwar years. And medical doctor John Boyd Orr was director of Scotland's Rowett Institute, which researched the relationships among soil nutrition, animal nutrition, and human nutrition. It was from this interconnected expertise that he pronounced forcefully in international fora the linked catastrophes of soil erosion, malnutrition, and world population growth. Speaking in Iowa as director-general of the new United Nations' Food and Agriculture Organization (FAO) in 1946, he proclaimed his credentials, just like Warren Thompson: "I am a farmer as well as a scientist."[6]

The work going on in all these locations extended the scientific tradition of agricultural chemistry that had long observed and sought to understand elemental conversions that took place between soil and atmosphere, what by the 1930s was known to be the nitrogen cycle. Soil breathed. It was a fundamental part of Vladimir Vernadsky's biosphere, the zone of life between lithosphere and atmosphere. In much of this scholarship and research practice, humans were conceptually integrated into the soil, as consumers of fuel and as producers of waste. This required expertise that comprehended land not just as mere territory or area, but also as "living soil"—the term became common—that could be enriched or wasted: its properties, its elements, were considered simultaneously microbial and biospheric. Soil was not just a matter of geopolitics, it was itself the matter of biopolitics.

The geopolitical trope of limited land provoked by world population growth and colonial expansion energized the long scientific tradition that sought to maximize the literal earth: agricultural chemists and early ecologically inclined botanists and zoologists were actively pursuing the reuse of waste in cyclical rather than net loss linear models. Thus, while soil would seem a reductive and even mundane matter, in fact, it was vastly expansive: conceptually, politically, and scientifically global. Here lies the tangled twentieth-century history of the later green revolution, of environmentalism and the population question, of mid-twentieth-century ecology's concern with soil and people. And here lies the origin of the (soil) conservation movement's uptake of population control and family planning.

Nitrogen and the Elements of Life

Delivering a presidential address to the British Association at Bristol in 1898, William Crookes presented "The World's Wheat Supply." Reprinted as *The Wheat Problem* in 1900, with further editions printed in 1905 and 1917, this address became a point of reference for early twentieth-century agriculturalists and economists concerned with the production and consumption of this key staple. The problem that Crookes put forward began with the British Isles, where at the end of the nineteenth century, only 25 percent of wheat was grown domestically and 75 percent was imported. The likelihood of growing more wheat in the United Kingdom was slim, given population growth; about 100 square miles of new wheat-growing land would need to be added each year, he calculated. How can the United Kingdom consider itself "safe from starvation" should wheat harvests fail or in the event of war, he asked. The problem was a global one. Crookes undertook a world audit, listing the limits of U.S. prairie land ("no land left for wheat without reducing the area for maize, hay and other necessary crops"), of Russia (where the "black earth" of the southern empire was vastly overrated), of Siberia (climate drastically limits yield), of the Canadian Northwest ("performance has lagged behind promise"), of Australasia and New Zealand (limited climatically, but "a potential contributor"), South Africa ("an importer of wheat"), and Argentina and Uruguay ("indifferent land"). Because of population growth, this added up to a crisis in food supply, a "life-and-death question for generations to come."[7]

Crookes's widely disseminated address initiated an avalanche of "wheat problem" research, writing, and international discussion that lasted right through the 1920s and 1930s, arguably into the "green revolution" work after World War II. It galvanized government, philanthropic, and commercial investment in agricultural research institutes across the United Kingdom, the United States, and beyond.[8] George Knibbs's book, *Shadow of the World's Future*, was reviewed as a descendant in the Malthusian tradition, "a younger and brighter brother of Crookes's *Wheat Problem*."[9]

From within the USDA, Oliver Edwin Baker wrote an important 1925 article in imitation of Crookes that swept across the world in a similar global audit of wheat-growing lands. Unlike Crookes, who had not mentioned Malthus, Baker explicitly contrasted his own global situation with that facing Malthus in the late eighteenth century. That world, Baker was at some pains to explain, had not yet witnessed the great global agricultural revolution that followed population growth, the revolution "which transformed the grass lands into grain lands and supplied the food and fibres that made the Industrial Revolution possible."[10] In Malthus's world, only small parts of the Russian prairies and steppes produced wheat, the plains of central and western North America had not yet been crossed by white men. Argentina was "an unmapped wilderness," and only the edges of the Australian continent had been explored, much less cultivated (figure 7.1).

Baker rehearsed the geopolitical trope of his own time: the great expansion of cultivation was drawing to a close. This, in combination with population numbers, created a looming "food problem of the future."[11] If the "wheat peoples of the world" continued to grow apace, even taking into account the reduction in that growth rate from the 1880s, argued Baker, there would necessarily be diminishing returns from the land: wheat production would be pushed back "into the cold, the wet, the dry, the hilly, the stony and sandy lands of the world." He offered the familiar discourse of agricultural limit, concluding that the world could probably grow three times its present population, he said, but not more.[12] Alternatively, and this is discussed in the next chapter, Baker thought that increased control of reproduction recommended itself in the light of this diminishing return.

William Crookes's earlier solution, however, had been neither reproductive nor territorial, but chemical. He was really interested in the element nitrogen: how to get more of it, or more correctly, how to make more of it usable, how to "fix" it to create fertilizer. Crookes was one of a long line of organic chemists who investigated the biological fixation of nitrogen, and he envisioned synthetic fixation as the great way forward for agriculture, for chemistry, and for the "Caucasian races" who were running out of wheat-growing land, because their numbers were growing so fast.

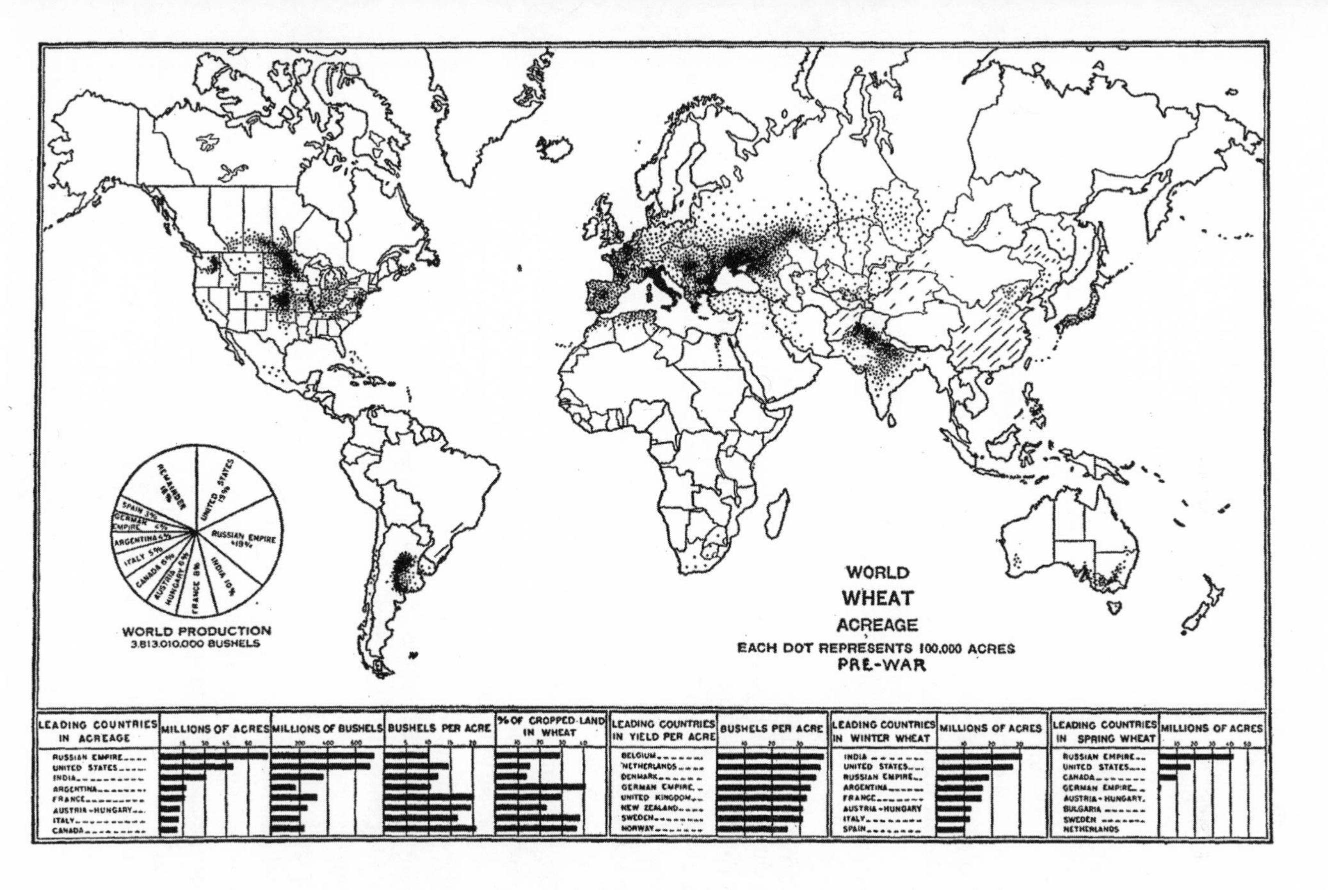

FIGURE 7.1 Global cultivation. Oliver Baker's point in this map was to show that with the exception of Italy and India, all the great wheat regions of the world had been converted from grasslands to grain lands over the nineteenth century. (From O. E. Baker, "The Potential Supply of Wheat," *Economic Geography* 1, no. 1 [1925]: 16. Reproduced with kind permission, Wiley.)

Current-day commentators still point to a relation between population growth and global capacity to fix nitrogen. Nitrogen's biospheric presence and use is unique, writes historian of energy Vaclav Smil. The element is critical for human life and survival, needed for almost every organic transformation. It is "abundant yet scarce, essential yet needed in rather minute amounts."[13] Nitrogen is abundant yet scarce due to the difficult process through which the tightly bound molecule is split into its atoms. Thus, although present in the atmosphere in large amounts (around 78 percent of Earth's atmosphere is nitrogen), either a natural or a synthetic fixing process needs to occur before nitrogen can be combined with other elements to become a usable fertilizer. This occurs naturally in the atmosphere as a result of lightning and was eventually achieved synthetically on an industrial scale through the use of electricity. Crookes's own experiments in 1892 had shown the possibility of synthetic fixation of nitrogen by atmospheric combustion, using electricity.[14] This is why he thought the real solution to the land–people problem was to be found in the atmosphere. In gleams of light, he wrote, nitrogen pressed down on "every square yard of the earth's surface." The wheat problem (the population problem, the race problem) "demands it *fixed*." Without an industrially viable process of atmospheric fixation of nitrogen, Crookes thought in the late nineteenth century, "the great Caucasian race will cease to be the foremost in the world, and will be squeezed out of existence by races to whom wheaten bread is not the staff of life."[15] Commercial application turned out to be impossible in Crookes's lifetime and, indeed, in Oliver Baker's lifetime, with large-scale industrial nitrogen available only from the 1950s.[16]

Part of the reason synthetic production was so pressing for Crookes was the marked depletion of organic sources of nitrogen and other elements for the vast nineteenth-century fertilizer industry. On this count, the prospects for increasing yield through fertilization on any kind of scale that matched the growing numbers of people were looking grim indeed. Yet there was another process of nitrogen fixation that was beginning to engage soil scientists and agricultural chemists toward the end of the nineteenth century. In 1887, Prussian chemists determined that agents (bacteria) on the roots of particular legumes fixed nitrogen from the atmosphere. This explained the empirically known beneficial effect of growing legumes in crop rotation, a practice that had been widespread in Europe for centuries.[17] Their work confirmed and explained the precise mechanism for agricultural chemist Jean-Baptiste Boussingault's earlier demonstration that the nutritional value of fertilizers was proportionate to their nitrogen content, and that legumes restored nitrogen to the soil. This was why, for example, when reporting on Cyrenaica for the Jewish Territorial Organisation, the explorer-geographer J. W. Gregory duly studied the amount of ammonia in the soil at every place his team stopped, converting it to a percentage of nitrogen.[18]

One fascinating aspect of agricultural chemistry, and the discipline's specific interest in nitrogen as an element of life, is that it pressed so many observers to think about the connection between atmosphere and soil in the search for the element's transformation. Soil–atmosphere exchanges prompted scientists to think on vastly different scales, the simultaneous action and effect of atomic and planetary transformations. It was the German chemist Justus Liebig who pursued the cyclical aspects of the transformation of nitrogen between atmosphere, soil, and plants, offering early renditions of global biospheric cycles. Writing of the elements that necessarily support animals and vegetables, he concluded that: "All the innumerable products of vitality resume, after death, the original form from which they sprung. And thus death—the complete dissolution of an existing generation—becomes the source of life for a new one."[19]

One of Liebig's most important students was Englishman Joseph Henry Gilbert, whose intricate and long-standing work with John Bennet Lawes at the Rothamsted experimental farm systematically demonstrated the value of nitrogen for agriculturalists as well as industrialists interested in its commercial application.[20] In 1847, Gilbert and Lawes commenced their joint field experiments, rotating turnips, wheat, beans, clover, fallow, barley, and grassland, chemically analyzing continuously. Their interventions were increasing yields of wheat by 200 percent, William Crookes later claimed, and so it is unsurprising that over time, the world's longest set of continuous field experiments dovetailed with the population problem.

Daniel Hall was Rothamsted's energetic director from 1902, greatly publicizing the farm's long-standing work in *The Book of the Rothamsted Experiments*.[21] For Hall, the detailed agricultural chemistry that was Rothamsted's trademark always played out in a larger political field. He upscaled his Rothamsted knowledge by connecting agriculture and fertilizer expertise with domestic and colonial statecraft. He was secretary and chief scientific advisor to the British Ministry of Agriculture and chaired the Kenya Agricultural Commission (1929), working with John Boyd Orr, who was then pursuing his famous African nutrition studies. As we shall see, this East African work became an important forum for internationalizing the questions of population, health, and agriculture. As Joseph Hodge summarizes the period and the "triumph of the expert," the war and the Great Depression "spawned a new realization of the need for greater integration of government and expertise, and growing enthusiasm in the interwar years for central planning as the basis for social and economic progress."[22] This was, Hodge explains, the prevailing logic of late colonial development, with a strong focus on new colonial agriculture initiatives, specifically tropical agriculture directed by the Colonial Advisory Council on Agriculture and Animal Health. Well might the experts have focused on the tropics, because as John McNeill has shown, in the late twentieth century most of the world's new croplands were converted to agriculture at the expense of tropical forests, grasslands having largely been converted during the nineteenth century.[23]

Like William Crookes, Rothamsted's Daniel Hall thought that future viability of population increase came down to the element nitrogen. He explained: "The earth's crust can have started with but a small stock of combined nitrogen, and every time organic material is burnt the nitrogen therein goes out of combination into the gaseous state . . . there is no source in sight for either the original stock of combined nitrogen in the world's soil and living organisms, nor any adequate means of repairing the losses it suffers by conversion into the free gas." Unlike Crookes's faith in synthetic nitrogen fixation, however, Hall looked to the organic processes of bacteria on the roots of legumes—"active agencies in the organization of life"—in which the Rothamsted experiments had specialized. He wrote that while nitrogen in the earth's crust was nearly exhausted, the "bacteria which live symbiotically in the nodules . . . found on the roots of leguminous plants" were theoretically renewable. As he put it, scientists needed to look to "recuperative agencies in plants and in the soil itself."[24] The symbiotic relationship between bacteria—various strains of *Bacillus radicicola*—and the host plant, modeled for Hall the ideal relationship between agriculturalist and soil.[25] This was part of the increasingly sophisticated ecological knowledge of the place of a single bacterium in a larger system of exchange, transformation, and adaptation. Agriculture that did not build from this symbiotic relationship, he argued, produced a net loss over time in a linear rather than cyclical system.

This imperative to think in terms of the capture, conversion, and reuse of energy within a necessarily limited system, rather than simply its use, dovetailed with Malthusian considerations of consumption and production that were increasingly not just economic but ecological. For Hall, failure to rotate crops with legumes was simply "wasteful mining into the resources of the soil," the main example being the continuous maize and wheat crops of early U.S. midwestern agriculture, as well as the shifting cultivation practiced in parts of Africa.[26] In Hall's mind, permanent cultivation, including crop rotation and fallow with legumes, would permit the land to absorb and support greater local populations. Hall's soil expertise was much cited, especially in international conversations on density and optimum population. At the 1933 Birth Control in Asia Conference, P. K. Wattal, author of *The Population Problem in India: A Census Study*,[27] drew on Hall to determine the average acreage needed per capita. India's governmental agricultural chemist had calculated a shortage of 500,000 tons of nitrogen. In all, this meant that India was producing food sufficient for only two-thirds of its population.[28] But it was also Hall's uptake of cycles, and in particular nitrogen cycles, that found its way into population talk. The connection between humans and earth was already alive.

Waste

The new knowledge of a nitrogen cycle, linking lithosphere, hydrosphere, and atmosphere, recommended the use, not the waste, of waste. The more the idea of geopolitical limit and containment gained traction, the more important the reuse of waste—and the political economy this implied—became. Harold Cox wrote that "[man] can make the ground itself more fertile, not only by tillage but also by feeding it with the waste materials that he or his animals have produced."[29] The whole idea of cyclical economies became quite literal.

It was in this context that Chinese agricultural practices came under close scrutiny in Anglophone texts. If British agriculture was massively dependent on imported fertilizer, Chinese techniques were local and functioned generally within a closed system, using both animal and human excreta.[30] Hall wrote extensively on the agricultural value of Chinese practices of manuring.[31] Just as Malthus had been fascinated with the apparent phenomenon of two crops per year, China modeled for Daniel Hall the possible use of waste to sustain denser populations, albeit at a lower standard of living. Although soil exhaustion was common all over the globe, he wrote, fertility had been maintained in China. There, an "astonishing density of population [is] supported wholly by the land." Compost was prepared through the use of soil, vegetable waste, and excreta, as well as crops grown for the purpose.[32]

The "fertility" excreted by human beings was conserved, not lost. Although the labor needed to produce the compost was itself significant and needed to be taken into account, for Hall Chinese practices added up to a sharp object lesson regarding the wastefulness of colonial agricultural policies of Great Britain and the United States, and by extension, world agricultural and population planning.

There were European traditions, too, that were mined, as agriculturalists and soil scientists considered how more people were to be fed and what that meant for economic and ecological systems of various scales. Historian Dana Simmons has elaborated the work of a group of anticapitalist French chemists in the 1840s, for whom a circular and replenishing relationship between humans and their environment (soil, atmosphere, animals, plants) was promoted both scientifically and politically. Seeking to balance input and output, Boussingault and others had experimented with the means to measure equilibrium between ingestion and excretion. Production and consumption were radically collapsed: "Every man is the producer—and even, precisely, the reproducer—of his own consumption," announced one. This was even reduced to the symbolic statement-slogan: "Make Bread with Human Excrement," an aspirational recycling universe where nothing was imported and all in the economic and environmental system was both closed and stable.[33] The recycling universes imagined by Boussingault had a particular anticapitalist, nongrowth politics behind them. As Simmons shows, the British scene was quite different: labor—of animals and humans—was always calculated as part of total energy bookkeeping. But in all of these traditions, agricultural chemistry was rarely just about agriculture; it was almost always about physiology too, as energies from sun, soil, plants, and animals entered, indeed became, human bodies that then expended energy through labor and as excreted waste. "The Relation of Nitrogenous Food to Work," for example, was one research agenda at Rothamsted. Animal manures, as well as nitrogen-fixing legumes, were the constant object of inquiry: animal feeding experiments were not just aimed at researching the best or most efficient meat for human consumption, but also for their production of different kinds of manures: "The Manure Value of Food."[34] The Rothamsted researchers looked at the relative value of nitrogenous and nonnitrogenous constituents of food, the relation of nitrogenous food to work, the source of fat in the animal body. As Harold Cox put it in the *Problem of Population:* "He is, himself but a product of the earth on which he lives. Not only the food that he eats but every material that he uses comes from the earth."[35] This kind of research brought the atmosphere into the biosphere, and via soil, into humans.

The natural and political economies of fertilizers raised spatial questions: Were fertilizers imported? Would or could waste be used where it was produced, as the French agricultural chemists had argued, or as the Chinese farmers practiced? British farming had been intensified over the nineteenth century, in part through the massive importation of various fertilizers—guano from the Pacific, bones from Europe and India, nitrates from South America. Critics considered this economy nonsustainable and wasteful. Hall quoted Liebig regarding the extent to which English importation of fertilizers literally wasted other lands:

> England is robbing all other countries of their fertility. Already in her eagerness for bones, she has turned up the battlefields of Leipzig, and Waterloo, and of the Crimea; already from the catacombs of Sicily she has carried away the skeletons of many successive generations. Annually she removes from the shores of other countries to her own the manorial equivalent of three million and a half of men, whom she takes from us the means of supporting, and squanders down her sewers to the sea. Like a vampire she hangs upon the neck of Europe, nay, of the whole world, and sucks the heart blood from nations without a thought of justice towards them, without a shadow of lasting advantage to herself![36]

Hall quoted Liebig extensively, because he also thought British agriculture unsustainable in this respect, and all the more reprehensible, because it had been so many generations since Liebig's original reproach. Later, ecologist Fairfield Osborn framed the spatial economy of fertilizers in just this way: "There is one steady movement of organic material to towns and great cities and industrial centres—there to be consumed or disposed of as waste but never to go back to the land of origin."[37] This was part of the plundering of the planet.

Human waste was accordingly to be considered within this economy. Just how might this be used to feed the planet's growing numbers? Hall himself calculated the possibilities in his book *Fertilizers and Manures*, reckoning the amount of nitrogen, phosphoric acid, and potash an average person would excrete each year.[38] The waste of potential fertilizer from human excreta in British and indeed in most Western agriculture, he thought vast. And yet it all went to the sea.[39] In fact, there were numerous public engineering and commercial attempts to do something with the sewage of large British towns and cities.[40] Crookes himself had been a director of the Native Guano Company that sought to convert London sewage into manure. He was certainly concerned with the use (and waste) of human waste, as Liebig had been before him. As more population writers picked up theories of carbon and nitrogen cycles over the 1930s, the more central the question of human waste became. Mukerjee, for example, detailed the significance of animal and human urine, feces, and other waste products, including dead bodies that might return nitrogenous compounds to the earth and be converted by the activity of other living organisms. Too often, he complained like Liebig, such substances are withdrawn from the biological cycle.[41]

In his 1935 Heath Clark Lecture at the London School of Hygiene and Tropical Medicine, Daniel Hall focused on the "Improvement of Native Agriculture in Relation to Population and Public Health," published the following year by Oxford. The lecture was framed by the ambition to relieve "the pressure of over-population." It advocated the reform of shifting cultivation in Africa and the recuperation of fertility and soil nitrogen through green manuring and composting in permanent agriculture. The cultivation of stationary plots should combine English-style crop rotation and Chinese-style fertilizer and composting techniques. Together, this would prevent desertification and assuage what he called "land-hunger," the urge to move into and cultivate ever-more-virgin land. This would be good for the health of Africans and for the conservation of land, he claimed, and together, these policies would solve "the increase of population in African tribes." Hall's plan was conceived within a mandate tradition that predated but morphed into post–World War II modernization and development: malnutrition held native people "at so low a level of physique and health that they cannot take their due place alongside the white man's civilization." As trustees, the British had a responsibility to improve this condition and were in Africa "neither to displace the natives nor to exploit them; actually we have to save them from themselves, for the intrusion of our Western civilization has a destructive effect upon the tribal cultures."[42] And yet historians have analyzed forced agricultural change as one of the most politically disruptive of all colonial interventions. As Henrietta Moore and Megan Vaughan have shown in their case study of northern Zambia, local government and economic policy rather more than population pressure itself encouraged increasing permanent cultivation, replacing "slash and burn" or citemene methods.[43]

If, by this view, mandated territories and colonies had to be assisted technologically, agriculturally, and financially to feed their growing populations, other parts of the world were increasingly taken with the idea of entirely closed and independent economic and agricultural systems whereby a nation could and should always be able to feed its own population. This would render it secure in a precarious international system. Autarky was a geopolitically derived concept of national self-sufficiency, whereby systems of production, consumption, and reuse of waste were self-sustaining. It was an idea honed by Karl Haushofer and implemented as policy in Mussolini's Italy and Hitler's Germany. But the idea had a purchase in Anglophone scholarship as well.

Agriculturalist O. W. Willcox set out the possibilities of a new systematic "agrobiology" that would enable nations to more effectively produce sufficient quantities of food within their own territories. There were nations that had over time shifted the basis of their subsistence "from its own agriculture to the products of soil controlled by others." Or as he put it elsewhere, nations that exchanged "the products of the surplus labor of its people in exchange for the products of other soils."[44] His study, significantly titled *Nations Can Live at Home*, was introduced with the familiar spatial question of density: "What is the maximum number of persons who can exist on the produce of one acre of arable land?" This question was pressing due to the new crowding, the restriction on global movement, and the lack of "outlets." "If it is true that pressure of population on the soil is one of the major predisposing causes of war, and if it is true that the new agrobiology offers a means of relieving this pressure, then it should be the first business of a world desperately in search of a formula for peace to square its agriculture and its agrarian polity with the new realities of science."[45] But this was ultimately an anti-Malthusian text, one that critiqued Edward East's "old style Malthusian prophecy" that there was a limit, that 2.5 acres was needed per person. Even a nation as "oversaturated" with people as Britain—"peoples beyond the threshold" Willcox called them—could become nationally self-sufficient by selecting the best agrotypes. The "agrobiologic dynamics" calculated much more finely the mineral content of particular plant species, the "maximum quantities of life that can exist in living plant species." The ultimate density of a population would best be finely calculated on the protein-yielding capacity (dependent on nitrogen) of any given species of plant that can be eaten. This should be put into place in overpopulated nations existing at what he called the Malthusian frontier, on his reckoning India, China, Germany, Italy, and Czechoslovakia. In all of these places, "population is pressed against its own soil." This pressure was compounded by immigration restriction, and so "no racial group can be reasonably sure of anything but its own home market and its own soil."[46] These were quintessentially 1930s statements. Willcox's ideas on agrobiology and autarky were considered by the Economic Section of the League of Nations.[47] The possibility that "nations can live at home" if their agriculture is carefully honed was an argument against the imperative of extended "living space."

Soil Conservation and Population

The politics of conservation and the politics of population were beginning to be linked through the 1930s in a way that only makes proper sense once the centrality of soil to both is properly understood. Soil depletion, deforestation, desiccation, and salination were only ever one intellectual step away from the population problem in these decades. Carelessness and overuse of forests in the past, wrote George Knibbs in 1928, made reforestation on large scales "an urgent necessity."[48] Warren Thompson pointed to insidious silting up of rivers as a result of deforestation.[49] Others understood the incursion of forestry and mining to produce food and wealth more in terms of preserving "natural wealth." The Peruvian engineer Pedro Paulet, for example, wondered at the World Population Conference "how to colonize the country without injuring or destroying the immense natural wealth perhaps unique in the world, of the still unexplored heart of South America."[50] And H. G. Wells, Julian Huxley, and G. P. Wells elaborated the world's reckless cutting without reforestation, in the same chapter in which planned births and populations were addressed, "Life Under Control."[51] This all followed a long history of soil and forest management in both exploitive and conserving modes,[52] but through the 1930s, it became a global issue, deeply connected to population increase and increased food needs in a world considered geopolitically closed, space- and resource-limited, and now in economic depression.

One of Radhakamal Mukerjee's central chapters in *The Political Economy of Population* was "The Theory of Conservation." In it, he explained that conservation meant nothing less than "a deliberate use of Nature as the ally of land-water culture for the continuity of human civilisation, a conscious planning of human food materials and energy circulation with due regard to the interests of unborn generations."[53] Soil fertility, water supply, nitrogen supply, and all sources of energy came under the purview of conservation. Overuse would bring about diminishing returns, he argued, with reference to Malthus. "The physical concept of the wheel of life and the ecological concept of the equilibrium of nature underlie the theory of conservation in biological economics for making the circulation of matter and energy in the region for human requirements as quick, efficient and wasteless as possible, and this for the longest time." The expansion of European people resulted in a process of soil exhaustion, desiccation, and denudation, he wrote, pointing first to North America. Large agricultural systems to enable greater populations would result in what he called abuse of land (soil exhaustion, single cropping without legumes, erosion), abuse of water (decline in the water table, meandering of rivers), and abuse of trees (deforestation, desiccation). This would return as "abuse of man and animal power": overcrowding, low standard of food and feed for both men and cattle, extermination of animals, and epidemic outbreaks.[54]

Fertilizer and agricultural practices were already globally political, then, by the time the soil started blowing off the U.S. plains in the early 1930s. The deep plowing of the American prairies and, some said, the lack of British-style crop rotation left hundreds of thousands of acres unprotected.[55] The Dust Bowl experience, hard on the heels of the Great Depression, had a major impact on the U.S. economy, on agricultural practices, and on infrastructure around soil conservation. The crisis prompted a sudden intensification of government concern about American soil. If Theodore Roosevelt's administration had galvanized the term "conservation" in an earlier era, it was Franklin D. Roosevelt's administration that took it to another policy and bureaucratic level. In 1933 he established the Soil Erosion Service, part of the Department of the Interior under secretary Harold Ickes (who later wrote forewords to Radhakamal Mukerjee's books). This became the Soil Conservation Service. The Dust Bowl sharpened and extended national U.S. policy and thought on planning for natural resources and dovetailed with a long-standing international politics of soil, yield, and density in which the USDA was already involved.

The Dust Bowl experience also brought the methods of ecology to the fore. The "conservation" philosophy stressed human use of resources, but also the human tendency to overuse and destroy natural resources. By the late 1930s, and through the question of soil, it was directly linked to the geopolitics of population: "the impossibility today of solving the problem of the exhaustion of resources by expanding to unoccupied lands or to those inhabited by primitive exploitable peoples."[56] Deserts were on the march, as ecologist Paul Sears famously put it in his 1935 conservation classic, and one effect was a reduction of net arable area.[57] At a time when so many were aiming to increase the global area available for human habitation, to extend the limits, the expansion of deserts due to soil erosion and ill-conceived agricultural practices effected the reverse. This added up to the problem of "vanishing lands."[58] Soil and people were considered mutually: "Soil conservation is a problem of human ecology."[59]

The desolate Dust Bowl, director of the FAO John Boyd Orr later wrote, jolted the people of America and Europe out of their complacency and into recognizing global destruction and the need for conservation.[60] He warned that the removal of forests and replacement with arable land losing its topsoil and the incursion of deserts represented the destruction of America's real wealth, and population growth would become unsustainable. But the ultimate significance for Boyd Orr was the global scale of the problem: the "wanton waste of the fertility of the earth, which nature has taken many thousands of years to create, is going on to a greater or lesser extent in all continents." As he put it, the age-long cycle of soil to plants to animals and man and back to the soil again, was broken.[61]

Food Security

Nineteenth-century economists and statesmen were constantly thinking about soil, food, and population, in part because of the great famines of that century—in Ireland, in India, in China—and partly because starvation was a constant in population talk. Malthus argued that want of food was the "ultimate check" to population growth, but also that it was rarely the immediate check "except in cases of actual famine."[62] In her own *Law of Population* (1880) Annie Besant had raised both Indian and Irish famines: "These checks may be 'natural' but they are not humane," she wrote, in her argument for "artificial" management of reproduction.[63] Yet even at the time, and certainly since, economists rarely understood famines as solely natural phenomena, but as responses to labor, market, and pricing systems, particularly within colonial economies. Both British and Indian writers established a strong intellectual tradition that connected famine, population, and statecraft, from a range of different political positions, evident most simply in the number of Indian Famine Commissions.[64] Some important figures emerged directly from this work, later to shape world population discussion.[65] Alexander Loveday, for example, the Cambridge author of *History and Economics of Indian Famines* (1914), went on to lead the Economics and Finance Section of the League of Nations, and it was under Loveday that the League established its Demographic Committee in the late 1930s, chaired by Carr-Saunders.[66] If food became part of twentieth-century national and international statecraft, perhaps this was initially less about early twentieth-century presence of the calorie, as Nick Cullather has argued, and more about nineteenth-century absence of food, counted via the metrics of starving and dead people, especially in data-rich, food-poor colonial India.[67]

After World War I, however, it was less India and Ireland than China with its intermittent famines that garnered particular interest. In 1920–1921, there was extensive famine in northern China, prompted by the dual action of drought and the substitution of opium production for grain. For postwar Anglophone Malthusians, famine was quite simply—and far too simplistically—evidence of overpopulation. Harold Cox cited the London *Times* reports on misery regrettably playing out: "They are selling their children for handfuls of coppers as the only way to preserve their own lives. . . . Those who cannot find buyers drown the children. . . . Large numbers of wells in the famine area are putrid from the number of dead bodies of children who have been thrown in, or of suicides."[68] For a commentator like Cox, the most successful preventive of famine was not infrastructural work of a famine relief committee, or even a rethinking of global markets for opium and grain, but local Chinese reduction of fertility.

Many population writers were made aware of food shortages, and of a changing global economy of food production and distribution by World War I itself and by its catastrophic aftermath. Europeans were also hungry. Margaret Sanger considered the prospect of world famine an argument for a "birth strike," as she put it in 1920, a term she would never use later in her career. Because the United States fed Europe—"we are no longer a nation to ourselves"—European hunger was impossible to ignore. It was inseparable from the North American economy, and for Sanger, inseparable from reproductive responsibilities.[69] This was an intimate economy of postwar reconstruction.

Many of the men of science with whom Sanger organized and corresponded throughout the 1920s, had come to the world population problem initially through their food-oriented wartime public positions. George Knibbs was member of a Food Supply in War Committee. Raymond Pearl headed the statistical division of Herbert Hoover's Food Administration Program between 1917 and 1919. He was asked to predict future food needs, and encountered the difficulty of doing so. "How much wheat can we spare for export?" "How much meat must be conserved to meet export demands and still not injure physiologically the home population?"[70] In this capacity, Pearl worked with Edward East, who was himself becoming suddenly aware of the small margin between the food supply and the world's increasing needs.[71] Reading Malthus in wartime heightened the geneticist's civic sensibilities. His significant pre-existing scientific expertise on genetics and crop production became strongly reframed by the process of considering the precarious ledger of food production and human reproduction.

By the late 1920s, it was Japan that began to receive the lion's share of economists' and political strategists' attention. In the brave new world of Pacific geopolitics, Japanese land, food, and population needs (and demands) dominated the agendas of many institutions. The Tokyo meeting of the Institute of Pacific Relations summarized that the problem behind population was food supply, and the problem behind food supply was land utilization and agriculture. Japan was understood to be at an economic turning point in the 1920s, emerging from a long period of isolationist self-sufficiency to being unable to supply sufficient food from domestic production. It was likened to Britain's economic position before the abolition of the Corn Laws.[72] In 1927, the shortage of rice in Japan was put at 55 million bushels.[73] In 1928, when rice had to be imported for the first time, one commentator described local panic and concern that was compounded by "the passing of the immigration law by the Senate of the United States [that] gave birth to the public feeling of national humiliation."[74] Food insecurity meant political insecurity, a situation that was rarely comprehended independently of population and its geopolitical implications.

The International History of Diet: Wheat, Rice, and Meat

When Edward East stated to his colleagues in Geneva that an average of 2.5 acres was needed to feed one individual, and that therefore the world population maximum was 5.2 billion people,[75] his presumptions were immediately exposed and attacked. Such a parochial formula, exclaimed Frenchman Henri Brenier, who had spent many years in China: "Everybody in the world need not feed on the celebrated Anglo-Saxon roast beef." Apart from the vegetarians of the world, half of humanity survives mainly on fish, he challenged, and they should not be omitted from the global equation. For Brenier, this fact alone meant that East's already-famous calculation of "cultivable" area was entirely faulty, because the productive oceans were omitted: where, for example, was his assessment of the area of the North Sea that was filled with edible fish?[76] In fact, East had dealt with oceans and seafood in *Mankind at the Crossroads*. As with land, he thought optimism about the productivity of the oceans quite fanciful.[77] George Knibbs was inclined to agree. As world population increased, he thought sea mammals and fish would be drawn upon to a much greater extent, both as food and as fertilizers. But overall, this would not significantly affect the world's maximum possible population.[78]

East and Knibbs might have minimized the prospects for seafood, but their very discussion of it indicates the extent to which the world population problem foregrounded, even created, a global politics of diet. This was all as much about what people ate, as how much they ate. Different diets—including fish and shellfish—would produce different scenarios of aggregate carrying capacity, many considered. Unsurprisingly, this had both divisive and uniting potential in the political realm.

William Crookes was certainly in the dividing camp. In his formulation of a global wheat problem, he had imagined the world to be fundamentally split into wheat-eaters and rice-eaters. The bread-eaters, as he sometimes called them—"the great Caucasian race"—were listed by Crookes as the people of Europe, the United States, Canada, whites in South Africa, Australasia, South America, and the white population of the European colonies. They had grown dangerously in number from 371 million in 1871 to 516.5 million as he spoke to the Royal Society in 1898. However, these "bread-eaters of the whole world" were in a precarious situation, because wheat-growing lands were stationary, and thus "England and all civilized nations stand in deadly peril of not having enough to eat."[79] "We are born wheat-eaters. Other races, vastly superior to us in numbers, but differing widely in material and intellectual progress, are eaters of Indian corn, rice, millet, and other grains; but none of these grains have the food value, the concentrated, health sustaining power of wheat."[80] For Crookes, white global power did not just derive from claim over land, but from the staple produced from that land. The world was fundamentally divided between the entitled white wheat-eater and those who ate rice. If he had his way, they should remain satisfied with rice, and with their place in the world, geographically, politically, nutritionally, and dietetically.

For others, equally blinkered, a rice-based diet was itself an indication of problematic overpopulation, something that the Japanese, for example, could and should aspire to change. Editors of the Malthusian *New Generation* thought that having to eat rice as a staple was itself the injustice, caused by pressure on land: "They hardly ever taste milk, butter, cheese, eggs, meat. . . . The only way to bring such doings to an end is universal birth control coupled with a reasonable division of the lands of the world."[81] But valuing rice and wheat in this hierarchical way was not solely a Western phenomenon. As Nick Cullather notes, there was some official incorporation of wheat into

Japanese military diets as noodles, breads, and batter, part of an aspiration to raise nutrition to Anglo-American standards.[82] And in Mexico, too, modernization was sometimes expressed in attempts to shape national cuisines, including changes toward wheat-based staples.[83]

If some British Malthusians were lamenting an Asian dependence on rice, M. K. Gandhi defended its virtues. Resistance to white rice, polished rice, was a key element of his conflated body politics and anticolonial nationalism. In essays collected as *Diet and Diet Reform*, Gandhi wrote on the relative value of polished and unpolished rice in a chapter nestled between "the moral basis of vegetarianism" and "waste to wealth."[84] The most famous vegetarian then alive, Gandhi politicized his diet in many ways, including argument for vegetarianism as well as fasting as a way to offset Indian food deficiencies, imagining the food saved being redirected within the body politic toward the starving. This represented, as Sunil Amrith has explained, Gandhi's commitment to the bio-moral value of food, as well as abstention from it.[85] The more processed the food, the less wholesome it was, pronounced Gandhi, and waste at the smallest level had political implications: "We need every ounce of the bran of wheat and rice if we are to become efficient instruments of production, not to be beaten on this earth by any race, and yet without the necessity of entering into killing competition or literally killing one another."[86]

Vegetarianism as local action for a global problem became a trope of popular environmentalism from the 1960s,[87] but formulation of its political economy in world terms derives from the population discussion of earlier decades. George Knibbs considered the scale of world meat production and consumption deeply problematic and thought that humankind would necessarily become more vegetarian. To do so would "better economise his energy and the earth-space per capita."[88] Following Malthus, Knibbs registered cattle and some other stock as large organisms that took up much valuable space. On his reckoning there were about the same number of animals as humans on the planet; these animals also had to be fed, even if eventually their carcasses were converted into energy for humans.[89] Similarly, Raymond Pearl had to calculate not just the consumption of human food, but the consumption of nutrients by domestic animals.[90]

The problematization of food, diet, and population by no means bypassed the wheat-eaters themselves, especially because they were the world's meat-eaters too. Oliver Baker, so interested in the wheat-problem literature flowing from Crookes's work, advised the Association of American Geographers in 1922 that a decrease in U.S. meat consumption would be beneficial for the long-term sustainability of U.S. land. It would make continuing population growth possible and would, all other benefits aside, be good for health. Baker estimated a maximum population in the United States of between 400 million and 500 million, but only if the population ate "a largely vegetarian and dairy diet, so long as the supply of fertilizers held out." This compared to his estimate of a maximum 250 million if "present tastes" were maintained.[91] For Benoy Kumar Sarkar, differential standards of living could be indexed by the consumption of different kinds of meat. There was an important intra-European standard of living that could be captured, he thought, by documenting consumption of beef, mutton, pork, veal or bacon. But outside Europe, this barely applied, because the standards of demand and culinary desires were simply too different. The Japanese eat rice because they like it, not because it is cheaper than beef.[92] Radhakamal Mukerjee (and the two had lived together for some time)[93] thought differently again, and identified meat-eating as the specific "stimulus and support of Western energy."[94] Far from the West's global leadership being physiologically reliant on wheat, as Crookes had it, Mukerjee put it down to meat and the nature of labor that it created: "The metabolic changes necessitate intense and intermittent spurts of energy."[95] Finally, some, far from thinking in terms of global vegetarianism, were advocating meat-only diets as the way of the future.

This was Vilhjalmur Stefansson's position, always unique. His advocacy of eating solely meat and fish derived from years living with Inuit people on his various Arctic excursions,[96] and he promoted the diet as not just perfectly possible but healthful. Just after his trip to Geneva, Stefansson embarked on a meat-only diet experiment that was monitored in a yearlong trial by a committee of scientists chaired by Raymond Pearl (and supported by contributions from the American Institute of Meat Packers), with results published in July 1929 as "The Effects on Human Beings of a Twelve Months' Exclusive Meat Diet: Based on Intensive Clinical and Laboratory Studies on Two Arctic Explorers Living Under Average Conditions in a New York Climate."[97] Prominent nutritionist John Harvey Kellogg called it Stefansson's "tallow-eating stunt."[98] A study of the effects of high protein, rather than of low carbohydrates, climate was as significant as diet, because doubters argued that it was the Arctic cold that produced high protein and fat tolerance. Stefansson was assessed in good health, even better health, at the end of the trial that had taken place over a sweltering New York summer. "White Men Can Live on Meats," declared one press report.[99]

The trial was important in this context, because Stefansson circled around the food, population, and land conversations constantly during the

1920s and 1930s, consistently arguing against the catastrophic projections of East and Knibbs. Writing in 1925, he estimated 1.8 billion mouths to feed, and after an elaborate discussion of how long it takes for populations to double, decided that conservatively global population would double once every hundred years. "If we are never going to starve, then how soon will our globe have to hang out the sign of 'Standing Room Only?'" But he saw several ways forward. First, don't waste. One outcome of the war, Stefansson advised, was the firsthand knowledge that a French family can live on what an average American family wastes. "We could save much direct waste that comes from our extravagant and slipshod ways." Second, extend the agricultural limits of global land use. His explorations in extreme climates were specifically intended to push back the apparent "limits of settlement" that so many geographers' careers were spent establishing. Stefansson, however, did not deal in limits, temperamentally or politically and posed all kinds of solutions. The Sahara and the Gobi deserts might be irrigated he argued in 1925.[100] Dry farming, a new "science and art" should be developed. Extreme geographies, beyond the apparent limits could be pressed to produce more food for humans; meat of course. In *The Friendly Arctic,* Stefansson proposed intensive reindeer farming that would not only turn wasteland into agricultural land, increasing food supply, but would also, thereby, secure territorial claims to the vast lower Arctic circle in the North American continent. The great northern zones—Greenland, Spitsbergen, Siberia, and Alaska—all offered grazing lands now wholly unused.[101] This sense of possibility for world population increase, understood in concert with food consumed and land cultivated, was captured in his 1946 book *Not by Bread Alone*, in some editions titled *The Fat of the Land.*[102] The geopolitical dimension of Stefansson's travels, his advocacy of a meat-only diet, and the physiological study of it (and him), encapsulates the scales of the food–population issue: at once a problem of cellular physiology and of global space.

Climate and Calories: The Physiology of Population

Political physiology had a lot to do with the calorie. Created in mid-nineteenth-century French physics as a unit of heat, the calorie transferred its meaning to physiology during the 1880s to signify a unit of human energy.[103] Once in this domain, the calculation of energy came to be meaningful in relation to labor on the one hand, and warfare on the other, part of the technocratic modernization of efficient national economies.[104] In imperial Germany, an Institute for the Physiology of Work—*Arbeitsphysiologie*—was established that measured humans doing different labor, with different fuel, in different conditions. In the United States, Wilbur O. Atwater's work on the calorie was widely popularized, as Nick Cullather has shown, and his techniques and data were picked up by the USDA and transferred to Britain, especially through Seebohm Rowntree's use of the "science of nutrition" to calculate poverty.[105]

The food needs of soldiers on the fronts and of civilians in blockades meant that nutrition science received a boost during World War I, even if nutrition itself did not.[106] Wartime physiologists studied "man as a machine," global nutritionist John Boyd Orr later criticized: calories for energy, and proteins to repair wear and tear to full laboring or fighting function.[107] Between 1915 and 1920, the Allied Sanitary Commission enquired into the physiological problem of a "normal" diet, one that would "suffice to keep a human organism in health without erring on the side of excess or want."[108] As Pearl put it, in the U.S. context, "The basis of any adequate survey of food resources must be essentially physiological, rather than one of commodities or trade."[109] The standard requirement of 3,000 calories per day was accepted by the commission, but influential Yale biochemist Russell Chittenden thought this too high.[110] He thought 2,500 calories per day was sufficient for a man doing heavy labor, or for a soldier in war, but less was needed for a "brain worker." This was accepted for some time, circulating within the global population discussion of the interwar years, with the League of Nations setting a standard of 2,600 calories per day per man.[111] By 1941, the U.S. National Research Council recommended 2,500 calories for a male sedentary worker, 4,500 for a male heavy worker, 3,000 for a nursing mother, 2,500 for a boy of ten to twelve years, and 1,200 for a child of one to three years.[112]

Geopolitics returned in this domain as well. The calorie—the fundamental unit of and for human life—was not about biology or physiology alone. When energy requirements were repositioned within a population problematic, calories were counted against land that produced that unit of energy agriculturally: both density and the variable arability of land again became the issue. In 1933, the USDA established what became known as the Stiebeling standard for food requirements. Chemist Hazel Stiebling related food requirements directly to the utilization of land.[113] Thus, for example, an emergency ration for a man could be obtained from 1.2 acres of cultivated land, an adequate ration from 1.8 acres, and a liberal ration would require 2.1 acres.[114] When Oliver Baker looked at the components of diet in the United States and Germany, "based on energy values," his comparison was made meaningful first in terms of how much crop land per capita was used to produce an adequate diet (Germany was more efficient than the United States), and second in terms of the relation between population growth and land available for wheat. This was to be read against a further map that showed the small land surface available, especially because corn, oats, hay, vegetables, and other crops also needed to be grown on land otherwise suitable for wheat.[115]

The calculation of energy required for life for one individual enabled the calculation of total food needs of an aggregate population—a nation or the world. It could be measured either side of the ledger: total world food production could determine maximum possible population (calculated against various definitions of need). Or the reverse: the world's current or projected population could be used to calculate total world food shortage or surplus (or maldistribution, as we shall see).

There was, as Cullather has shown, a new food inventory being pursued at many levels. He understands this as preceding the "population bomb" discourse of the Cold War period.[116] Likewise, Megan Vaughan in her groundbreaking *The Story of an African Famine* periodizes the food supply and population question after World War II, compounded by popular ecology of the 1960s onward.[117] Yet once we understand the extent to which the "population bomb" discourse already loomed after World War I, we can see how interrelated the food inventory, the calorie, and world population "carrying capacity" already were, generations earlier. In other words, calorie counting and food inventories were meaningful in the 1920s, partly because population growth was already perceived to be such a problem.

In the food and population debate, climate-specific calorie needs were a regular part of the discussion: this generation of experts was trained to take climate into account. Thermal energy research on the calorie here met medical geography's long-standing research on climate and human constitutions. Mukerjee did not accept East's 2.5 acres per man benchmark, precisely because food requirements varied so much by climate: energy needs were less in India because of the warmth, he argued at the 1936 All-India Population Conference at Lucknow, the meeting that also produced the new Institute of Population Research. Mukerjee's book *Food Planning for Four Hundred Millions* was the first large piece of research to emerge from that conference, and in it he found two theoretical standards useful for forecasting shortages according to population numbers, one for northern India, and another for Bengal and southern India.[118]

The food question was also pursued at the 1938 All-India Population and First Family Hygiene Conference in Bombay. There, I. B. Adarkar and her husband, B. P. Adarkar, brought climate, vegetarian diet, energy needs, and population together in a key paper, "The Problem of Nutrition in India." Basing their paper on Mukerjee's work, they thought that vegetarian diet was better suited to Indians than "a heavy cereal or a meat diet."[119] And like Mukerjee, they considered problematic the practice of presuming European food standards to be indicative of Indian requirements; the vastly different climates in India needed to be taken into account. Certainly, they thought the League of Nations' calorie criteria should not be universalized, and they even disagreed with nutritionists working within India. Such requirements indicated for them the impossibility of "universal application to both temperate and tropical countries." The "expenditure of heat energy" needs to be considered, they wrote: "The metabolism of the human organism in the Punjab cannot be the same as in South India."[120] As Mukerjee had done with respect to Indians and the tropics, the Adarkars also argued that body surface area affected expenditure of heat energy, and national/racial differences in average weight, volume, and surface area should therefore be taken into account.[121] The fact, they said, that in many parts of India, the outside temperature is higher than body temperature, meant that body temperature had to be kept down by artificial means, by "cold bath, sleep, laziness, etc." This was the reverse of the need to supply food to the human body for warmth. If the object in Europe is to calculate food needs for both nourishment and radiation, in India "nourishment and replacement rather than heat energy should be the object." Calories should be decreased, and "life-giving ingredients such as vitamins and mineral salts" increased. Given the sun energy absorbed by Indian bodies, a "cooler diet" is sufficient. In all, and taking climate properly into account, they calculated as adequate 1,400 to 1,600 calories per man per day.

The consequences of rethinking calorie requirements were that maximum and optimal population could be projected from that base and in relation to food produced. Mukerjee calculated that 250 billion calories were available for consumption in 1931. Working on 2,800 calories per day, he could conclude that food supply was inadequate for India, and one could thus demonstrate and substantiate a state of "overpopulation." But if the Adarkars' standard 1,400 to 1,600 calories per day was accepted, there was currently a surplus of food in India. None of this was put forward naively or apolitically. The Adarkars were the first to argue that distribution within India and globally was so unequal that those below the "subsistence line" could not derive even necessary nourishment and "heat energy" for life.[122]

This kind of climate-based determination of nutritional need that recognized human difference came up against the politics of the single measure for all humans, insisted on by progressive cosmopolitans working within the League and later agencies of the UN. One of John Boyd Orr's many contributions was the firm dictum that there was to be one calorie standard for all humans of the world. "Nutritional needs are known to be the same for all races." No science was to be pursued under his auspices on differential climate or race-based standards. "In treating all humans as strictly alike, the assumption is that every family in every country is going to have the basic necessities of life." This, Boyd Orr wrote, was a measure for health—the only one reasonable, but not yet applied "generally to the human race."[123] A single standard, on which he based world food needs, was to be sufficient not just for bare life but for full health; not a minimal, but an optimal benchmark of nutrition.

World Food Plans and the League of Nations

Nutrition science had an important origin in calories—the calculation of energy—but the real innovation from the 1920s was the newly discovered "protective foods." The Adarkars told their Bombay audience that the poorer classes eat too many grains and not enough food containing minerals, vitamins, salts, and proteins.[124] "Calorific value" had to be balanced with these essential elements. In drawing attention to the protective foods, the Adarkars were picking up on the great post–World War I shift in the sciences of food and life, from the concept of undernutrition to the concept of malnutrition. The political economy of food and population came to be less about calorie needs, famine, and starvation, and more about the ingredients that made up a total diet, the multitude of minerals and vitamins now known to be necessary, sometimes in very small quantities, for health. Their absence gave rise to "deficiency diseases." As Boyd Orr recounted it, the revolutionary idea contained within "deficiency diseases" was that absence of a substance could cause disease, displacing the idea that disease was caused by an active presence—a microorganism or toxin.[125]

Importantly, Boyd Orr was one of two nutrition scientists who featured centrally in the scientific establishment of the deficiency diseases in the 1920s, and who went on to be instrumental in the formulation of a global politics of food and population within the League of Nations and later in the UN's FAO. He was accompanied along the way by Wallace Ruddell Aykroyd. Both Aykroyd and Boyd Orr were medically trained, and through early field-based research on deficiency diseases, became interested in agriculture in relation to both health and political economy, first in local contexts and then in national, imperial, and global frames.

It was Aykroyd, the Irish nutritionist who had made the early discoveries about beriberi, the poor Newfoundlander's disease of malnutrition, endemic also in south and east Asia. Aykroyd's Newfoundland studies linked beriberi to the use of white flour rather than wholemeal flour in winter months, a problem with a solution that Aykroyd interpreted explicitly as economic rather than medical.[126] He also researched rice at the Lister Institute between 1928 and 1931. Aykroyd showed that parboiling rice before it was milled protected it against the loss of vitamin B1. After a period with the League of Nations, working closely with Boyd Orr in the early to mid-1930s, Aykroyd moved to south India as the director of research laboratories in Coonoor (1935–1946). There, he worked closely on the nutritive value of Indian foods, on extensive dietary surveys, and on clinical assessment of nutritional status.[127] During this period, the chief scientific work of the laboratories was an inquiry into the rice problem, in which he reworked his earlier material on beriberi.[128] After World War II, Aykroyd rejoined Boyd Orr as a member of the latter's senior team in the new FAO. In the last decades of his life, he published widely and increasingly popularly on nutrition and health for the Freedom from Hunger organization.

Similarly, Boyd Orr had long understood nutrition to be a political issue. He was early and controversially interested in malnutrition in England[129] and was especially well known as one-half of the Orr and Gilks study that compared the diet of the meat-eating Massai and the vegetarian Kikuyu in Kenya during 1926 and 1927.[130] In the tradition of Rothamsted agricultural experimentation, he constantly thought in terms of cycles of energy and nutrition transfer between soil, pasture, animals, and humans. Boyd Orr had researched in east Africa with Daniel Hall and had a similar background in agricultural research and its links with animal and human physiology. At the Rowett Institute for Animal Nutrition in Scotland. later the Rowett Institute for Nutrition and Health, he directed research on the relation between the chemical composition of soils and plants, as well as "the more subtle influence," as he put it, of humus and various chemical fertilizers.[131] In the 1940s, he went international with *Food: The Foundation of World Unity* (1948).[132] It was from this base that later, as first director of FAO, he envisioned and attempted to implement a world food plan and to promote a global awareness of the linked problems of "food and people" with Julian Huxley. In 1949, he won the Nobel Prize for Peace.

Boyd Orr was fundamentally interested in, and learned about, both the problems of distribution and the significance of diet changes. Not just food security but nutritional equity was his ambition, his early class-based studies informing his later global "development" work. In the United Kingdom, he noted a more equitable distribution manifesting as a decrease in the consumption of more expensive foods in the richer classes and an increase of these foods in the poorer classes. The health of the lower-income group in the United Kingdom had improved with this diet, even though it did not reach the standard needed for full health established by the League.[133] Nonetheless, this was the equalizing trajectory he thought possible to implement on a world scale.

Much of this discussion about world food took place during the Great Depression and in its wake. Many understood the Depression itself to be an indicator of global overpopulation. In 1932, the World Monetary and Economic Conference was convened by the League, at which it was proposed that grain production be restricted, scarcity created, and prices driven up. But some were appalled by this formal world response. Led by Stanley Bruce representing Australia, delegates of the League of Nations Assembly expressed both dissent and outrage that foodstuffs were destroyed while millions starved.[134] He argued that in a poverty-stricken world, those without the "primary necessities of life" would turn to fascism and Communism. A better international response was imperative. Bruce lobbied for a change of policy to rapid expansion of agriculture and industry, and an end to the aspirations of national self-sufficiency.[135] Famously, he called for a marriage of health and agriculture.[136] In 1935, there was a three-day discussion on nutrition by the full assembly of the League of Nations.[137] A Technical Commission on Nutrition was appointed to work out a world food policy based on global human need, rather than on economic strategy.[138] A Committee of Physiologists, including Boyd Orr, drew up a statement of the kind of diets needed for health and helped produce an influential series of reports and papers, each of which stressed a balance between "energy" foods and "protective" foods.[139] At a time when the League was coming under increasing attack, Boyd Orr strongly defended its pursuit of improved international nutrition.[140]

In the global political economy of food and population, the proportion of energy-yielding food to "protective" food was increasingly taken as an index of comparative standards of living and of economic development. Rothamsted agriculturalist Daniel Hall wrote that the poorer the community, the more they would be forced to live on cereals, the cheapest source of energy: "the foods which call for least expenditure of labour in relation to their power of maintaining life."[141] The greater the proportion of protective foods in a national diet (milk, eggs, meat, fruit, vegetables) to energy-yielding foods (cereals, rice, potatoes) the richer the nation. John Boyd Orr mapped this distinction in an arresting table of the "Proportion of Energy Yielding to Protective Foods in the Diets of Different Nations," published in 1943 (figure 7.2). It showed that 64.2 percent of the world's population

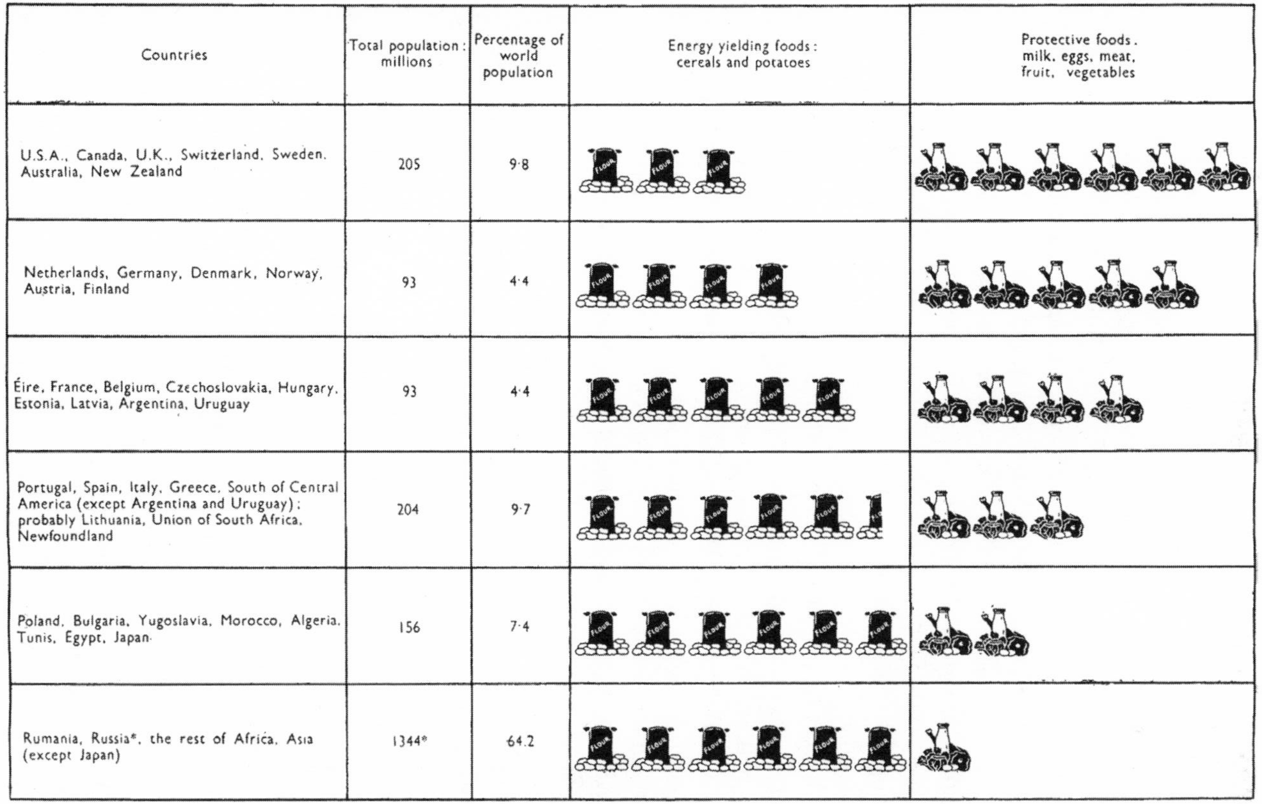

Countries	Total population: millions	Percentage of world population	Energy yielding foods: cereals and potatoes	Protective foods. milk, eggs, meat, fruit, vegetables
U.S.A., Canada, U.K., Switzerland, Sweden, Australia, New Zealand	205	9·8		
Netherlands, Germany, Denmark, Norway, Austria, Finland	93	4·4		
Éire, France, Belgium, Czechoslovakia, Hungary, Estonia, Latvia, Argentina, Uruguay	93	4·4		
Portugal, Spain, Italy, Greece, South of Central America (except Argentina and Uruguay); probably Lithuania, Union of South Africa, Newfoundland	204	9·7		
Poland, Bulgaria, Yugoslavia, Morocco, Algeria, Tunis, Egypt, Japan	156	7·4		
Rumania, Russia*, the rest of Africa, Asia (except Japan)	1344*	64.2		

FIGURE 7.2 Nutritional map of the world: protective foods and energy foods. John Boyd Orr found different ways to present global disparities at a glance. (From John Boyd Orr, *Food and the People* [London: Pilot, 1943], 44.)

(Romania, Russia, Sub-Sahran Africa, and Asia, except Japan) only ate one part protective foods to six parts carbohydrates. North America, the United Kingdom, Switzerland, Sweden, Australia, and New Zealand, at the top of the table (9.8 percent of the world's population) ate six parts protective food to three parts carbohydrates.[142] The table was designed as a conspicuous illustration of discrepancy; the gap between the richer and poorer nations that development discourse would soon seek to bridge. This was a new way of "mapping" the political economy of world population and food.

At a particular point in the web of meaning and the web of life, the geopolitics of land, soil, and fertilizer crossed over into the biopolitics of food, diet, nutrition, and human health. Soil science became nutrition science. Food was part of the biology, the geography and the political economy of world population, the matter between the productive Earth and the fueling of humans. Determining global energy needs created by population growth required thinking of soil and food jointly, each part of the earth and each part of life. Their management formed a significant element in twentieth-century national and international planning. At the end of the nineteenth century, Crookes had viewed the world and perceived it as divided on the axis of entitled wheat-eaters and others. Fifty years later, John Boyd Orr saw not justified division but great inequity. For him, the nations of the world were divided into those that met minimum nutritional standards across their populations, and those that did not. Such a global demarcation was the precursor to the Third World and First World distinction soon to be devised by French demographer Alfred Sauvy with reference to the hungry third estate.

8

Sex

The Geopolitics of Birth Control

> Unless sexual science is incorporated as an integral part of world-statesmanship, all efforts to create a new world and a new civilization are foredoomed to failure.
>
> MARGARET SANGER, *THE PIVOT OF CIVILIZATION* (1922)

It would not have been Margaret Sanger's preference to leave birth control off the official agenda at the first World Population Conference. Nonetheless, she strategically conceded and was certainly not the only one disappointed. In February 1927, before the decision to make the meeting "scientific" and not "applied," John Maynard Keynes suggested that "it would fill a gap in the programme as sketched out if you could have a paper on the actual progress of contraceptive methods. How widely are they in fact employed? Have they already affected the birth rate?"[1] For Keynes, birth control in all its aspects was a necessary, even a central, part of economic planning, whether or not one was an advocate of it on other grounds. In fact, almost all of the Anglophone contributors to the Geneva conference specifically, and in the conversation on the world population problem generally, were advocates of the new methods of contraception. This includes Julian Huxley, Edward East, H. G. Wells, Carlos Paton Blacker, George Knibbs, Warren Thompson, Oliver Baker, Radhakamal Mukerjee, Alexander Carr-Saunders, and others. Even Raymond Pearl, who thought that a population equilibrium would automatically transpire once a certain density was reached, came around to accepting the significance of active and planned regulation of fertility. His new advocacy of birth control was set out in *The Natural History of Population* (1939).[2]

Historians often presume or imply birth control to have been mainly or even solely women's political and intellectual business through the 1920s and 1930s.[3] Plainly this was not the case. Given the prominence of many of these men in framing national and international polities and policies in these years—Keynes was the most significant economist of the early twentieth century, Mukerjee was member of the Indian National Congress's population committee, Baker was a long-standing expert within the U.S. Department of Agriculture, Knibbs was chief statistician, Huxley was the first director-general of the United Nations Educational, Scientific and Cultural Organization—there is even a case for suggesting that the steady uptake of birth control in the public sphere was at least as much due to their advocacy as to the far better-known women's lobbying for birth control. But this raises a more basic question: Why were they all such strong proponents? Not, in general, because they were concerned about women's health or reproductive autonomy. Many of these men had little if any interest in the feminist and health-based arguments then being put forward by Sanger and others who shared her approach. Rather, in their schemas, the regulation of fertility was a means by which food scarcity might be ameliorated, war averted, and global security achieved. As a rule, geopolitics not gender politics energized these prominent men.

Some, however, were more aligned with the strand of neo-Malthusianism that grafted women's health, alleviation of poverty, and geopolitical arguments together. Margaret Sanger herself should be understood in that tradition. As this chapter shows, a few perhaps unlikely men combined advocacy of birth control with remarkably strong critiques of masculine power in the private and sexual spheres. None of this sits comfortably with a tradition of women's historiography that claims the twentieth-century birth control debate as its own special domain. And yet someone like Edward Alsworth Ross should almost be admitted to that historiography, because he railed against the era in which men dominated and women bore the cost. He himself called this the "he-epoch."[4]

By the same token, histories of international relations that traditionally presume birth control or sex to have been a private sphere matter, the domain of feminine or feminist politics and history, need to register that population (and therefore sex, reproduction, men, and women) was a major business of state.[5] The much-heralded Cold War uptake of population control as U.S. foreign policy was a fairly late manifestation, if an internationally important one, of a long-standing geopolitics of birth control. International relations and intimate relations would seem to be worlds apart, but sex and reproduction have long been part of international history through the economics of population and the politics of food security.

Economic historians who have ignored gender, and feminist scholars who have ignored political economy might both have looked to Thomas Robert Malthus's own work. There is, in the *Essay*, a certain interest in the history of the social status of women vis-à-vis men, and in their different bodily and social relations to reproduction, birth, and death. This was a key factor for Malthus in formulating the relation between population checks (control) and successive economic stages. These were "civilizational" stages in which societies across the world developed or stayed stationary. Any number of demographers in the 1920s and 1930s recapitulated this as a world history of population and ultimately as a world demographic transition, always incorporating comparative standards of living. Savagery and civilization, tradition and modernity, the Orient and the Occident, the East and the West, each divided the world across both time and place. Here lies the remote as well as the proximate antecedent to the great post–World War II issue of economic development, including its tight links to women, gender, fertility, and sex.

Intimate Relations and International Relations

In the 1920s, agricultural economist Oliver Baker found one set of solutions to food insecurity in land and soil: might the tropics be exploited, as the grasslands of the temperate zones had been during the nineteenth century? Might fertilizers increase yield? Another solution for him was sexual and reproductive: "Will the decline in religious authority and development of materialism diminish the birth rate?" he asked hopefully in 1925. Like any number of fellow agricultural experts Baker thought through population within a discourse of global resource limits, specifically capacity to produce food. That led him to the advocacy of birth control. The slowing birthrates that were demonstrably under way were as desirable for Baker as they had been for geographer Ernst Georg Ravenstein in the 1890s. The trend in the United States, Canada, and Australia was "encouraging," Baker wrote, and all other countries should follow. The driving imperative was to equalize European and Asian fertility and mortality trends: "There is left to the white man a century of grace" to get global birthrates in order, he wrote in this article that was on the face of it neither about population generally nor birth control specifically, but about the world's supply of wheat.[6]

Conceptually and politically, for most of these population commentators, food came before, not after, sex. This is reflected not least in the organization of their multiple studies. Food as the international problem and birth control as the corresponding solution literally structured their thoughts. George Knibbs foregrounded food security in his chapter titles as "Man's Agriculture, Forestal and Animal Needs" and "The World's Cereal and Food-Crops and its Mineral Needs." His interest in birth control as "New Malthusianism," derived from these problems.[7] For economist Harold Wright, "Food and Raw Materials" and "International Population Problems" both prefaced a chapter on birth control. In the conclusion to his study it was figured as "the way out." The recent decline in growth rates "gives rise to the hope that mankind may in time assume the conscious control of one of the greatest forces by which the richness or poverty, the happiness or misery, of his life on the earth is determined—the power of population."[8] Harold Cox's *Problem of Population* moved similarly from problem to solution, from "The Arithmetic of the Problem" (where he cited Benjamin Franklin as much as Malthus), to "The Economics of the Problem," to a final chapter on "The Ethics of Birth Control."[9] The food problem and birth control solution was a constant in Anglophone Indian literature as well. At the 1938 All-India Population and Family Hygiene Conference, S. P. Srivastava argued for birth control, because 60 percent of the population suffered from malnutrition.[10] And Radhakamal Mukerjee advocated birth control for similar economic and ecological reasons in *Food Planning for Four Hundred Millions.*[11] Birth control was supported by population writers across the political spectrum, separated both geographically and in terms of disciplinary expertise. But food security, not women's health or autonomy, almost exclusively drove their convictions.

The derivative issue was war, extending the classic late nineteenth-century neo-Malthusian position that "our views alone can secure lasting peace to the nations."[12] In an underrecognized way, birth control was a central part of postwar internationalism and cosmopolitanism. As early as 1926, Carlos Paton Blacker, psychiatrist and leader of organized British eugenics for many decades, worked from the prewar Malthusian interest in pacifism through "unification" and federations of various kinds. Notwithstanding economic unification, he suggested there had been little "ethical unification" of the human race, and restriction of human fertility—"biologically good"—would be one of the most significant steps forward to this end.[13] His Oxford teacher, Julian Huxley, thought the same. Throughout the late 1920s, Julian Huxley wrote constantly in support of birth control, as we have seen, one method of which might be sterilization "after a certain number of children had been born."[14] In his ecological framing of human reproductive trends, the significance of birth control derived not just from resource limits always in operation, but also from the new geopolitical circumstances: "The age of migration is passing, as most of the more empty countries are closing their doors, and when we look at the impossibility of any considerable extension in such saturated areas as India, China, and Belgium, we are faced with the fact that the world is filling up. . . . Wherever there is no birth control there will be an outwards population pressure which clearly indicates the danger of war."[15] By the end of the 1920s, this was standard population talk. Birth control was being written into the very idea of international relations in the decade after World War I: sex and geopolitics were entwined, a matter of high politics. This is what constituted "population" as a world problem. And so, when zoologist Wilfred Agar delivered a luncheon address, "Birth Control as a World Problem" to the International Relations Society in Melbourne in 1934, he said that he welcomed a decline in the world's birthrates as the "only way of avoiding an appalling catastrophe." He told his audience that such a decline was already under way, at once "a physical necessity for the human species" and an international relations problem because of the unevenness of the global response. In this sense, he thought, the reduction in population was akin to reduction in armaments. Who shall begin? Fortunately, the signs were good, he reassured his audience. Birth control was already a "mass movement" in Japan and in several provinces in India.[16] Here, birth control was discussed as nothing less than a foreign relations matter. This prefaced and prepared later national and international policy when it became precisely that. It is no coincidence that a demographer and politician of a younger generation, Sripati Chandrasekhar, considered population control to be "demographic disarmament."[17]

Some of these authors, as we shall see below, thought through birth control and birthrates in terms of gender, some even through feminist conceptions of gender. But others, and this is possibly the more important point, pressed forward birth control without considering women or gender, let alone contemporary feminism at all. Warren Thompson, the demographer most often credited with formulating the earliest versions of fertility transition theories, is notable in this respect. He strongly politicized population growth, exploitive colonialism, and birth control without considering women or gender in the least. There is no discussion of women in *Danger Spots in World Population*, no entry for women in the index, not even under "mortality," when it was well understood that maternal mortality, alongside infant mortality, was critical for projecting demographic trends. At the same time, this book made one of the strongest pitches for birth control of all the comparable texts. Indeed, Thompson thought birth control one of the great technological developments of humankind, as significant in the history of human affairs as fire, printing, or electricity: "In time it will change the entire course of history." The medium-term relief of congestion through redistribution of crowded peoples to new lands would itself improve economic conditions, and that was "the most effective means of spreading the knowledge and practice of birth control that can possibly be devised." It was birth control that was the long-term solution, however. "Birth Control the Only Remedy for Over-Crowding," was how he concluded his book, a claim that became widely endorsed in this milieu. Indeed, the final sentence of Thompson's study stated that birth control constituted a "new and better system of international relations."[18]

In many ways the postwar spokeswoman for an integrated Malthusianism, Sanger was well aware that many of her colleagues argued for birth control without necessarily considering or referring to women's health or feminism. Though she strategically applauded any argument for birth control—those like Warren Thompson were her natural allies, not her political enemies—she was also upfront about nominating what she saw as their serious conceptual limitations: "They treat the world of men as if it were purely a hunger world instead of a hunger-sex world. Yet there is no phase of human society, no question of politics, economics, or industry that is not tied up in almost equal measure with the expression of both of these primordial impulses."[19] The point about Sanger is less that she had an alternative rationale for birth control than that she had an additional one: she, too, argued for birth control in terms of food security.

Accordingly, the various arguments for birth control should not be seen as aligned to the gender of historical actors. This was not a neat case of national and international geopolitics (e.g., arguments for food security) being masculine intellectual territory, and intimate familial and sexual politics (e.g., reproductive choice) being feminine or feminist intellectual territory. Indeed, given her Malthusian intellectual pedigree, it should be unsurprising that Sanger argued for birth control on the basis of global security, as much as did her international relations colleagues. Sometimes she functioned as their effective mouthpiece. For example, her editorial "When the Earth Is 'Full Up' Shall We Starve to Death?" reiterated Edward East's work on the minimum requirement of 2.5 acres per person, along with his "fatal maximum figure" for world population.[20] In fact, geopolitical argument was standard in Sanger's repertoire throughout the 1920s and 1930s. When she lectured on birth control in China and Japan at Carnegie Hall in New York, the entire discussion was framed by peace, population, and war.[21] And by 1938, when the expansionist ambitions of Hitler were evident to all, she wrote: "New horizons stretch before us as a wider understanding of population problems awakens those who look for international peace to consider the population factors underlying national policies that lead to wars and national aggressions."[22] It was all restated with fresh urgency for the new atomic age after World War II. Her audiences in each of these decades would have understood the international and global scale that linked economic processes, the population problem, and world war.

Sanger set up population as a "world problem" in her early book, *The Pivot of Civilization*. She saw sex and reproduction as the critical elements of human activity that connected, as she put it, political, industrial, religious, and ethical factors. It was not that sex was not recognized as a public issue, she said—statesmen, economists, and churchmen respectively foregrounded it all the time—but they did so independent of one another. Her point was that political, industrial, religious, and ethical issues all "pivot" around reproduction: over that practice and that idea. For Sanger (and even moreso for her associate Edith How-Martyn), birth control was an issue of power in the private domain: sex, reproduction, birth, birth control, and population were never *not* about sexual politics in the private sphere. But it is also the case that these issues were never *solely* about the private sphere: "Birth Control is not merely a problem of the individual woman, it is not merely a national question. It is now a world problem." Always expediently name-dropping, this time she referred to Keynes, who thought population the "greatest of all political questions." Yet it was more due to her own intervention than Keynes's that the problem manifested simultaneously on multiple scales: for men and women locally, as global security and for "the welfare of the biological species."[23] Some of these new-generation Malthusians combined feminist, economic, and health arguments for birth control, often at the individual and family scale, but rarely limited to that scale. Their arguments for birth control typically ranged from those based on the health needs of individual women ("this universal movement for maternal health," as Anna Chou of the Shanghai Birth Control League put it),[24] to the global imperative to avert famine and war on a planet of limited size.

This internationalism shaped feminist ideas on birth control centrally. Edith How-Martyn and Margaret Sanger's Birth Control International Information Centre (BCIIC) appeal asked: "Why Should Birth Control be

World Wide?" The answer: "Increases of population without expansion of territory lead to friction between nations and ultimately may lead to war."[25] The BCIIC utilized the small profit from the World Population Conference and became a network of women physicians, social workers, and birth control activists who aimed for the dissemination of applied knowledge on contraception.[26] The term "applied" pointedly countered the "pure scientific" agenda of the conference itself, and purportedly of Raymond Pearl's initiative, the International Union of Scientific Investigation of Population. The BCIIC headquarters were in London, but How-Martyn constantly sought to relocate to Geneva (her first choice was India). Administering from Geneva "would help to show that the movement is truly international and its educational work worldwide."[27] The point, of course, was to enter birth control into the work of the League of Nations. Of particular interest here are the terms on which this inclusion was sought. Both women and men throughout the 1920s and 1930s argued that the League of Nations should take up the question of birth control under the terms of its covenant, "to promote international co-operation and to achieve international peace and security by the acceptance of obligations not to resort to war."[28] When Viola Kaufman, who funded Sanger's work, wrote to the secretary-general in 1930, she pressed that birth control should be on its agenda as an international issue precisely because of its connection to war. "The League of Nations ignores birth control—the only thing that can permanently abolish war."[29]

Their bids failed. Even the slightest reference to birth control was picked up by vigilant Catholics, who carefully watched the League of Nations and its agencies. In a draft of the Report on Maternal Welfare and Infant Welfare (1932), for example, headed by the physician Dame Janet Campbell, contraception had been presented as desirable in certain cases for the woman's ongoing health. This was stepped on quickly by Catholic lobbyists from several countries and by Catholic member states in the assembly, not to mention Secretary-General Eric Drummond. The offending section of the report concerned the possible need for women who were already ill to avoid pregnancy, but it also implicitly acknowledged that contraception was widely in use for other reasons. The context in the report was the health implications of abortion and related maternal mortality rates.[30]

The League was never going to endorse birth control, and the work of the BCIIC continued to be centered in London. The BCIIC did hold the International Birth Control Conference in Zurich in 1930, however. It was everything the 1927 conference was not. A technical meeting about sexual and applied matters, its papers detailed all the different contraceptives then available: the occlusive pessary; the cervical cap; sponges; sheaths; the use and consequences of the intrauterine pessary; the composition and use of various chemical contraceptives used as douches, powders, and jellies. New methods and experiments were reported: sera, spermatotoxins, X-rays. These were the "discoveries" that impressed commentators like Thompson and prompted him to locate "birth control" within a progressive history of technological advancement. All the while, however, fertility decline appears to have been actually effected by the entirely nontechnological practice of coitus interruptus, and in some locations, later marriage and abstinence from sex. The millions of individuals and couples who were actually practicing some form of fertility regulation, whose collective change in conduct registered as the reduction in fertility growth rates from the late nineteenth century, were not necessarily imagining themselves politically at all.[31] But the proponents of birth control could not have been more political in their comprehension of what it was all about. Alongside all this technical and medical elaboration, the framing problematic was still one of international relations, peace and war.[32]

The BCIIC coordinated international activities, talks, and tours for visiting birth control proponents in London.[33] In 1933, it hosted the Birth Control in Asia conference held at the London School of Hygiene and Tropical Medicine, a meeting that foregrounded links between a global political economy, ecology, and birth control. The meeting combined the economic approach (the key question of the standard of living in Asia) and the technological approach ("the practical problem of contraception in the East").[34] At least some of the British organizers were well aware of the politics of doing so. "It would be presumptuous to offer advice to Easterners," cautioned Carr-Saunders.[35]

Sanger and How-Martyn were imagining a world space, not just an international space, for action. Indeed in the first report of the BCIIC, "international" was struck through by How-Martyn, to become "Birth Control Worldwide."[36] The words "birth control" have circled the globe, Sanger wrote, calling their various tours "Round the World for Birth Control."[37] But of course birth control was already part of "the world." When Sanger and How-Martyn visited India on several tours during the 1930s, birth control talk and birth control clinics had long been under way, led in the main by men schooled in neo-Malthusian economics.[38]

Political economies of population in India were long-standing, and "overpopulation" formed an explicit element of late colonial British renditions. As David Arnold has shown, the official British response to censuses in the

1920s and 1930s was already in favor of birth control—population control—measures.[39] And at the opening address of the 1938 All-India Population and Family Hygiene Conference, the practice of family limitation was presented as a solution to the rapid increase in population and the pressure on land, wherein the latter was perceived actively to retard the prospects of self-government. Galton was invoked by the nationalist president of that conference, B. G. Kher, but it was a Galton who ventriloquized Malthus: "According to Sir F. Galton the effect of a few years' delay in marriage on the population is considerable."[40] Radhakamal Mukerjee thought the small family system essential for India, deploying Carr-Saunders's term. And, switching to Ross's terminology, he argued that "'an adaptive fertility' will relieve the present heavy population pressure." For him, this was all part of Indian modernization and Indian independence.[41]

These were quite late interventions, however. In Madras, a Hindu neo-Malthusian organization had been established as early as 1882, re-emerging in 1928 as the Madras Neo-Malthusian League, run by elites and publishing in English.[42] It aimed for "human welfare through birth control."[43] Biology professor Gopalji Ahluwalia created the Indian Eugenic Society in Lahore in 1921, and one year later established the Indian Birth Control Society in Delhi.[44] Elsewhere, the Sholapur Eugenics Education Society ran a clinic for married women, headed by medical doctor Alyappin Padmanabbha Pillay. It combined health, eugenic, and neo-Malthusian interventions into women's reproduction, and for two hours every Tuesday and Friday, men were advised as well. Birth control can "as easily be popularised in India as elsewhere," wrote Pillay in 1930.[45]

On their tours, Sanger and How-Martyn grafted onto any number of pre-existing organizations and individuals, and the benefit of "internationalization" was typically mutual. In 1937, How-Martyn toured India at the invitation of A. P. Pillay. She spoke at meetings for women and for men, for doctors and for nurses, and toured villages, towns, and cities, addressing in her estimation about 5,000 people directly. She also broadcast her message and translated her talks into vernaculars. She spoke to the Indian Institute on Population on "World Population" in a session chaired by Radhakhamal Mukerjee.[46] How-Martyn made much of the fact that she gained entry to talk about birth control more easily in India, Ceylon, Burma, Japan, and the Philippines, than the United States, where in 1936, her luggage with birth control exhibits and literature was confiscated. "The Far East Welcomed her Teachings" was the useful headline.[47] When Margaret Sanger toured Singapore, she became especially interested in the spatial issues on the tiny island;

in the hands of local Malthusians, as well as outsiders like Sanger, it was a microcosm of the crowded planet. An advice pamphlet for family planning noted, "Over 700 babies are being born in Singapore every week. Houses cannot possibly be built fast enough for such a rapid increase in population." Accompanied by an image titled "No More Houses to Let!" it warned: "The population of the world is increasing by twenty millions every year. Already there is not enough food in some countries."[48] The density issue became the real estate issue.

How-Martyn held a thoroughly gendered understanding of the politics of birth, death, and war. She broadcast on her Indian tour: "Motherhood knows no frontiers. Mothers give, War takes, life. Yet a fraction of the amount being spent by the world on armaments for the destruction of life would, if spent on making motherhood safe, bring health and happiness to millions of mothers."[49] This was an international feminism that comprehended gendered power and gendered relations as a major transnational force. Edith How-Martyn's pursuit of birth control was, to some considerable extent, unusual for British feminism, a movement that typically, almost traditionally, gave the topic a wide berth. Nonetheless, her early militancy with prewar British suffragist movements, like Sanger's socialism, imbued her with a sophisticated set of ideas about the public politics of sex.[50]

While Sanger and How-Martyn's analysis of a world population problem consistently folded together the politics of private and public spheres, fewer of the men of the period engaged with what might be called the sexual politics of world population. One such was Indian nationalist leader M. K. Gandhi, another was general secretary of the Eugenics Society Carlos Paton Blacker. They serve as alternative instances to Warren Thompson's seeming incapacity to think about women in relation to population.

Sanger met Gandhi on one of her Indian tours in December 1935 and had a long and well-publicized conversation about sex, reproduction, gender, politics, and world affairs.[51] They were completely opposed to each other's positions on the use of contraceptive devices, the dissemination of which was Sanger's mission. Gandhi, however, thought that prevention of pregnancy should be achieved through abstention from sex, "self restraint versus self indulgence," deeply informed by Brahmacharya philosophy of spiritual purity as embodied practice.[52] Sanger and Gandhi met, talked, and parted, respectful and interested, but neither was swayed on the question of the means by which fertility might be controlled. Notwithstanding this divergence, on some matters Gandhi's views were arguably closer to Sanger's than those of her fellow birth control advocates in the United States, with whom she worked very closely indeed. Unlike many of these men, Gandhi was acutely aware that sex, reproduction, and birth control spanned intimate politics and international politics.[53] They disagreed, but the reach of the question from private to public, from sexual to global, exercised both of them: they shared an understanding that population was always, in every domain, about the power of gender and sex difference.

Sanger and Gandhi spent two days talking, in their different ways both theorists of freedom. Like Gandhi, Sanger connected political freedom and personal freedom. Unwanted maternity is a form of enslavement, she wrote, extending several generations of feminist thought that had conceptually developed the links between women, slavery, and freedom (as well as Malthus and John Stuart Mill's standard analogy to slavery). For her, individual unfreedom manifested as repeated unwanted pregnancies, in both the West and the East. Birth control represented progress for individual women and, aggregated, for nations and the world as a whole. For Indian women, then, birth control built a stronger and healthier population "for the climb up the ladder of progress that they are envisioning for their country." The "personal freedom" granted by birth control was the corollary to the "political freedom" women sought as Indian independence.[54] Annie Besant's early presidency of the Indian National Congress was not incidental.

Gandhi was also a theorist of women and freedom, of sorts: "I tried to show them [women] they were not slaves either of their husbands or parents, that they had such right to resist their husbands as their parents, not only in the political field but in the domestic as well." This is why, Gandhi argued, women should insist on abstaining from sex with their husbands, as the only method of preventing conception. Both Sanger and Gandhi agreed that this might be considered an act of "legal and marital disobedience." They also agreed: "We have to start with the individual. You feel that the beginning is with the individuals' control of sex. There is no argument there."[55] But they disagreed with respect to what was possible for women, in intimate power dynamics. Sanger thought the expectation that women could demand abstention unrealistic and ultimately dangerous. Gandhi says he knows women, Sanger dismissed. She scoffed at his claim to be regarded as "half a woman because I have completely identified myself with them." He has "not the faintest flimmering," she wrote, because he advises women to "resist" or even to "leave" their husbands in order to control the size of their families. But is this possible for economically dependent women? And what about both the men's and the women's "thwarted longings," she asked.

While Gandhi advocated and practiced ascetic chastity, Sanger advocated and practiced the need for women to "express their love sexually." Gandhi, she wrote, "cannot conceive of this force (i.e., sex) being transmuted from lust into beauty."[56] This cross-cultural discussion also crossed gender lines. The Indian Malthusian men who edited the *Madras Birth Control Bulletin*, the organ of the Madras Neo-Malthusian League, supported Sanger unequivocally, and opposed Gandhi: "Gandhiji's advocacy of continence is not practicable." Conversely, some Indian women opposed Sanger. At a meeting in Travancore, for example, they stated: "Brahmins of ancient India had small families, but they did not use artificial means. . . . Spiritual strength was acquired by self-control and not indulgence." Following Gandhi, they linked the private and public sphere differently to Sanger: "National discipline was impossible with self-indulgence."[57]

In another context altogether, Carlos Paton Blacker argued not dissimilarly across multiple scales, from the personal to the national and international public spheres. Organizationally close to British birth control work, Blacker appreciated something of the sexual politics that drove population politics. His capacity to see the connections among women's desire for birth control, an economic rationale for birth control, and an international agenda for birth control added the individual layer to the whole question that most other men involved in contemporary population politics ignored. Well short of Edith How-Martyn's radical feminism, he nonetheless attempted seriously to address many women's desire to avoid pregnancy, to limit fertility for their own sake. This fitted his eugenic agenda as well. The point, though, is less to argue simply for an expedient use of birth control on the part of a key eugenicist like Blacker, and more to show how easily eugenics and birth control fitted together: so much so that the Eugenics Society offices in London doubled as the offices for various birth control organizations. Understanding the scope of his early work helps explain how birth control and eugenics were linked from the beginning, de-exceptionalizing the connections between eugenics and population control after World War II.

Blacker mapped arguments for and against birth control in his book *Birth Control and the State* (1926). On both sides, these could be divided into matters for the individual and matters for the "race." He designated as "International" his first argument for birth control: "The connexion between over-population and war is nowadays fairly obvious," he opened, developing the cases of Germany and Japan, and importantly raising as a problem the "restrictions imposed by America upon Japanese immigration." He was alive to the international relations aspect of the issue. The medical profession should unite to call for the state to step in, where small private groups were stoically proceeding, but at too small a scale (in London, the Malthusian League's clinic or Marie Stopes's clinic, for example). But birth control was also deeply personal. Contraceptive information, Blacker thought, should be disseminated to the "poorer classes . . . in the interests of the mother and child" simply because so many infants were "unwanted." He detailed research at Guy's Hospital in 1924. Of seventy-eight mothers asked, forty-seven said they definitely did not want their newborns. This raised (and explained, he said) the prevalence of abortion. "The damage done to the health of many poor women by such practices is enormous, and might largely be avoided by a judicious instruction in Birth Control."[58] Blacker had every awareness of the unequal status of men and women in questions of sex, reproduction, and population. And the simultaneously local and international implications of birth control were already firmly part of his repertoire in the 1920s, decades before his instrumental involvement in establishing International Planned Parenthood Federation at the 1952 family-planning conference in Bombay.

The combined insights of Sanger, a growing feminist lobby group for birth control, some men like Blacker, and the ever-vigilant Charles Vickery Drysdale into the different implications of the population question for men and women gained traction over the 1930s. They nonetheless remained the exceptions. By the middle decades of the century, as first national then international agencies created population policies, it was sex as a solution rather than gender as a problem that drove it all forward. It was food security that put health and reproduction into the international public sphere. And ultimately, it was intractable geopolitics that put women and gender so squarely, if expediently, into "population and development."

Civilization: The Global History of Population Control

When Warren Thompson forgot to think about women, he was forgetting his Malthusian roots. Certainly no protofeminist,[59] Malthus nonetheless recognized, however imperfectly, that "population" in practice had different implications for men and women. To put it in late twentieth-century terms, population was always gendered. This mattered to him, not in the least because he imagined a Mary Wollstonecraft–style vindication of women, let alone the sexual and marital libertarianism she also represented. Nothing could be further from the case. The place of women and reproduction mattered, because it related to stadial economic and social progression. Women were key players in matters of life and death, and this was part of "political economy" due to the circumstances of perpetual limits to land and space on which to grow or gather food.

In the savage state, as Malthus put it, particular hardships fell on women. And so, lack of civilization—the state of savagery—was most manifest in and as the status of women in those societies. "The women are represented as much more completely in a state of slavery to the men than the poor are to the rich in civilized countries. One half the nation appears to act as Helots to the other half, and the misery that checks population falls chiefly, as it always must do, upon that part whose condition is lowest in the scale of society." The exertion of "preparing every thing for the reception of their tyrannic lords" caused miscarriages, he thought, and prevented any but the most robust infant from surviving.[60] Malthus was always interested in the poorest in any given stage of civilization as he saw it, and posited that the less civilized, the more that lowest group would be represented by women. John Stuart Mill put forward similar ideas, but more politically; that is, not to describe things as they were, but to prescribe changes that should occur. He put the gendered aspect of reproduction thus: "a degrading slavery to a brute instinct in one of the persons concerned, and most commonly, in the other, helpless submission to a revolting abuse of power." Mill also rendered sex and gender a major locus for civilization and progress. This included a lingering savagery in his mid-nineteenth-century British present. Civilization, he argued, "in every one of its aspects is a struggle against the animal instincts." And social organization is possible, because "man is not necessarily a brute."[61] By planning and managing this instinct (sex), and its effects (pregnancy), humans would become more civilized.

The second and subsequent editions of Malthus's *Essay* incorporated a large amount of new empirical content that included ruminations on the South Seas, Africa, Siberia, Turkey, Indostan, Tibet, China, and Japan.[62] John Toye suggests that this places Malthus at least as much in the intellectual genealogy of "development studies" as that of political economy.[63] The detail with which Malthus explained the stadial development of civilizations also places him in a tradition of writing world history. This was a process by which analysis of Pacific, American, European, Asian, and African worlds were brought into a global past. It should also be considered early ethnography and anthropology of sex and gender,[64] investigating, however spuriously, causes of female sterility, effects of nomadic existence on fertility and infant mortality, prolonged breast-feeding, abstinence, castration, late marriages, practices of polygamy, the fertility of one wife and not another, ritual killings, and the status and treatment of women. The real significance of this material (for him) was comparative, as he moved from describing preagricultural economies, to agrarian economies of various kinds, across the world. For Malthus, all this went some way toward explaining the "thinness of the population."[65] Indeed, the low density of indigenous societies remained a constant point of reference in population literature from Malthus to Carr-Saunders to demographic and economic historians, such as E. A. Wrigley and beyond.[66]

Malthus considered past stages manifest in his present as "savage" societies, but not biologically so; he was rather less determinist than his later nineteenth- and early twentieth-century intellectual descendants, who recast this in evolutionary terms. But already, in Malthus's *Essay*, the twinned regional and temporal axes of "development" can be discerned: some peoples moved forward in history, into civilization, and from hunter–gatherer to pastoral to agricultural and commercial economies. Others—the savage, the undeveloped—stayed stationary. They might be brought into contact with civilization, a process marked on the one hand by changing economies, and on the other by changing cultural practices around sex, birth, and death; alternatively, as he worried, such contact would manifest as starvation. It was the lack of correlation between population needs and food supply that drove civilizations forward: "Had population and food increased in the same ratio, it is probable that man might never have emerged from the savage state."[67] This is what energized the development of human societies from one stage to the next. This was a world history of both "improvement" and "development."

Alexander Carr-Saunders conceptualized his *Population Problem* as a modern *Essay*. He was himself, perhaps, an up-to-the-minute Malthus. His studies were also world histories of population, though written after industrialization, after the facts of Darwin and Mendel, and after a century of anthropology. Extending Malthus's interest in stages of civilization, Carr-Saunders's "primitive" was similarly evident in the present world as in the deep human past. "Hunting and fishing races" were of interest to him in terms of prepuberty intercourse, prolonged lactation, postponement of marriage, abstention from intercourse, infanticide, and the killing of the old and sick. He drew here on studies of Tasmanians, "Eskimos," "Bushmen," Native Americans on the Pacific coast, and Fuegians. In "Primitive Agricultural Races," he looked at other Native Americans, Africans, Oceanians, and to societies in some parts of Asia. And then came his examination of the "historical races," in which he also investigated practices of abortion, infanticide, celibacy, contraceptive practices, and venereal disease.

Carr-Saunders synthesized a mass of contemporary anthropology, much of it dealing with sex and reproduction. In parts of Africa, Carr-Saunders explained, people abstain from intercourse for several years after the birth of a child. This also took place in parts of Polynesia, he claimed, "where in addition abortion and infanticide are practiced." Among Native Americans infanticide and abortion were "fairly common." Indeed, he considered that the practice of infanticide most acutely represented the primitive mode of controlling population. Notably, though, he assessed this in neutral terms, as a matter of anthropological fact, not a moral judgment. Carr-Saunders cited such practices on the Pacific Island of Funafuti, where every mother was, apparently, permitted to keep alternative newborns only. While in Tikopia, he reported that all children above the number of four were killed.[68] Some Islanders committed infanticide to such an extent, he claimed, that the population had recently been brought near to extinction—recall Friedrich Ratzel's work on just this problem.[69] Other scholars, far more knowledgeable about the Pacific than either Ratzel or Carr-Saunders, were analyzing the introduced diseases as well as the psychological issues that in fact caused mortality and depopulation in Melanesia. But Carr-Saunders's point was that in general, native peoples limited population increase by such means.[70] These customs, more than war, famine, or disease, he thought, kept numbers down in such "primitive" societies, except where those customs had been destroyed or minimized by European influence. And if the West was responsible for reducing infanticide through the health measures that accompanied colonial rule, it was also then responsible for introducing alternative methods for the relief of population pressure.

Carr-Saunders thought that the power of multiplication, in world history terms, had been limited primarily by infanticide, abortion, and restraint from sexual intercourse.[71] In his view, then, infanticide and abortion constituted "what may fairly be called a small family system."[72] If the practices of abortion and infanticide were endorsed in Asiatic, Pacific, and Oriental peoples up to living memory (as he would categorize them), Carr-Saunders also indicated their sanction in Europe up to the beginning of the Christian era. His point was that abortion and infanticide were universally accepted, with the exception of post-Christian Europe, where, significantly, abortion and infanticide came to be illegal under secular and canon law.[73] These practices were largely replaced with a system of postponement of marriage in premodern Europe, which had proven only apparently effective in keeping numbers down: the real reason that there had been a stationary European population for so long, he clarified, was the high death rate.

Certainly for Carr-Saunders, the means by which different human societies (over time and place) kept numbers down were more and less preferable and marked human populations and societies as more or less civilized, more or less modern. Despite the differences and preferences of various methods, for Carr-Saunders, this was all of a piece, a historical or civilizational continuum whereby human populations limited their multiplication. This constituted part of a long-standing world (temporal) and global (spatial) history of the means by which population control had been achieved.[74] Typically optimistic about human progress, Carr-Saunders thought that humans were always capable of manufacturing and controlling their own conditions of reproduction and always had, more and less effectively. Someone like George Knibbs might predict catastrophic futures, but for Carr-Saunders "Nothing is more certain than the fact that it lies within the power of the human race to dictate its own future."[75] This was why there was little evidence in human history that population ever increased up to the point of starvation. For him, then, the new epoch was not really marked by a "discovery" of birth control, as some of his contemporaries insisted on putting it, because measures of fertility restriction were and always had been present in human populations. Rather, the key difference, the index of civilization, was the *means* of birth control or fertility adaptation. This idea apparently buoyed the beleaguered Marie Stopes, who told Carr-Saunders she kept a copy of his book by her bed, instead of the Bible.[76]

Population Control as Civilization

While Carr-Saunders treated infanticide as part of a population control continuum, for others infanticide was the practice that most distinguished civilization from savagery.[77] This was the more typical neo-Malthusian position. Malthus had written of the "barbarous practice" of exposing infants in China, "a custom that thus violates the most natural principle of the human heart," but one that was an index of economic distress, rather than an essential trait that marked China as uncivilized.[78] In the late nineteenth century, Annie Besant wrote similarly of infanticide as "life-destroying, brutal, anti-human," and definitely savage, not civilized.[79] In 1892, Charles Robert Drysdale thought that the practice of infanticide was identifiable in the primitive, the non-European, and ancient social domains. In cases of famine and extreme want, the Chinese kill their children, he wrote, as do the inhabitants of New Holland, and as was the practice in ancient Athens.[80] (Here was a twentieth-century version of Malthus's "lowest stages and past times.") But for Drysdale, significantly, infanticide was also a feature of his own neo-Malthusian backyard: infanticide was on the increase in London.[81] And although abortion is now understood on a continuum with contraception, then it was understood on a continuum with infanticide. Edward Alsworth Ross went a step further and wrote about abortion only in the present, and as part of his immediate world: most abortions in the United States were undertaken on and by married women, he told his readers.[82]

For the Victorian neo-Malthusians, infanticide and abortion wherever and whenever they were practiced, were barbarous methods of limiting births and were precisely what contraceptive technologies would render both civilized and infinitely more effective. These were Sanger's early twentieth-century arguments as well. Contraceptive practices and technologies would and should replace all other checks, including infanticide and abortion, as civilized and humane, as life preventing, as Sanger put it, following Besant, not life destroying.[83] Or in Radhakamal Mukerjee's terms: "The selection which is left to nature is by killing and devitalization, which are both relentless and haphazard, and these should be replaced by man's deliberately controlled and preferential reproduction, which allows the best types to evolve under most favorable conditions." The connection between these neo-Malthusians and eugenics is clear, as are the progressive politics to which they typically adhered. Mukerjee continued: "As civilization progresses man will control both natality and mortality more and more effectively."[84] The new means of birth control came to be heralded as an index of civilization itself.

For this generation of commentators, "civilization" was indicated not just by the prevalence of infanticide or abortion, but also by a complex of biopolitical practices around marriage, sex, health, and death. That these practices were always deeply embroiled in gendered power relations was a fact more interesting and more apparent for some commentators than others. For Alexander Carr-Saunders, the default actor of "population" was the man/husband deciding whether or not to marry, whether or not to have more children. Women entered the field less as individual actors with particular motivations or capacity to make decisions, than as a group whose social status was changing for many reasons, and Carr-Saunders was interested in them all. It was in that capacity that women affected aggregate rates of pregnancy and birth. The rising status of women was one of the chief elements of social evolution, he considered, signaled by education and greater independence. He thought this correlated to an acceptance of family limitation. Carr-Saunders's most consistently cited instance was not cross-cultural, or even cross-class, however, but cross-occupational. He often compared cotton-manufacturing districts in Britain, where women were typically employed, to mining districts, where women were more dependent on workingmen. The birthrate was far higher in the latter than the former.[85] Given that Carr-Saunders approved of "the small family system," there was also an implicit approval of such shifts toward women's financial independence, however small. He was deeply interested in the world trends with respect to birthrate decline, "all races of European descent are treading the same path, though at different intervals."[86] And the more other economies changed, the more they would follow that demographic path.

Compared to his contemporaries, however, Carr-Saunders was understated about the significance of such changes: his work was typically descriptive not prescriptive. Others wrote about the changing status of women far more politically and polemically, seeing it, like the uptake of birth control, as epochal and heralding major global trends. It was certainly part of the long-standing Malthusian argument that "higher standards of living are correlated with a deepening sense of individual and social responsibility toward women and children."[87] There were approving and disapproving versions of this, but the positive rendition was not solely the domain of feminist thinkers of the period.

Progressivist Edward Alsworth Ross might have been an anti-immigrationist, but he was also a strident supporter of birth control, and one who saw it firmly as part of the "woman question." He considered the growing control over fertility as "nothing less than a revolution in the vital history of Western Europe, the United States, Canada, and Australasia." This was as beneficial, he projected, as the discovery of the germ theory of disease, indeed "even more momentous in the life of humanity."[88] He was also a proponent of a "universal" raising of women's social status, and accordingly an opponent of "female subjection," one manifestation of which was intolerably high maternal mortality. "The best brake on this murderous prolificacy is the elevation of girls and women in the home, the state, the church, the university, industry, the professions, social life; not in their own eyes only, but in the eyes of the other sex as well." For him this was a major aspect of Western and Eastern world difference. In "progressive peoples," he considered, the last half-century had been marked by a correlating rise of a woman's movement, fall in maternal mortality, and fall in fertility.[89] The physiological cost of childbearing falls on the mother. But when "males dominate" or as he put it rather remarkably, "in the *he*-epoch," this sex-specific cost is deemed inconsequential. He cited Luther and Napoleon's views as characterizing this "he-epoch," as much as peasant Slavs or Aboriginal "primitives." But for fifty years, he said, "This male domination has been well rapped. The universal woman's movement is breaking the fetters on the wife's mind and causing the heavy cost of motherhood to be more considered." This was the "emancipation of women," wherein "emancipation" simultaneously signified the slavery parallel and an imprisonment parallel, in that Ross referred to "*penal* maternity."[90] Emancipation of women, for him desirable in itself, would also contribute to the desirable decline in birthrates, relief of population pressure, and a more secure world.

In *Standing Room Only?* the birth control imperative was a domestic issue as well as a world issue. Absence of knowledge about modern birth control or lack of access to contraceptives was an irrational and unjust "embargo" that chiefly affected the poor; monied women could access both information and materials privately through their physicians. It is, he wrote, "a class privilege." Ross was appalled at the illegality of contraceptive literature in states of the United States. At the same time, although he claimed to retain an interest solely on the quantitative population problem and to bypass the "quality" question, in fact the latter was entirely evident in *Standing Room Only?* A driving problem for him was domestic differential fertility on class, cultural, and race axes. The readers of his book will already be responsibly limiting their offspring in all likelihood, he commiserated, while the message should really be getting to those who will never read the book, "the dullards" who have been kept from education.

Standing Room Only? was framed by a tradition–modernity continuum that both racialized and spatialized civilization in relation to gender. "It is the undue subordination of the wife to the husband, of the female sex collectively to the will of the dominant males, which has brought upon Asia a condition of overpopulation."[91] Equivalent arguments were put by Indian population writers. "Indian women have been treated as slaves for centuries past," argued Prabbu Dutt Shastri at the New York International Malthusian meeting in 1925. The higher classes were already practicing birth control (presented as a way out of such "slavery"), but the poor, without this knowledge, were "embarrassed" with large families.[92]

Across much Anglophone argument, high infant and maternal mortality rates and high fertility rates marked a less progressive society, as both cause and index. For Ross, but also for Indian economists like Mukerjee, birth control offered a new economy of reproduction. This did not necessarily translate into similar positions on national or international policies. Mukerjee was deeply opposed to immigration restriction, for example. By contrast, Ross's spatial solution to high fertility, as we have seen, was the "Great Barrier." The immigration restriction acts should stay in place not as a permanent fixture, Ross thought, but until "civilization" (a lower birthrate) was achieved. This was about gender too. Civilized societies required "no subjection of women, no pre-puberty marriages, no propagation from physically immature females, no infanticide or abortion, no high infant mortality, no unschooled ignorant masses, no congested labor market."[93] Such statements entailed obviously problematic presumptions that the West led the way in gendered civilization and took upon itself to set the standards of it. Such assessments were part of a colonial discourse of tradition and modernity and of a problematic evaluation of class and race difference. However, they cannot simply be reduced to that. Ross's arguments were as feminist as Margaret Sanger's, though marginally more Eurocentric.

Malthus's intellectual descendants received multiple versions of "civilization." One strong strand saw birthrate decline as threatening a kind of "over-civilization." Charles Darwin's *Descent of Man* opened this idea.[94] Herbert Spencer's derivative thesis declared that declining fecundity correlated with progressive civilization.[95] It was reincarnated in the 1920s as Corrado Gini's theory of the senility and "old age" of Western civilization, a version of degeneration.[96] It is also the case that planned limitation of births seemed to announce a major new progressive era, a new stage of civilization. Thus, birth control did not mean "race suicide," for many. On the contrary, it was itself the great mark of progress, a key material condition for and expression of advanced civilization. As such, many wrote, the new methods of limitation of births should gradually spread across the world, part of the great human move away from nature and from savagery, and toward more complex and developed economic systems.

Keynes saw birth control as heralding just such a new age: "A great transition in human history will have begun when civilized man endeavors to assume conscious control in his own hands, away from the blind instinct of mere predominant survival."[97] It was about "control" and for economists, "planning." Birth control signaled a modern rationality, a new "governmentality": the capacity to plan into an abstract future, as well as capacity to make sense of the deep past, both as a species and as individuals.[98] Of special importance was the fact that birth "control" and family "planning" fit the whole idea of planned economies up to but not including Marxist states that persistently refused Malthusian population-based explanations or solutions for economic well-being (at least until the late twentieth-century turnaround in China).

As civilization progressed, Mukerjee wrote, natality and mortality would become both more controlled and more *effectively* controlled toward an optimal population. This would be assisted by good government policy and economic planning: precisely his role in the Indian National Congress. In less advanced societies, the average expectation of life is low and "there is great waste of people." By contrast, "in advanced and well-organized societies, the agencies of social control effectively keep down numbers as well as the death-rate, resulting in better economy of reproduction and higher physiological well-being." Bringing the ecological rationale into civilizational development, he asserted that "the physiological waste is reduced." In a repetition of the argument put forward in *Science of Life*, it was this efficiency that defined the most civilized societies, biologically, economically, ecologically, politically, and socially.[99]

"Civilization," though, was a complicated concept. Witness Sanger and Wells's attempts to apply civilization both to their domestic contexts and to their aspirational world plans in 1922. They used "civilization" to signal old and new generational approaches, the 1920s moderns responding to the aged conservatives. Wells wrote, in the foreword to Sanger's book, *The Pivot of Civilization*:

> The New Civilization is saying to the Old now: "We cannot go on making power for you to spend upon international conflict. You must stop waving flags and bandying insults. You must organize the Peace of the World; you must subdue yourselves to the Federation of all mankind. And we cannot go on giving you health, freedom, enlargement, limitless wealth, if all our gifts to you are to be swamped by an indiscriminate torrent of progeny. We want fewer and better children who can be reared up to their full possibilities in unencumbered homes, and we cannot make the social life and the world-peace we are determined to make, with the ill-bred, ill-trained swarms of inferior citizens that you inflict upon us."[100]

"Civilization," here, was temporal: the new generation (who responsibly planned their reproduction and planned their future) chastised the old generation, the old civilization, that did not. But the potential for this idea to shift from a temporal axis to a regional/racial axis across global geography, is plain. "After my eight months tour of the world," Sanger wrote,

"I am glad to agree with H. G. Wells when he says that the whole world at present is swarming with cramped, dreary, meaningless lives, lives which amount to nothing and which use up the resources and surplus energies of the world."[101] The familiar Malthusian overcrowded masses of nineteenth-century London were in the process of being relocated to twentieth-century Bombay. Lives seemingly wasted in poverty in the neo-Malthusian tradition were by implication, and sometimes by directive, lives that were themselves waste.

The Universal League of Low Birth Rate Nations

Comparative anthropologies of sexual conduct and gender were both a reference point and a sustained axis of analysis. In early twentieth-century versions, more than later twentieth-century versions, gender and sexuality of the peoples who had once "swarmed" but were no longer "swarming"—Europeans, Euro-Americans, "Europeans Overseas" as it was sometimes put—were of particular interest. What was their motivation for contraceptive use, and how could this be translated from one context to another? The studies that Pearl, Blacker, Sanger, and others were undertaking of sexual practices and use of contraceptives in London or Baltimore were as significant as the contemporary studies of sex and primitivism. As Enid Charles put it: "The sexual behaviour of Manchester or Montana is of more moment than that of Melanesia. At present we know rather less about it."[102]

Birthrates were certainly declining in Manchester, Montana, and Melanesia but for utterly different reasons. The Melanesian islands were not being "naturally" depopulated as a result of the kind of land pressure that Malthus and later Ratzel had theorized of such islands, or as a result of economic changes toward industrialization. They were depopulating as a result of their modern colonization by diseased British and French. But in other respects, Charles's focus on the industrialized United Kingdom and United States is to the point. In the mid- to late 1930s, for a second moment in the history of the Anglophone West, birthrate declines grabbed considerable attention. For some in Britain, the decline seemed to mark a shift from leading a global trend back into a potential domestic problem. Even Carr-Saunders momentarily changed his signature "small family system" into "the small family problem."[103] There was a re-emergence of the anxiety about birthrate decline in the late 1930s in the United States as well, led by the work of Frank Lorimer.[104] The fall seemed to be commensurate with the

Great Depression. A decade after toasting Annie Besant at the Malthusian League dinner, Keynes was delivering a Galton Lecture to the Eugenics Society. "Some Economic Consequences of a Declining Population" may bring difficulties, the economist conceded, but in the end a stationary or slowly declining population might also "enable us to raise the standard of life to what it should be."[105]

Enid Charles was theorizing the "declining phase" in population epochs based on Robert René Kuczynski's methodological innovations, which distinguished between fertility rates, gross reproduction rates, and net reproduction rates.[106] Fertility rates were calculated according to the number of reproductive humans—that is, women between fifteen and fifty: the number of children born in one year to 1,000 average women of childbearing age. Kuczynski narrowed this first to the gross reproduction rate: the number of girls born to one average woman throughout her reproductive life. His most important contribution was even more specific: the net reproduction rate. This took into consideration the number of female children who would survive to be reproductive, that is, taking into account the likelihood of death before these women reproduced themselves. "From the standpoint of calculating changes in population, girl children are more important than boy children, because it is from its women that the community replaces itself."[107] This was all very different compared with Knibbs's standard statistical formulation of the "masculinity of the population" in his 1917 census report and was one of the drivers for the increasing focus on women, as national and international policy makers (in the United States, in India, and especially in Sweden) moved closer toward the incorporation of birth control based on demographic transition ideas.

These new statistical considerations and concerns were the brief of a British enterprise called the Population Investigation Committee that went on to publish (and still does) *Population Studies* out of the London School of Economics. The initial committee included Lord Horder, Julian Huxley, Blacker, Carr-Saunders, Kuczynski, and David Glass.[108] In 1939, it published the popular *Population and Fertility*, addressing the fall in the British birthrate and its significance for the British Commonwealth and colonies. And yet, it did not recommend incentives to raise fertility rates, and in this sense was nothing like turn-of-the-century pronatalist inquiries or 1930s German and Italian pronatalism. Indeed, following Keynes, it still did not regard as "necessarily undesirable a decline in our numbers." Rather, the committee sought to investigate and understand the cause of this decline and to clarify commonplace misunderstandings of the much-discussed phenomenon.

Ultimately, the group was driven by the mission to universalize fertility decline, stating that the most desirable global future would be for fertility to fall uniformly throughout the world. If this took place, the prospects of peace would be increased, the committee wrote in 1939.[109]

When Carlos Paton Blacker developed his own version of demographic transition as modernization, it involved five phases: the High Stationary Phase (high birthrates and high death rates); the Early Expanding Phase (high birthrates and declining death rates); the Late Expanding Phase (declining birthrates and death rates, with the death rates consistently lower than the birthrates, thus yielding annual increases in population); the Low Stationary Phase (low birth and death rates). And finally, in recognition of Charles's work and the British demographic context, he added a final, but not automatically following, "Declining Phase."[110] Charles's observations and Blacker's nomination of "decline" seemed to indicate a problem for Britain, but it did not gain real traction, largely because the declining local trend reversed again after World War II. Rather more abiding was Blacker's proposition that fertility and mortality declines were a social and biological good, linked to political security, but only so if pursued internationally, or, in the increasingly used term, "universally."

The question is, then, how this "universal" ambition for demographic transition sat with a history so deeply shaped by racialized and spatialized geographies of "civilization." We have seen that one expression of this global geography was the League of Low Birth Rate Nations, a precursor of the Third World/First World division. The core of such a "league" was always imagined to be the West, with its "occidental birth rates" and higher standards of living; industrialized societies in which declining birthrates and declining infant mortality rates seemed to be stabilizing as the norm. The global geography would seem to double neatly as the settler-colonial "white man's countries," where high-fertility "orientals" were being excluded by immigration restriction acts. Yet simply to conflate the racism of immigration restriction with the emerging geopolitics of demographic transition is to miss the core point of "transition" (and to ignore the demographic opponents of immigration restriction).[111]

First, high fertility and high mortality were not correlated automatically with societies in the "Orient" or in the "Far East." It was not just about "race" but also about "civilization" read as economic development; this was why Japan was so often singled out, for example. Scholars also looked at other parts of the world altogether. George Knibbs thought Russian demography a major problem, wherein high fertility and high mortality characterized its preindustrial civilizational/economic stage and kept its standard of living low. Poland was another such case. How to educate peasants in such economies, Knibbs wondered? If agricultural methods were improved, he thought the short-term effect would be high fertility and increasing net population, but the longer-term effect would be to raise the standard of living, reduce fertility and mortality, and eventually reduce density, and desirably so.[112] This was a quandary for his "New Malthusianism," anticipating the twists and turns that demographic transition theory would later take with respect to active population control vis-à-vis industrialization.

Second, the aspirational League of Low Birth Rate Nations often aligned with "white men's countries," but did not necessarily do so. Blacker, for example, imagined an internationalized birth control in the 1920s that would be a coalition of organizations in the United Kingdom, the United States, Germany, Japan, Italy, and India. This league would be a "first step in the direction of an international control of population, and would thereby lay the basis for a genuine and permanent world peace."[113] Third, the whole point for those who increasingly advocated a world demographic transition was that any "league" was expansive: more and more national populations would enter it by virtue of declining birth and death rates. The differential mortality and fertility rates across cultures and global regions would ideally be minimized, just as the difference between class and occupational groups within Western countries was being slowly minimized. Carr-Saunders, for one, was comfortable predicting in 1935: "It cannot well be doubted that things are moving or will move in this direction in all Western countries." Just as he predicted differential class fertility to pass away, so too would differences between regional peoples.[114] Presenting birth control within a world history, Carr-Saunders normalized the controversial practice—the fact of limitation was always present and is "no new thing."[115] This trend toward a uniform "small family system" is what many aspired to globally, and some already predicted to be one of the great transitions of world history.

The idea of demographic transition and modernization emerged, then, from the old idea of civilization.[116] The strong eighteenth- and nineteenth-century intellectual provenance was still current throughout the 1920s and 1930s, although stadial economic concepts were beginning to be replaced by more neutral "phases" or "stages" of high and low mortality and fertility in various combinations. The world was discursively regionalized by its demographic status, its relational and comparative birth and death rates. This was global "modernization" in which the aspiration was for the whole world to become "the West," with its higher standard of living. "Westernization" also meant that the world's women might have, perhaps, three children on average, not eight, all of whom would survive to reproduce themselves at this neat, low rate. Add global ecology, and the entire complex turned into waste *versus* efficiency. This is what Mukerjee meant by "the new economy of reproduction." And when Kingsley Davis sat down to sort out his own famous ideas about all this, "the world demographic transition," he did so in just those terms.[117]

9

The Species

Human Difference and Global Eugenics

It is clear that what man can do with wheat and maize may be done with every living species in the world—including his own.

H. G. WELLS, JULIAN HUXLEY, AND G. P. WELLS, *THE SCIENCE OF LIFE* (1931)

Edward Alsworth Ross ended his book *Standing Room Only?* with an entitled set of stipulations about the conditions under which the West might, at its pleasure and discretion, lift the "Great Barrier," its system of immigration acts. If the social and economic criteria that marked a civilized state were achieved, "the Barrier should be removed." But this was his penultimate comment only. Ross's long study concluded with an afterthought: "Unless, to be sure, the future should demonstrate that interchanges of population between racially unlike peoples—which leads to crossing—is biologically undesirable."[1] Ross's preference, clearly, would be permanent racial difference, manifested and maintained spatially. Constantly pressing for new international systems, George Knibbs was also quite clear about forcing responsibility for the mentally and physically "inferior" back onto the national populations that reproduced them: "It is inimical to the world-future's interest that inferior sections of humanity should be transferred even to relatively empty countries." Rather, it is in the interests of the human race, he wrote, that each nation should retain, learn to deal with, and figure out ways to prevent the reproduction of "its degenerates or defectives, its derelicts, and its poor."[2] This, he said, was the *New* Malthusianism.

Conceptions of human difference were core to the biopolitics and the geopolitics of world population. Yet difference was manifest on many axes.

Race—the limit for otherwise progressive Ross—was one kind of human difference. Another was that of mental and physical ability and disability, capacity and incapacity, calculated and diagnosed a thousand ways in the era of eugenics. The apparently heritable characters of "race" and "disability"—sometimes intersecting, but not necessarily so—were affected by fertility and mortality at population levels. The intricate technicalities of this proposition energized any number of brilliant minds over the 1920s and 1930s. Human difference became the business of emerging demography, of eugenics, and of the new population genetics, none of which could quite be reduced to the other.

None of the historical actors who considered themselves advocates of eugenics would disagree with Knibbs's broad claim that the reproduction of the mentally "unfit" should be managed, even if they did argue over the means by which this might reasonably be done. Mental and physical dis/ability was eugenics' core problem. On the question of race, however, "eugenicists" presented any number of scientific and political faces, from strident segregationism and insistence on hierarchized difference, to equally strident absorptionism that would obliterate difference, to public and passionate antiracism with its own complicated and unpredictable relationship to the new genetics. The antiracism of some of the key players in the Anglophone biopopulation establishment—Julian Huxley or Radhakamal Mukerjee, for example—by no means precluded a sharp valuation of human difference on other axes, especially the classic and persistent eugenic issues of intelligence, disability, and class. Their positions were always complicated. Warren Thompson, for example, made a career in active opposition to presumptions of superiority built into both eugenics and various race sciences that were consolidating in this period, even as he actively wrote out indigenous people from any geopolitical solution to overpopulation. Nothing in the linked population–race–eugenics triad was predictable or can be presumed, so multiple were the agendas, the politics, and the sciences of population, so polyvalent the actual practices of reproductive management.[3]

How did eugenics intersect with or form part of the world population problem? In what ways was eugenics international or global or both? To answer these questions, we need to recognize the strong connection between Malthusian ideas and eugenics; Malthusianism not in contrast to eugenics, or reducible to eugenics, but as formative of it. While over time and place not all eugenicists aligned with Malthusian ideas, proponents of the social application of "natural" selection, as well as the new Mendelian genetics, shared an interest with neo-Malthusians in the possibilities of the management of human reproduction, sometimes at world scales and at the species level.[4] Understanding this connection allows us to properly grasp some of the twentieth century's key turning points in a politically engaged science, including the famous Geneticists' Manifesto (1939) that was to be extended and challenged after the war in the successive United Nations Educational, Scientific and Cultural Organization (UNESCO) statements on race.

Quality and Quantity: Eugenics and the Neo-Malthusians

Because so many advocates of eugenics were troubled by differential fertility rates within national contexts (along class, race, ethnic, or national axes of difference), historians often presume them to be more separate and separable from Malthusian ideas than was necessarily the case. Especially for those who came to population questions via political economy and Thomas Robert Malthus, it was not difficult to admit Francis Galton's original eugenic ideas as well, because they both rested squarely on the claim that more organisms reproduced than could survive. For both Galton and the neo-Malthusians, natural selection with respect to humans was incompatible with civilized modern sensibilities and aspirations, involving as it did, high infant mortality. Human-directed selection—eugenics—was thus figured as far preferable and even humanitarian. At least this was Galton's argument, and one that carried much currency and influence. As it was put at the sixth Neo-Malthusian conference in New York, eugenics should replace nature's "cruel and inexorable methods of eliminating" with birth control as way to check the greater fecundity of "the inferior."[5] This is partly how and why eugenics became the domain of progressives.[6]

The Malthusian idea that was at the heart of eugenic selection is not often recognized. "Natural Selection rests upon excessive production and wholesale destruction," wrote Galton. Eugenics was a solution to this natural violence, he thought, based on "bringing no more individuals into the world than can be properly cared for." As Galton would have it, the birth of the "unfit" would be prevented, and this would prevent also their likely infant death, because they were "doomed in large numbers to perish prematurely."[7] Although the Malthusian League in Britain often had territorial disputes with various eugenic organizations, they constantly cited the intellectual common ground, Charles Darwin's original inspiration by Malthus.[8] At the 1910 International Neo-Malthusian conference, alternative organizational names were put forward: "Eugenics" suggested one delegate; "Race Control," suggested another. But they were troubled that "control" could not be successfully rendered into French. Perhaps "Régéneration," or "Procreation Consciente" [*sic*], or "Ligue Eugénique," usefully equivalent to the German *Rassenverbesserung*. This, they agreed, might indicate that "the Neo-Malthusian movement dealt with the quality as well as the quantity of children."[9] As president of the Malthusian League, and as a member of the British Eugenics Society until 1961, Charles Vickery Drysdale wrote voluminously and thought carefully about neo-Malthusianism and eugenics as complementary. Most importantly, he often understood the two population movements as successive; the reduction in the birthrate was a necessary precondition for successful eugenics.[10] Far from being oppositional, in this view, the "quality" and "quantity" aspects of population problems and population laws were interdependent. *Ned Quantitas sed Qualitas* emerged as the tagline for British Malthusian journals. Eugenics was, in other words, phase two of the temporal program of population improvement; quantity would be dealt with first, then quality.

Keynes also thought that quantity projects should and would merge into quality projects: eugenics.[11] Keynes was interested enough in the mathematical application of biology to join early Cambridge eugenics initiatives, and he kept up these connections. Although Keynes scholars tend to worry over his eugenics activity,[12] they do so for the wrong reasons. Keynes's interest in eugenics is explained less squarely by "race prejudice" than by his interest in Malthus and population itself; the phenomenon of too many organisms (people). If population growth was to be reduced, the eugenic question was derivative: Which kind of person might be reproductively restrained or encouraged? British eugenicists of his generation sometimes thought about this question in racial terms. But they were far more likely to think of it in terms of mental and physical ability and disability of various kinds.

Drysdale wrote that in most parts of the world, quantity was still the issue—the birthrates and death rates were still too high—but that there were some instances where this had been dealt with adequately, such that "quality" became the population problem of first importance. New Zealand and Australia were his primary examples, where both fertility rates and infant mortality rates had slowed markedly. It was, by contrast, useless to think of eugenics in Russia, India, or China, he thought, or anywhere where the infant death rate remained higher than twenty per 1,000. There, infant mortality needed to be addressed first, not only, but not least, by reducing birthrates. In societies with such high levels of mortality, "the struggle for existence due to over-population" trumped the application of "rational eugenic principles on any important scale."[13] Radhakamal Mukerjee also subscribed to a continuum of reproductive planning from quantitative selection to qualitative selection, leading toward the net improvement of the "quality of his species by preferential breeding of the best types."[14]

For George Knibbs, too, neo-Malthusianism quite simply *was* eugenics. His 1926 *Scientia* article was a manifesto for the New Malthusianism. Humanity should now work toward "ameliorative action based on true eugenics and a finer sense of the claims of those who are to be."[15] Working from the premise that the rate of global population increase cannot continue ("it *must* diminish"), some measure of control of births was therefore inevitable, and so population managers of various kinds, especially those considering birth control, must take the laws of inheritance into account. What he called "Constructive Birth Control," meant selective breeding or eugenics, and "true eugenics" was ameliorative for Knibbs, following Galton: New Malthusianism would so regulate the conditions of life that the newborn would enter a far healthier world.[16] Margaret Sanger agreed, revealing the sharp edge to which apparently ameliorative humanitarianism could be honed. Her "Plan for Peace" included the diminution of death rates, maintenance of immigration restriction of the mentally and physically "unfit," sterilization and segregation of the unfit, and their relocation to farm colonies to live out their years as nonreproductive citizens.[17]

From this theoretical and conceptual starting point, neo-Malthusianism and eugenics became virtually indistinguishable in some contexts. Indeed, Malthusianism was occasionally the original frame of reference through which earliest discussion and dissemination of eugenic ideas took place. William Schneider noted that in early twentieth-century France, before the policy and legal pronatalist crackdown on contraception, *Le Malthusien* was one of the only journals even to use the word "eugenics," and indeed, changed its title to *Le Malthusien: Revue eugéniste* after the 1912 International Eugenics Congress.[18] In China, birth control was described plainly as a "method of eugenics."[19] And in India, it was neo-Malthusian organizations and personnel who first discussed eugenics, as subsidiary to economically driven rationales for the dissemination of birth control information.

In the United States and the United Kingdom, as in India, eugenics, neo-Malthusianism, and birth control advocacy elided over time. In the United States, eugenicists were certainly anxious about differential fertility on both class and race grounds, yet even there, neo-Malthusianism as "birth control" was ultimately incorporated. At the early Population Association of

America meetings, Harry Laughlin of the Eugenics Record Office spoke of the need for both quantitative and qualitative population research, distinguishing birth control from eugenics along these lines. Any "population" association should deal with quantitative research only, according to Laughlin, but less because it was unconnected to eugenics, than as an assertion of his Office's prior territorial claim over "quality." Also at these formative meetings, Henry Pratt Fairchild thought a population association should consolidate all this activity, while Frank Lorimer thought the quantitative and qualitative aspects entirely inseparable.[20] By 1934, Fairchild was convinced that quantitative and qualitative aspects of population "improvement" could no longer be regarded as distinct.[21] In 1940, he announced, "These two great movements [eugenics and birth control] have now come to such a thorough understanding and have drawn so close together as to be almost indistinguishable."[22] Ultimately, as Dennis Hodgson has so carefully shown, the association combined population problems in both its quantitative and qualitative aspects.[23] He is absolutely correct to name some eugenic advocates "biological Malthusians."[24] At the same time, the Birth Control International Information Centre indicated that it addressed global population distinctions not eugenically but economically.[25]

The marked demographic change toward Carr-Saunders's "small family system," which he and most others were quite sure had started among the wealthier, was by 1935 asserted to be a trend among the poorer classes in northwestern Europe. The small family was the way of the future, and many eugenicists who privileged quality understood that they had to factor in this clear demographic trend. Thus, although eugenicists are often understood to be discordant with neo-Malthusians, because of their belief that the more responsible and fit bred less, and the thoughtless and less fit bred more, by the 1930s, the typical eugenic position was in favor of the neo-Malthusian agenda to limit fertility across class, region, race, and globe. United States eugenics organizer Frederick Osborn, imagining a reformed U.S. eugenics in the late 1930s,[26] saw the public relations benefits as well as the genetic benefits of promoting the "universal" small family system as part of reproductive choice in a kind of population free market. Because "eugenic philosophy dealing only with extremes would be limited in its effect, as well as socially controversial," conditions should be freed up to make the choice of the "small family system" available to all. If a society could be so organized that parents were relatively free to determine the size of their families, "it is likely that on the whole the trend of births would be eugenic." Ideally, this would accord with freedom, not coercion.[27]

The best eugenics, most agreed, was when the bulk of the population made eugenic choices as individuals and saw that reproductive choice as beneficial for themselves and as part of good citizenship for a larger polity—the race, the nation, even the species. The idea of "voluntarism" was a key element of reforming Anglophone eugenics, so critical, they all ultimately knew, to eugenics' success, palatability, and apparent legitimacy. This manifested strongly in the debate about sterilization.[28] The British liberals were concerned that sterilization be made legal and be voluntary, if at all possible.[29] "Freedom" was the point, even if that involved the strongest persuasion: Huxley and Wells said at one point that "these low types might be bribed or otherwise persuaded to accept voluntary sterilization."[30] There is no doubt that entitlement to "freedom" lay on a sliding scale according to fitness, for such as Huxley and Wells. It is also to the point that their British sensibilities, along with mainstay sterilization campaigner Carlos Paton Blacker, could not stomach legal compulsion, in explicit contrast to North American state and provincial laws, and later, Nazi laws. Often enough, eugenics functioned not through the legal coercion of authoritarian states, but in open contradistinction to such measures. The intellectual and political context that linked freedom and reproduction had a prior, perhaps an original, history within versions of feminism and was sustained within the neo-Malthusian feminist agenda that Sanger and others articulated. Aligned with eugenics, Sanger nonetheless distinguished her own population politics: "Eugenists imply or insist that a woman's first duty is to the state; we contend that her duty to herself is her first duty to the state."[31]

All the key Anglophone writers on world population over the 1920s and 1930s had to deal with eugenics one way or another. But there was certainly no smooth alignment. Raymond Pearl's apparent objection to "the biology of superiority" is the well-known instance.[32] Warren Thompson is also important in this respect, given his influence on the emergence of demographic transition models, the framing of a world population problem, and his reworking of Malthus for the postwar world. At the end of his 1915 economic study of population, wages, labor, and food production, Thompson concluded that a greater control over the growth of population was essential. But if population must increase more slowly, he thought this left open the problem of which sector of any given population should increase and at the expense of which other sector. Thus, the final point of *Population: A Study in Malthusianism* was the open question of eugenics. This study, he concluded, is "merely a necessary preliminary to a further study. . . . If the population is increasing more rapidly than it can survive, then the questions about which we are particularly concerned are questions of selection." The ambitious young demographer ended his doctoral thesis by indicating that he hoped to make this study himself, some time in the future.[33] And yet, when Thompson picked up his publishing momentum after the war, he was critical, not supportive, of this process of active human selection.

In the great mix that was population thought, Thompson was one of the few who actively resisted the self-appointed racial and civilizational superiority that his colleagues in general presumed. He considered this one of the great blocks "in trying to establish a greater equality of resources on the earth." The Anglo-Saxon belief that they were the "chosen people" needed addressing, perhaps directing the critique more to Americans than to the British. "So long as we believe in our inherent superiority, we cannot give fair consideration to the claims of peoples of other races to share in the goods of this world."[34] Thompson's criticism of race prejudice was one of the bases of his critique of eugenics, but not the only one. He was unimpressed with many of his colleagues' concerns about differential fertility rates, their concern, essentially, about their own aggregate birthrate. Put simply, the presumption that old stock was good stock should be up for questioning. Moreover, the conceit of the economically successful was equally spurious, as was any premise that those superior in intellect were also superior in social value.[35] Thompson's case rested on the inflated claims of Mendelian geneticists; in his view, thinking about human society was strictly the domain of the sociologist and the economist.

Immigration Restriction and World Eugenics

Writing his strong critique in 1924, Warren Thompson had not just eugenics in his sight but the problematic (to him) recent U.S. Immigration Act. The 1924 Johnson-Reed Act has rightly been analyzed as the high-water mark of U.S. eugenics.[36] Prescott F. Hall, founder of the New England–based Immigration Restriction League had earlier written of "immigration restriction and world eugenics" in that internationally resonant year, 1919. Hall also headed the Immigration Committee in the Eugenics Section of the American Breeders Association, and this partly explains the location of his article in *The Journal of Heredity*. It appeared on the same page as more standard "breeding" intellectual fare, "Two Striking Color Variations in the Green Frog."[37] The genetics of variation, just then stunning the biologists of the world, did not interest Hall technically, but the biology of the relation between fertility rates and density did. This was a biology, as we have seen, that derived from the spatial history of Malthusianism. Fit North Americans should be given the continental room necessary to increase their own fertility (Benjamin Franklin would be pleased), and at the same time the less-fit southern Europeans should be kept in their own crowded homelands. Fertility limitation would thereby by forced upon them. "Just as we isolate bacterial invasions, and starve out the bacteria by limiting the area and amount of their food supply, so we can compel an inferior race to remain in its native habitat, where its own multiplication in a limited area will, as with all organisms, eventually limit its numbers." Pearl would surely be satisfied with this leap from microorganisms to humans. "On the other hand," Hall continued, "the superior races, more self-limiting than the others, with the benefits of more space and nourishment will tend to still higher levels."[38] Given limited space, both in the North American continent, and on the surface of the earth as a whole, "world eugenics" was a process by which the space necessary to promote increased birthrates was safeguarded for the purportedly superior. Eugenics, it turns out, was globally spatial too.

Charles Davenport, also involved in the various immigration acts, though not quite so intimately as his talented protégé Harry Laughlin, identified several phases of immigration regulation in North America. The phase following the 1891 Immigration Act, he called "hygienic": medical examinations were required of all immigrants, and epileptics and "defectives" were barred. The next phase, culminating in the 1924 Immigration Act, he called "eugenic," because immigrants were barred in the light of their ethnic or racial identity, again framed in terms of the density of the United States—"no land remained for settlement."[39] One of Margaret Sanger's regular arguments was a twist on this idea. Birth control, she argued, should be thought of as the "internal" or domestic equivalent to "external" immigration restriction, whereby exclusion was primarily figured in eugenic or "quality" terms. "Surely immigration from inside, namely births, are equivalent to immigration from without, and the same principle should be applied to both." Approaches to the "quality" aspect of the population problem were one-sided, she argued; undesirables were kept out, but there was no attempt to discourage the reproduction of undesirables within the nation, either "aliens or natives," she said, like Prescott Hall meaning immigrants or native-born white Americans.[40]

In addition to the well-documented racial eugenics of immigration, almost all of the immigration acts then proliferating in multiple settler-colonial contexts encoded the powers to exclude individuals on the grounds of physical, mental, and sometimes moral unfitness, irrespective of race or nationality. Thus, in thinking about what constituted "world eugenics," there needs to be an extension of understanding of the eugenics of immigration restriction beyond racial politics alone. Indeed the immigration acts were rarely, if ever, solely mechanisms to keep out "colored aliens"; they also functioned to exclude a range of other undesirable or unfit entrants. Many New World nations were hurriedly writing and rewriting the exclusion of the Old World "feeble-minded" into statutes, policies, and regulations as a means by which their own national populations might be improved. The "unfit" individual of the same "race" was to be excluded. More strictly eugenic than any race-based exclusion, then, were the clauses that sought to screen out the white physically and mentally "unfit" who would otherwise be permitted to enter: the feeble-minded, syphilitic, criminally inclined, or alcoholic.

This is the less familiar legacy of eugenics on the international bioregulation of global movement. In the U.S. 1917 Immigration Act, for example, section 3 prohibited "all idiots, imbeciles, feeble-minded persons, epileptics, insane persons; persons who have had one or more attacks of insanity at any time previously; persons of constitutional psychopathic inferiority, persons with chronic alcoholism."[41] The earlier Canadian laws were similar, the 1910 Immigration Act prohibiting "idiots, imbeciles, feeble-minded persons, epileptics, insane persons."[42] Even earlier again, the 1901 Australian Immigration Restriction Act had effectively ceased Chinese, Indian, Japanese, or Islander entry. This meant that the powers were far more commonly implemented to prevent entry of the eugenically suspect from the most populous country of migration: the United Kingdom.

For some in this period, individualized "eugenic" exclusion was considered in explicit distinction from the generic exclusion of a population on the basis of race or ethnicity. Geographer and race theorist Griffith Taylor, for example, thought the obsession with blanket race-based exclusions should sensibly be dropped in favor of health and fitness criteria of entrants as individuals. "Eugenics rather than nationality," he wrote, "is the best criterion."[43] Kiyo Sue Inui of Tokyo University of Commerce held a similar view. The best principle for international migration regulation, Inui argued at the Geneva World Population Conference should not be the exclusion of whole national or racial populations, but rather the "classification" of individuals: the former was applicable only "vertically" to a social group, the latter, more appropriately, was applicable horizontally and in a more individualized and refined way. He invoked two subsidiary principles. First, no nation should send emigrants where they are not wanted. Second, invitations to migrate or admission to settle should be issued to "all peoples alike."[44] The barely hidden subtext here was the international decision in 1919 not to accept the Japanese-proposed racial equality clause.[45] "Every nation has the right to protect itself from deterioration by racial intermixture," wrote geographer J. W. Gregory. Thus, "Australia is biologically well supported in its claim to restrict immigration to the white race," whatever the political objections might be.[46]

This was on Prescott Hall's mind too. In proposing in 1919 that immigration acts might function as a kind of world eugenics, Prescott Hall was urging his compatriots then in Versailles to hold the line on sovereign rights to determine who can enter a nation; the right to determine any criteria on who shall be admitted not into its territory, but its "life," as he tellingly put it.[47] This was the final message of his manifesto for world eugenics, though Woodrow Wilson was hardly likely to be working his way through the *Journal of Heredity* at the time. As it transpired, nothing in the constitution of the League did limit this right; quite the reverse. On Prescott Hall's terms, and in this sense, the League upheld "world eugenics."

However, when Charles Davenport tried to internationalize eugenics by having it raised formally onto the League of Nations agenda one year later, he failed. It was specifically on the grounds of the global significance of human migration and its regulation that Charles Davenport was convinced eugenics should be recognized as an international issue. He wrote to the secretariat in 1920: "In view of the fact that racial differences are now recognized as matters of the greatest possible concern in a world organization, in view of the fact that they played so important a part in the Peace Conference and in the delimitation of countries, and in view of the fact that they form so important a consideration in matters of immigration, it is thought that the progress of the world would be advanced by having a definite sub-section of the Health Section."[48] Davenport was one of many who sought this formal internationalization of eugenics. None were successful. The League's secretariat consistently put a diplomatic arm's length between itself and any idea of world eugenics. When Davenport tried again, later in 1920, his rationale rested more specifically on race difference and its significance in world affairs. He linked eugenics this time to peace (resolving race tensions, as he saw it) through immigration regulation and through the constitutions of nations. In this case, the bid to argue eugenics onto the League's agenda through race was foiled by Nitobe Inazō: "In view of the historical fact that races of all colors and grades have freely mingled all through the ages, I cannot share Dr. Davenport's view that the progress of the world would be advanced by accentuating race differences. German scientists under the lead of Gobineau . . . tried to find scientific basis to demonstrate the absolute superiority of the 'Hun.' I hope America will not follow the German example."[49]

Despite Davenport's arguments that eugenics promoted the "cause of the comity of nations and international good," the flagship international organization would not enter it. Yet this did not stop the honorary secretary of the International Eugenics Congress from trying again, this time pressing Dame Rachel Crowdy, then secretary of the Social Questions and Opium Traffic Section of the League of Nations, on whether the Health Section had yet considered a eugenics subsection:

> The fundamental importance of this subject is now fully recognised, and the close relationship between Eugenics and a progressive Health policy necessitates all progress in our knowledge of human heredity, of miscegenation, and vital statistics, being disseminated as rapidly as possible to the Health Services of all countries. The great Powers with their Colonial responsibilities cannot afford to neglect any opportunity of increasing the knowledge of such practical eugenic questions.[50]

For the British, eugenic aspects of population were world affairs due to the global character and obligations of the British Empire and Commonwealth. This idea retained real purchase well into the mid-twentieth century: "Our imperial and world responsibilities impose on us a special obligation to foster the good inborn qualities of our population," the Eugenics Society reported to the UK Royal Commission on Population in 1944.[51] As Karl Ittmann has detailed, British arguments often became "world" arguments on the back of perceived imperatives to pursue imperial population quality and fitness, especially after Robert René Kuczynski's massive studies of colonial population during the 1930s.[52] But in this instance, even when read by countrywoman Dame Rachel Crowdy, the ideas of British eugenicists, like those of the Americans, were rejected in Geneva.[53]

It was not under the logic of race, race-mixing, or immigration regulation, but of infant health and protection, that eugenics came closest to consideration by the League. As a result of a resolution put by the Cuban delegation to the League's governing assembly in 1926, the Health Organization was asked to what extent it could enlarge its scheme of work on the protection of infants toward eugenics. The secretariat's file, originally titled "Protection l'enfant" was struck through to become "l'Eugénisme: questions générale."

D. F. Ramos, who had established the Pan American Eugenics Committee, represented the Cuban Ministry of Health and Welfare and presented eugenics to the League's secretariat as "homiculture," the French-influenced brand of hygiene and improvement.[54] But the recipients remained ever-reluctant: "Avoiding all questions of a purely national character [the Health Organization would consider] only those problems which deserve international consideration."[55] Although eugenics was without doubt considered "international" on any number of other criteria—its expression within a global system of immigration regulation, its international organizations and conferences, its link to differential global fertility rates, its self-definition as "the science for the betterment of the human species"—to the League of Nations, eugenics was clearly associated with national policy, as well as being closely connected with birth control and sterilization, and so part of Roman Catholic objection. Prescott Hall's "world eugenics" was, of course, in the end nationalist eugenics. While this nationalist quality had led the Paris peacemakers to uphold the race criteria of immigration regulation in 1919, the same quality kept eugenics out of subsequent discussion by the League of Nations, though not, as we see in the following chapters, off the agenda of agencies of the UN after World War II.

World Population and Assimilation

Charles Davenport's eugenics involved asserting, investigating, and legislating for human difference. This also made him interested, one might say obsessed, with human sameness, fusion, mixing, and intermarriage. Part of the problem for Davenport was the increasingly verified argument that there was no pure difference of race, nor had there ever been. Even more of a problem for Davenport and all he represented was the interest some scholars—even eugenicists—were showing in policies of assimilation through "racial" intermarriage. A global amalgamation and obliteration of the geographies and biologies of difference was both possible and even desirable, claimed some of his biology colleagues.

This dovetailed with the political tradition of cosmopolitanism that so energized the neo-Malthusians. On the one hand, cosmopolitanism and holism could sit quite comfortably with irreducible and hierarchized difference, as Jan Smuts, who authored both the preamble of the UN charter and South African apartheid, knew better than most.[56] On the other hand, the "one world" discourse also offered the possibility of a promising and progressive human singularity. F. James Dawson, in *Aggression and Population*, pronounced: "Humanity's very long-term destiny in this small world is fusion, which involves intermarriage, between all races."[57] Dawson wrote this in 1946, at a high point of world federalism and when a politics of human identity, not difference, was very much on the international agenda, notwithstanding Smuts. Yet such ideas had a history in the earlier decades of the century, even a history within eugenics, the field otherwise built on purported hierarchies of biological difference.

From the earliest part of the twentieth century, the idea of world overpopulation lent both to a race-based competitive model of the future and to conceptions of humans on the planet emerging from cosmopolitan political traditions. The latter was certainly not free of racialized discourse on human difference and capacity, and colonial discourse on the right of certain populations to dominate and govern. Nonetheless, part of the interwar "retreat of scientific racism" that Elazar Barkan has traced,[58] and the antinationalism and anticolonialism of Mukerjee, Thompson, and to some extent Knibbs, lay in this cosmopolitan desire to think about humans as a species, rather than as racially or nationally divided populations.

One aspect derived from the universal applicability of evolution by natural selection. At the 1912 international eugenics conference in London, Leonard Darwin framed eugenics as his generation's work, which extended "the practically universal acceptance of the principle of evolution in all fields of knowledge in the nineteenth century." This was, for him, the great international achievement of the Victorian period.[59] Once eugenics was accepted as part of a larger evolutionary principle, it would and should be understood to govern humans universally. In hands other than strident patriot Leonard Darwin, this line of inquiry was sometimes used as scientific ground on which eugenics would become not just an international but a cosmopolitan science, applicable to all humans. Legal writer C. E. A. Bedwell pursued this aspect of science and the new world order when he presented "Eugenics in International Affairs" at the 1921 New York meeting. Scientists know, he argued, that national boundaries do not limit research, that there is an "international character of knowledge" that needs to be incorporated. Bedwell approvingly quoted jurist Sir John Macdonell's 1916 essay in the *Eugenics Review*, which raised the possibility that a dispassionate eugenic science might show that "unions between certain races" are possible, even "desirable and propitious." It might find that "certain stocks would be enriched and strengthened," and humans might thus, in his opinion, become "citizens of a better world." For Macdonell writing in the middle of World War I, and for Bedwell writing in its aftermath, "eugenics in international affairs" could potentially lead the way by showing the "unity of humanity" a "rational *jus connubii* as yet undreamed of."[60]

Such statements were certainly not mainstream eugenics. Indeed Bedwell's audience in New York was made up of the architects of the 1924 Immigration Act, whose eyebrows and ire would have been raised by his arguments. This was an alternative tradition of "world eugenics" that derived from the global population problem itself and the cosmopolitanism that was an early part of neo-Malthusian orthodoxy. For example, Edward Isaacson wrote in precisely these terms in 1913: population growth was a problem that stimulated "united action by the whole human race . . . thorough and intelligent co-operation of the whole race to control the forces of Nature for the best good of the whole." This was, in his terms, a "universal consideration."[61]

From the beginning of the twentieth century, then, some eugenicists participated in the politico-scientific project of "species," the biological version of cosmopolitan world citizens, wherein difference might progressively be erased through fusion. Warren Thompson thought that world population pressure might be reduced by shifting Indians to Portuguese-influenced zones, specifically because of the racial integration that would likely take place: the Portuguese "have never shown the same repugnance to the social equality of races that the northern Europeans have." He suggested that in Portuguese Africa, and possibly French Madagascar, there would be little difficulty assimilating the populations. This would simply repeat the intermarriage of the Portuguese and Indians in Goa, where "the two races have become one people."[62] In other global instances, however, such assimilation was imagined genocidally. In the first issue of *Population,* G.H.L.F. Pitt-Rivers also talked about the point "when two races meet," an ecological moment, not so much in terms of biological convergence, as elimination through replacement. "Disequilibrium," he wrote, experimenting with the new ecological vocabulary, would always force a re-equilibrium, either extinction or modification of the "weaker culture" and some kind of adaptation by the stronger. Extinction could be brought about directly by violence, gradually by "substitution through differential birth and survival rates," or by "selective elimination of less adaptable characters," that is natural selection in operation when different races "mix freely."[63] This kind of assimilation was both biopolitical and geopolitical, and often the homogeneity it offered was addressed as a solution in a world where difference was difficult.

Also thinking ecologically, Radhakamal Mukerjee maintained the need for recognition of racial difference, but within his aspirational "oneness of mankind." For him, difference was part of a permanent natural scheme related to environmental and planetary location. "Nature has decreed that human evolution should include a colour scheme which is an essential factor in the adaptation of man to his surroundings, and which is an outward and visible indication of his fitness for life under certain geographical conditions. And the time has come for international legislation and supervision to ratify the judgment of nature."[64] Mukerjee was arguing for irreducible difference to be admitted into the global racial economy, where difference and equality were not just possible, but actively pursued, even enforced. Well might he argue thus, since national legislation had, for several generations already, upheld the inverse of Mukerjee's argument, that racial difference and racial belonging justified racial exclusions. This was the principle endorsed in 1919.

George Knibbs called such statements ill-based "racial and national vanities."[65] He was perplexed by the global questions of population heterogeneity and homogeneity. The world was obviously living with difference, filled with peoples "living in all stages of development," but he also imagined a tendency toward homogeneity that would be biologically produced as absorption. He agreed fairly easily with Pitt-Rivers, as to the indigenous people of his own Australia: "Whenever a people in a more advanced stage become part possessors of the territory of a people in a less advanced stage, the latter either become merged in the former, and pass into their stage of development, or, if not, die out or vanish." Such naturalizing of racial homogeneity presumed a spurious white genetic dominance (that apparently accompanied dominant white civilization). But Knibbs was far less comfortable with presumed racial superiority over Chinese, Japanese, or Indians. In "The Fundamental Elements of the Problem of Population and Migration," published in *Eugenics Review*, Knibbs asked his reader, and indeed himself, whether comprehension of human difference was founded on "baseless prejudice" or an "instinctive recognition of some fundamental incompatibility." If the former, it may and should be corrected by education. Knibbs was cautious about popular and scientific opposition to miscegenation, and the apparent biological problem of "crossing." Much popular opinion "is quite ill-founded," and so scientific work needed to be approached with particular care. Moreover, the nationalist politics of the problem needed actively to be kept out of research, replaced less by a politically neutral approach than an international (by which Knibbs would also mean a pacifist) one: "Doubtless this question ought to be reviewed from a world-wide standpoint."[66] Knibbs carefully distinguished between racial, linguistic, "sociologic," political, and economic axes of difference. He was most concerned about the undesirable and unfit as an economic and physical underclass, sometimes conflated with nonwhiteness, but not necessarily so.[67] This, rather more than a racial axis of human difference per se, is how and why Knibbs the Malthusian, uncomfortable with race-based immigration restriction, was also a firm advocate of world eugenics.

If racial difference had these thinkers in disagreement, they typically aligned smoothly and virtually with one voice, over the issue of "defectives." Thompson might question presumptions of racial superiority and the new vogue of intelligence tests, but he had no quarrel with "arrangements to eliminate those who have been proved defective."[68] Alexander Carr-Saunders problematized "race," but he also had an eliminatory agenda (cessation of reproduction) when it came to the mentally and physically disabled. They were, over generations, to be eliminated as a human type. This was also, without a moment's hesitation, Margaret Sanger's position. In this respect, too, Knibbs was as keen an international eugenicist, as he was an active international neo-Malthusian. Advising the Norwegian Jon Alfred Mjøen on the urgency of race hygiene, Knibbs put all his population expertise together into one package of advice: "The rapidity of the world's increase of population reveals the urgent necessity of the consideration of question of Race-Hygiene." Segregation was necessary for the feeble-minded, epileptics, and the physically and mentally "crippled." Indeed, it should be obligatory for those whose willful behavior foreclosed citizenship: drunkards; habitual criminals; professional beggars, who refuse to work; "and it is desirable for cases of incurable tuberculosis." But race difference? Knibbs paused here, in his otherwise relentless advice to the Norwegian and his Consultative Eugenics Committee. "Crossing between races needs to be immediately studied with an open mind. I am not satisfied that the popular opinion is valid. The question of miscegenation generally ought to be reviewed without delay."[69] This challenge was issued in 1924, and clearly Knibbs was already sensing the shifting ground.

Eugenics and (World) Population Genetics

Throughout the 1920s and 1930s, George Knibbs's discipline—mathematical statistics—met the genetics of Mendel in its second generation. The "population" element of the world population problem was being recast, even as demography was establishing itself professionally and institutionally. In other words, part of what formed a "world population problem" in these years was population genetics—the great but difficult synthesis of Darwinism and Mendelism.

Alexander Carr-Saunders followed up his enormously successful studies on population with *Eugenics*, a small book that appeared in a series edited by Oxford historian H. A. L. Fisher and Julian Huxley. It was the opportunity for him to return to biology, to integrate genetics, and to fill in some detail on the "quality" question that he had nominally put to one side in his original study. Carr-Saunders was worried about the direction eugenics was taking in Britain, he confided to Huxley, pressing his friend to consider succeeding the aging Leonard Darwin as president of the Eugenics Society. "It is of real importance that the Society, and with it eugenics as whole, should not fall into the hands of political propagandists."[70] The science of eugenics—the part played by inheritance in human affairs—was to be clearly distinguished from applied eugenics; a most important point for Carr-Saunders, who always asked: "What do we actually know?"

The science of eugenics concerned the changes in human genetic material from one generation to another, at a population level. Carr-Saunders typically called these "racial changes." "Race," here, was a synonym for genotype, the genetic material of a population. When Carr-Saunders wrote that "proposals for racial improvement are proposals to take our evolution into our own hands," he meant changes to the "germinal constitution" of the species. Some of these genetic changes were brought about via mutations, he explained, but these were relatively unimportant at a population level. Far more significant were "effective fertilities," by which Carr-Saunders meant the relationship between birthrate and infant mortality rate—"children surviving per 100 families." Changing effective fertilities of populations would directly impact the distribution of any particular character between one generation and the next, between one subpopulation and another. This was how quantity and quality were connected; indeed, in this view, it was impossible to understand them as separate factors. Carr-Saunders's first example in this popular book was apposite: the genes that control the character of skin color. If one group, "say the yellow," has a greater effective fertility than another, "say the white," then in the next generation there would be a larger proportion of genes that gave rise to yellow or white characters, "and racial change will have thus been brought about." But here, again, by "race" Carr-Saunders was not referring to "yellow" or "white," but to the fact of population-level alteration in the characters produced by that particular gene. Carr-Saunders deferred political discussion, and instead asked why and how such changes in fertility and mortality rates take place, a problem both "intricate and obscure," and one that obviously exercised him deeply for the rest of his life as a biologist-turned-social scientist. In this instance at least, "skin colour" was an exemplar of a principle—emergent population genetics—rather than the substantive point.[71]

Carr-Saunders sought to explain to his readers the genetic variation that occurred within whole populations and the differences between genotype and phenotype. To what are these differences among persons of the same ancestry due? "The work of Mr. R. A. Fisher has solved this difficulty," he announced.[72] As early as 1911, at the Cambridge University Eugenics Society that Fisher had formed (with John Maynard Keynes as treasurer), the mathematically talented student was bringing together classical Darwinian natural selection with the newer theories of Mendelian inheritance.[73] In a paper published in 1918, and authoritatively in his book *Genetical Theory of Natural Selection* (1930), Fisher rethought the whole of Darwin's theory in terms of genetics.[74] And like so many of the biologists interested in both Mendelism and quantitative study, he saw that these ideas applied to human populations, and might be molded to a particular end: racial improvement. Arguably the most important of the original generation of population geneticists, Fisher had developed this line of inquiry under the patronage of Major Leonard Darwin. As a sixty-year-old, Darwin became President of the Eugenic Education Society in 1909, succeeding Francis Galton,[75] and encouraged Fisher to review many of the biological works that were submitted to the *Eugenics Review*, around two hundred in all. For a substantial amount of time, Fisher worked as a statistician at Rothamsted Agricultural Research Station, under E. J. Russell's directorship, and extensively researched plant variations and published on bacteria in soil. It was in that capacity that he attended the World Population Conference in Geneva, where he disagreed with Pearl's logistic curve and law of population growth.[76] Fisher succeeded Karl Pearson as Galton Chair of Eugenics, in the process changing the subtitle of the *Annals of Eugenics* from "the scientific study of racial problems" to "the genetic study of human populations."[77] The significance of this change was less "race" to "population"—because that was essentially what Pearson already meant. Rather, the key change was the insertion of "genetics." In the 1940s, he directed the Galton Laboratory at the Rothamsted Agricultural Research Station, with a team of geneticists, anthropologists, serologists, and statisticians.[78]

The *Genetical Theory of Natural Selection* was important to demography, as well as population genetics, highlighting the age structure of any given population; age-specific birth and death rates were significant, and different age groups were weighted as having different reproductive value. Alfred Lotka at Johns Hopkins had a similar idea, and demographers based at the London School of Economics—especially Robert René Kuczynski—applied this demographically, critically adding the sex of the differently aged humans: women were more important for reproduction than men.[79] Fisher's ideas were also important to demography, because he discussed the inheritance of fertility itself. Apparently confirming generations of eugenic investigation, he inversely correlated higher birthrate and class. And, venturing well past his own skill set, he suggested that this had led to the decline of civilizations in the past. He concluded his book with controversial ways to counter "the social promotion of infertility."[80] Fisher sought to apply his population genetics politically, but by 1930, his conservative politics were sitting less well with a reforming English eugenics movement than they had in the Leonard Darwin years.

Carr-Saunders saw birth control as critical precisely because of the significance of effective fertilities for population genetics. Eugenics was thus linked to fertility rates in very technical ways—clarified by Fisher's work—which led Carr-Saunders to conclude in his 1935 Galton Lecture, "Eugenics in the Light of Population Trends," that "whether we take long or short views, voluntary parenthood occupies the centre of the field."[81] This population focus of a new eugenics was also being presented in the United States, driven by Frederick Osborn and Frank Lorimer and shaping what they called "social eugenics" as well as "social demography."[82]

Zoologist Wilfred Agar put the basics clearly in a 1937 talk, "The Eugenics Outlook for the Future." The premises of eugenics were first that humans exhibit heritable differences, and second, that "certain types procure more or fewer children than other types." All eugenics proposals must therefore involve alteration of either the birthrate or the death rate of various subpopulations or "types." Agar then became more specific: "Since I imagine no one contemplates doing anything with the death rate except to reduce it to the lowest figure possible, practical eugenics is concerned with birth rates."[83] In fact, by 1937, there was a range of possible and actual methods of managing population-level human difference through death. Agar's Australian colleagues, for example, were in that very year discussing the possible policies for the management of Aboriginal people in the nation's Northern Territory. One option, though "repugnant," was nonetheless raised: a laissez-faire policy whereby Aboriginal people would not be treated medically: Aboriginal women "would become sterilized by gonorrhoea, many would die of disease, and some of starvation." The race would "become extinct within 50 years." The policy that was actively recommended, however, involved birth, not death, a policy of absorbing the Aboriginal into the white population, "and doing so with haste, lest the reverse take place."[84] In Nazi Germany, death, not life, became a "solution" to the kind of problem that Agar put forward. Active nontreatment and active killing of people with disabilities, and later under racial categories of unfitness, characterized that particular nation's "eugenic outlook for the future."

It was in the light of fascist dogma on race and racial purity that Carr-Saunders became involved in a project with Julian Huxley and anthropologist Alfred Cort Haddon: *We Europeans: A Survey of "Racial" Problems.* This was a moment at which the international issue of Jewish refugees from Germany was also exercising the Political and Economic Planning Group, based at the London School of Economics.[85] The driving mission of their joint work was to lay bare the vagueness of the term "race," as well as the expedient use of this vagueness in fascist Europe. Rather than use "race" to mean "genetic" or "germinal," as Carr-Saunders had done in 1926, the book sought to question all uses of the word and to undermine "a vast pseudo-science of 'racial biology.'"[86] The trio, along with the two undeclared Jewish authors, historian of science Charles Singer and ethnologist Charles Seligman,[87] pronounced that despite great and admitted scientific ignorance, they lived in times of increasingly dogmatic "proclamations of certitude" about racial difference. As Elazar Barkan notes, this was a turnaround for Huxley in particular. From a series of articles in the mid-1920s that revealed a frankly Southern position on "the Negro Problem," Huxley's volte face was sharp but committed. Nationalism and Nordicism were the joint targets, not just Hitler but also Madison Grant, connected by a shared presumption of superiority. One could make a shrewd guess at the complexion of the author of *The Passing of the Great Race*, they noted.[88] The idea of "race" had long been unclear, but it was being harnessed to a "violent nationalism"; in the first instance, this book was an intra-European study, as its title makes clear, primarily about the distinctions made between kinds of Europeans. Intermittently, it gestured to worldwide, species-level significance as well.

In undermining the popular and political usage of race, the authors' main point was that nothing in the nature of a "pure race" can be said to exist or ever to have existed. Charles Davenport, hardly the most cutting-edge geneticist at the best of times, was being conceptually, as well as politically and institutionally, retired. There had been "intercrossing for tens of thousands of years." All modern nations were the result of extensive "amalgamation," many waves of migration, constant divergence and convergence, such that the typical "tree" image and metaphor of evolutionary change was not just an inaccurate, but a positively misleading way to represent *Homo sapiens*. One aspect of humanity's uniqueness, vis-à-vis other animals, they put down to the tendency, or even instinct, to constantly migrate. This, they claimed, gave rise to a species vastly more differentiated than other organisms, both within its own "geographical races" (better called subspecies, they suggested), and between them.[89] Partly in recognition of the significance of continental migration, Carr-Saunders was brought on board to write the final chapter of *We Europeans*, "Europeans Overseas."

Famously, Huxley, Haddon, and Carr-Saunders argued for the replacement of "race" with "ethnic group." This presumed the more popular usage of "race" as "national" or "linguistic" group that was especially problematic for them, not the usage of "race" that Carr-Saunders had presented in his earlier *Eugenics:* race as the genotype of the species. But *We Europeans* was not in the least a book that sought to replace biological thought with cultural thought. It did not seek to render human difference a nonbiological matter, even though the authors stressed that social and economic factors were typically far more important. Huxley's and Carr-Saunders's biological and zoological training trumped cultural anthropologist Haddon's expertise in this respect. The book was written in "a new biological era," so defined because of the post-1900 rediscovery, dissemination, verification, and extension of Mendel's laws of inheritance. Mendelian genetics offered great clarity about what was known, what could be known, and possibly most importantly, what remained unknown, with respect to the biology of human difference. Genes mattered more than ever, but in a completely different way.

First, after Mendel, it became clear that "there is no such thing as blending," genetic recombinations do not disappear gradually after "crossing." Unlike the earlier Darwinians, who expected blending and a diminution of variation, the Mendelian geneticists expected that human groups would possess a large range of internal variation. There are geographical trends in these variations, the authors quickly qualified, but these trends can only be identified by statistical methods at population levels. Population genetics shaped this picture. Second, there were connecting links even between the most apparently different of the geographically differentiated subspecies. That is, they explained, there is no pure difference at a population level.

This had always been the case, but the connections of the modern world augmented this variation and gradation.[90]

With the great synthesizer Julian Huxley taking the conceptual lead, the book stressed again and again that the study of genes and the study of "characters" that genes give rise to are comprehensible and meaningful only at a population level. The method of Karl Pearson's biometrics is an important corrective to the currently grossly inadequate data collected by field anthropologists, he wrote. The only possible way to retain a scientifically viable biological usage of "race," would be through quantitative measurement of all exhibited characters or actual genes. Either way, it would only be by quantitative study—population genetics—that differences could possibly be identified and defined.[91] And the early signs of this project were looking inconclusive, they stressed. The book included a series of world maps that displayed the global distribution of various standard characteristics of "race." The distributions did not correlate, from one characteristic to another, one map to another. This was the point, underscoring the growing meaninglessness of the idea of genetic racial difference. This was a key moment, of course, for the "retreat of scientific racism." Haddon was personally retreating from the whole endeavor, "tired of the race concept," he confessed to Huxley in 1936, when asked to deliver the Huxley Lecture.[92] But "retreat" was not a sudden turnaround, as Elazar Barkan shows.[93]

Race science and eugenics never aligned perfectly or predictably. If Huxley utilized the new Mendelian genetics to argue *against* any straightforward global distinctions of race, but *for* the usefulness of eugenics, the leading population geneticist of the era, Ronald Fisher, was doing the reverse. Withdrawing from eugenics, Fisher retained his belief in "innate capacity for intellectual and emotional development," as he later put it in his dissenting position on the UNESCO Statement of Race. At a time when Huxley was urging a world population conference and creating UNESCO with both eugenics *and* antiracism as its core values, Fisher was also expanding his genetics to the international stage. For Fisher, the key "practical international problem is that of learning to share the resources of this planet amicably with persons of materially different nature, and that this problem is being obscured by entirely well intentioned efforts to minimize the real differences that exist."[94] Fisher gradually distanced himself from British eugenics because of the turn to a focus on birth control, because he thought the society was becoming unscientific, and because of the reforming and liberal politics that were dominating under the strong leadership of Carlos Paton Blacker. For his part, Blacker had long understood eugenics and birth control to be linked, both as solutions to and because of world-level population growth. By 1942, when Fisher resigned from the Eugenics Society, Blacker was insisting even more strongly on "global eugenics" as world population control, but this was no sudden postwar turn. Rather, world population was part of eugenics from the earliest decades of the twentieth century, and conversely, eugenics was part of Malthusian population control.

Most prominently, the post–World War II statements on race were set in train by claims put forward in the so-called Geneticists' Manifesto. This was a massively influential joint statement by some of the world's leading biologists, led by the British left-wing circle: F. A. E. Crew, Julian Huxley, J. B. S. Haldane, Lancelot Hogben, along with American Hermann Muller, who had been colleague of Huxley's at Rice University, Texas, just before the outbreak of World War I.[95] They were all gathered in Edinburgh, in the high summer of 1939, for the Seventh International Congress on Genetics, hosted by Crew and his department of animal genetics. But the six hundred attendees were reading the daily papers as much as they were reading one another's abstracts. On August 30, the congress ended and the war began. The Poles were defending Warsaw against the Germans when the editors of *Nature* published this remarkable document on September 16, 1939. Those who had mobilized so strongly after World War I to declare the world politics of science gathered again to press science into the service of "mankind at large."

The manifesto stands as one of the world's most important statements about genetics and race, and it was to be an intellectual battleground after the war was over. The extent to which its signatories were or were not "eugenicists" often monopolizes discussion. And it is typically framed in and by the immediate context of the war: the manifesto's key significance is understood to be its implicit opposition to Nazi race theories.[96] The extent to which the Geneticists' Manifesto was also about world population is barely recognized at all. In fact, most of the document only makes sense in the light of broadly Malthusian argument for birth control, notwithstanding the Marxism of some of its key signatories.

The biologists were replying to the Washington's Science Service's question: "How could the world's population be improved most effectively genetically?" The scientific respondents showed their collective left-wing hand immediately. Economic and political leveling needed to be implemented before any other kind of "improvement" can be appropriately implemented:

"There can be no valid basis for estimating and comparing the intrinsic worth of different individuals, without economic and social conditions which provide approximately equal opportunities for all members of society instead of stratifying them from birth into classes with widely different privileges." And the second great hindrance to world population improvement, they insisted, was ongoing racial prejudice based on unscientific doctrines that presumed good or bad genes to be the monopoly of particular peoples. That idea needed eliminating, they said. To do so required "some effective sort of federation of the whole world, based on the common interests of all its peoples."[97] This was signature Huxley. But it also signaled the socialist uptake of Malthusianism and world federation after the style of H. G. Wells. This was the tradition that appealed to Haldane and Hogben, for example. Here, some of the world's leading biologists were reiterating the old Malthusian interest in federation, the cosmopolitics that had never quite materialized within the League of Nations. But cosmopolitics was on the cusp of finding an institutional home, if a fleeting one, in UNESCO under Julian Huxley.

The Malthusian lineage is quite plain in the Geneticists' Manifesto. Their answer to how "the world's population could be improved" was birth control. The biologists argued that the improvement of future generations required the economic security of parents, that women should be offered "special protection" so that reproduction would not interfere with their larger social participation. It called directly for the legalization, universal dissemination, and scientific investigation of fertility regulation: voluntary temporary or permanent sterilization, contraception, abortion ("as a third line of defence"), control of fertility and the "sexual cycle," and artificial insemination. All this would be "always voluntary," they insisted, in what was to become the important pattern for postwar eugenics. The famous Geneticists' Manifesto, then, was also a manifesto for world-scale birth control. Indeed given that the descriptor "Geneticists' Manifesto" was only ever retrospectively applied, it might well be retitled the "Malthusians' Manifesto."

After World War I, eugenicists had recognized that the past, present, and future of the human species, its difference or its sameness, was playing out with deep political consequences in the new world order and in the "new biological era" that Mendel's laws had ushered in.[98] The politics of this global population scene were highly unpredictable: Anglophone advocates of eugenics could be, and were, racist geneticists, antiracist and cosmopolitan geneticists, or anticolonial social scientists who nonetheless ignored indigenous people. Eugenicists could be entitled and benevolent white colonials or Indian anticolonial nationalists. Many, like Davenport, were strident nationalists. Perhaps more sustained in influence and argument were the "internationalists" of one shade or another. They were the intellectual survivors into the next new world, when the events of World War II foregrounded the great question of human sameness and difference—at once raw and immensely complex. Human multiplication, human extermination, and the very concepts of "race" and biological "assimilation" were up for questioning.

Part IV

Between One World and Three Worlds, 1940s to 1968

10

Food and Freedom

A New World of Plenty?

> Those who have never known hunger are rightly interested in political freedom. The ill-fed masses are much more interested in the freedom from want promised in that forgotten document "The Atlantic Charter."
>
> JOHN BOYD ORR, *SOIL FERTILITY* (1948)

In 1943, agriculturalists, economists, physicians, and international policy makers met in Hot Springs, Virginia, a gathering that pre-empted what was to become the first agency of the new United Nations (UN): the Food and Agriculture Organization (FAO). The sinister backdrop was devastating famine in Bengal and in parts of China and millions hungry in the Soviet Union.[1] A world food crisis was declared, and food and population were immediately linked in grand plans for the postwar era. This manifested as national politics as much as international politics over the turbulent late 1940s. As the British withdrew from the subcontinent, independent governments in India, Pakistan, and Ceylon put food, population, and agricultural reform at the center of economic and political planning. In civil-war China, "Anti-Hunger" became one political rallying point.[2] In Japan, the occupying U.S. forces, as well as the new Japanese government, puzzled over demographic projections for the country in relation to food production: Was a transition to low birthrates and low mortality rates already taking place? And if so, how much food would be needed?

As the world war turned into the Cold War, Europe was also gripped by a postwar food crisis, its solution seen by many to be the core imperative of political and social recovery. John Boyd Orr, director-general of the new FAO, wrote about his experiences in Prague, Autumn 1946. He reported bad harvests reducing diet to starvation levels, hunger and discontent contributing to the downfall of the government. Meanwhile, the Greek government had just requested the FAO to assist in reconstruction and development, specifically to address the "impoverishment of land" where forests, it was claimed, had been reduced from 60 percent to 5 percent of the total area.[3] Like many others, Boyd Orr was watching Communist and anti-Communist dynamics in Europe very carefully indeed. This was part of his new brief, not least since U.S. initiatives for food aid were quickly escalating with the Marshall Plan, later extending into its Food for Peace program.

Food policy increasingly accompanied U.S. bids for simultaneous world peace and global domination, key to its pursuit of a stable non-Communist world future.[4] Actual famine, and fear of the political ramifications of threatened famine, drove vast amounts of research, laboratory, and field development of agricultural sciences. New high-yielding varieties of wheat, maize, and rice; irrigation systems; pesticides; and herbicides became the "green revolution," a descriptor first used in 1968, but nominating technological developments in place from the 1940s. In Mexico, India, the Philippines, and beyond, investments by national governments, the Rockefeller and Ford foundations, and the new World Bank manifested as a doubling of cereal production.[5] For the hungry, this was about food, while for statesmen it was about food security, the political stability that food both brought and bought. For ecologists and Malthusians, it was unsustainable. Behind the food problem and its accompanying anti-Communist politics, was major discussion about global resources and energy needs. What could be produced from the earth, and was it enough?

It was in this context that population control eventually became a formal element of U.S. foreign policy, part of Cold War attempts to contain the spread of Communism.[6] Food/political security was the ends for which birth control was the means. As Mohan Rao correctly observes, the work of Kingsley Davis and Frank Notestein on demographic transition eventually constituted a Cold War resistance to the defection of the Third World to the Communist Second World. The "Third World became the arena of this competition" between the U.S. and the Soviet Union. Indeed, Kingsley Davis himself described the nonaligned third of the world as a prize to be won in the struggle between Communist and "free" polities.[7] This prize might be won through food, ostensibly minimizing social and political unrest, the new manifestation of an idea already strongly established in the 1920s. When some of America's richest businessmen wrote to John D. Rockefeller III in 1954, it was all patent: "We are not primarily interested in the sociological or humanitarian aspects of birth control. We *are* interested in the use which the Communists make of hungry people in their drive to conquer the earth."[8] To expose population control as political in this way is really to do nothing more than repeat what the key actors were themselves saying, publicly and explicitly. This is all the well-known postwar history in which world population and demographic transition became part of U.S. foreign policy and deeply part of the institutions and logic of "development" and "modernization."

While population thought in the twentieth century was a strong element in the history of a new global imaginary of "one world," it was foundational also to the invention of three worlds. In 1952, it was French demographer Alfred Sauvy who re-invented the relations between East and West as First and Third Worlds, recasting and recalling the politics of bread from another era: "The Third World, like the Third Estate, wants to become something too," he wrote in "Trois Mondes, Une Planète."[9] It was the demographic profile that differentiated First World and Third World, the East and the West, and later the global "north" and "south." The very idea of developed and developing worlds built on apparent demographic transitions from high fertility and mortality to low fertility and mortality.[10] The old League of Low Birth Rate Nations was on its way to becoming the First World, its identity increasingly, but not originally, shaped by anti-Communism.

Food and hunger were also rendered global problems by those whose politics was opposed to the U.S.-led anti-Communism, and in ways that recognized a recent colonial and anticolonial past as much as the Communist and anti-Communist present. In other words, while global food, hunger, and population were key elements in the discourse of anti-Communism, they were key elements also to postwar anticolonialism. Brazilian doctor, anti-Malthusian, and sometime chair of the FAO's Executive Council, Josué de Castro argued that the "geography of hunger" matched the global geography of colonial rule.[11] But concern over a lingering colonial geopolitics of hunger was not aligned solely with de Castro–style opposition to Malthusianism. Demographer Sripati Chandrasekhar agreed that "the world's great areas of endemic hunger are exactly the colonial areas." Deeply driven by Malthusian conviction, this was also part of his postcolonial critique: "The myth of Western military and social invulnerability has been exploded and the day of Western domination is over."[12] Thus the geopolitics of a "free world," a Communist world, and a "Third World" that did not have enough to eat, emerged not just as a U.S.-led neocolonialism but also as a critical geography of the global north and south. The capacity to achieve "freedom from hunger" was related to global calculations of food production, consumption, and distribution, but also more surprisingly it was tied to calls for "freedom to move," rehearsing interwar conversations for the next new world.

Population and the United Nations

When war broke out, the business of the League's freshly-minted Demographic Committee under Alexander Loveday and Alexander Carr-Saunders was transferred from the Palais des Nations in Geneva to the Princeton Office of Population Research, to be headed by U.S. demographer Frank Notestein. This was part of the wartime transfer of the entire Economic, Financial and Transit Organisation.[13] Even though the League generally eschewed population studies, once this move was made, several major population works were produced at the eleventh hour under its imprimatur. Notable was *The Future Population of Europe and the Soviet Union*, authored by Notestein and his team.[14] And it was essentially this group—propelled by the juggernaut of United States demography—that formed the core of the UN's new Population Division, part of the Department of Social Affairs, from 1946.[15]

Trained as an economist at Cornell University, Notestein had researched at the Milbank Memorial Fund during the 1930s, focusing largely on differential fertility. The Office of Population Research was itself financed by that fund. While the League had actively sidestepped the "reproductive health" implications of population (because of eugenics, birth control, and pronatalism) in favor of the economics of population (surplus labor, resources, raw materials), the institutional relocation to Princeton served to move "population" into an intellectual domain more connected to health, fertility, and the emerging study of reproductive choices and conduct.[16] At the same time, it was squarely connected to international relations, since the Office of Population Research itself was located within the Woodrow Wilson School of Public and International Affairs. In this sense, the Office of Population Research extended the interwar work on population, territory, density, and security.[17] Notestein himself addressed food supply and food security as a standard element of his work, contributing to the publication *Food for the World*, for example.[18] Princeton's Office of Population Research had a relatively open hand in terms of research agendas: they did not have to avoid the study of birth control, for example. But the new United Nations Division of

Population, also led by Notestein, did not have this freedom. In this curious institutional arrangement, the Population Division was straitened, required to focus on compiling and processing population data only and to strictly avoid the implementation or advocacy of population limitation.[19] It consequently experienced difficulty distinguishing itself from the Statistical Commission of the UN (under the Economic Affairs Department).

The Population Division itself was not the only branch of the new UN that dealt with population questions, however. With an active Julian Huxley at the helm, world population came under both the "science" and the "education" brief of United Nations Educational, Scientific and Cultural Organization (UNESCO). He argued in his manifesto for UNESCO for a decrease of the global population growth rate through sponsored educational programs of fertility control. And he immediately began pressing for a UN-sponsored world population conference. The still controversial birth control argument fell into abeyance, and the conference did not materialize until 1954 (Rome).[20] But UNESCO successfully incorporated other elements of the population question, inherited directly from its predecessor, the International Institute of Intellectual Cooperation, and its 1937 meetings on Population and Peace. This is where the "tension" vocabulary that tended to dominate UNESCO's early discussion of population came from. At the First General Conference of UNESCO held in 1946, a group for the Study of Tensions Crucial to Peace considered problems "relating to population."[21] A further meeting held in Beirut in 1948 resolved to promote enquiries into "population problems affecting international understanding, including the cultural assimilation of immigrants."[22] And by 1951, the Social Tensions project within UNESCO's Social Sciences Program settled on four platforms: the studies of "tensions" arising from population questions fell alongside studies of national character, studies of technology, and studies of racial and ethnic "tensions."[23] Continuing the interwar logic, then, UNESCO took on the study of cultural, scientific, and educational aspects of "the pressure of populations" through, and as part of, its mandate to contribute to peace.

The FAO also dealt with population, not least from its particular concern with the welfare of rural, that is to say agricultural, populations. FAO's brief covered "the internationalism of population and food supply," as well as the assembling and preparation of population data.[24] Indeed it was through and within FAO that population was discussed most politically in the earliest years of the United Nations, largely licensed by the politics and experience of its first director-general, John Boyd Orr. He was beginning to argue that increased levels of nutrition in and of itself would work to diminish fertility rates.

Boyd Orr had been invited to Washington in 1942 to take part in discussions on a world food plan led by Frank Boudreau, previously director of the Health Section of the League of Nations and subsequently director of the Milbank Memorial Fund.[25] This gathering was the immediate precursor to the May 1943 meeting at Hot Springs, Virginia, where John Boyd Orr was joined by Wallace Ruddell Aykroyd representing India, only too aware that in his absence millions were suffering in the Bengal famine. International delegates were asked to consider ways of intensifying world agriculture to produce "food for all" at the end of hostilities.[26] Some, including Boyd Orr presented expressly political findings, including the statement that the first cause of hunger and malnutrition was poverty, and that "resources and knowledge are sufficient to enable the food required for the health of the whole world population to be produced."[27] A circumscribed FAO emerged from these meetings, however, and Boyd Orr's directorship was both a reluctant appointment and a grudging acceptance: his world food plan was seriously diminished and FAO itself reduced to an agency that would gather data, publish documents, and promote research, much as the International Institute of Agriculture in Rome had been doing since 1906 and much as the Population Division were asked to do on the demographic front.[28] Nonetheless, Boyd Orr pressed on, appointing Aykroyd with his Indian experience as director of the Nutrition Division.[29]

The initial efforts of the FAO were to gather all data on the current global food situation, which revealed a much more acute situation than many governments thought.[30] The fundamental question for John Boyd Orr was: "How much food does the world need?" That depended on how many people there were, how much and precisely what they would, could, and should eat. At the end of the war, Boyd Orr thought it would take another five years for the world to produce grain at prewar levels; nothing happened quickly and much was weather dependent, he warned. By contrast, populations could and did increase rapidly: globally about 20 million a year, he estimated in 1945. "It will go on increasing until the peoples in countries with a high birth rate have the standard of living and the education that we enjoy. Then it will begin to fall." But that fall would not occur for about twenty-five years, he predicted in a talk entitled "The Road Through Plenty to Peace." In the meantime, there would be "another 500,000,000 people in the world."[31] His prediction, it turned out, was remarkably accurate, the slowing of world growth rates beginning in 1963.

Boyd Orr placed this whole demographic circumstance within a long modern history, wherein the most important change was less the shifting geography of population growth from Europe to Asia, and more the great global constriction of land. Accelerating population growth is not a new phenomenon, he was quick to explain, but European increase in the past had been enabled by "exploitation of virgin lands in North and South America and Australasia." Can the earth feed its teeming millions? Yes, he answered, but only if governments, the Economic and Social Council of the UN, and the World Bank cooperated on a (his) world food plan. The British Commonwealth of Nations might lead the way—"and among that I include the great new Dominions of Asia"—by creating a commonwealth and colonial food production plan. With its unrivaled experience in international affairs, he thought Britain could "lead the nations in the march to the new world of plenty which science has made possible."[32] In this way, although the UN recognized "population" by virtue of its new division, the long tradition of politicizing population pressure in global terms was expressed rather more openly via the food problem in its other sectors and specialized agencies.

Boyd Orr and Julian Huxley: Food and People

Respectively leading FAO and UNESCO, John Boyd Orr and Julian Huxley were by now elder scientific statesmen in world politics. Huxley was thinking about world problems from his new UNESCO office in the Paris Hotel Majestic, the same hotel that had hosted delegates at the Peace Conference in 1919. Then he had been thirty-two; now he was fifty-eight. Boyd Orr was now sixty-five. They would seem to suit the new climate of global possibilities perfectly, but they were too impatient, too left-wing, and too outspoken for their new jobs. Their terms at the top of the UN were short. In many ways, neither remained aspirational enough, seeking to put ambitious plans into operation too quickly.

Huxley had supported the need to bring "Life under Control" since the early 1920s, and he moved much more squarely into the space of world population control after World War II, actively forging its terms and reception in the international public sphere. His early biological training in ecology, his Malthusian–eugenic interest in pursuing human quality, and his conservationist politics and concern about the effects on the natural world of human population growth, were all of a piece. While Huxley conceded that new land needed to be brought into cultivation, there still would not be enough food, and so "we must have birth control."[33] Most importantly, he thought, the UN must take an official interest.[34] "I would say categorically that the control of population, birth-control applied on a large scale, is a prerequisite for anything that you can call progress and advance in human evolution . . . We want the United Nations to have a population policy."[35] The newly co-operating agencies of the UN, possibly with the assembly itself, Huxley thought, could and even should implement such policy alongside a world food policy based on structures of equitable distribution. In his years with UNESCO, he traveled to member countries constantly, appraising among other factors the links between illiteracy and high fertility.[36] Huxley declared that "population is really a world problem, involving potentialities of good or evil for the whole human species."[37] And although such statements have been received by historians as novel, situating Huxley "ahead of his time,"[38] they were hardly visionary for his wider intellectual milieu. He was, rather, developing links between global population studies, ecology, and internationalism that were already several generations old. He was reworking *The Science of Life.*

In the late 1940s, it was still controversial for a UN agency to publicly support birth control. It was to be Huxley's institutional undoing, but not before he could set several programs in train. One outcome of his long connection with Boyd Orr was "Food and People," a joint UNESCO/FAO project.[39] This was a public education campaign that commissioned a series of pamphlets by well-known writers. Any number of social and natural scientists were invited to contribute their studies, translated into French, Spanish, Dutch, Portuguese, and German. The titles in the series distilled many of the elements and implications of the world population question. Anthropologist Margaret Mead was commissioned to write *The Family's Food.* Sociologist and later chair of UNESCO's social science section Alva Myrdal produced *Are We Too Many?*[40] Charles E. Kellogg, who had been director of the Soil Survey Division of the U.S. Department of Agriculture (USDA), wrote of the "vital relationships among people and countries" in his contribution, *Need We Go Hungry?*[41] He explained: "Population trends and food production are no longer the exclusively specialist domains of the demographer and the farmer, but have been made the object of international planning." There might not be much empty land left, but new technologies of agriculture, food storage, and transport could match or even surpass the needs of a growing population. It was all an exciting project, he concluded, contributing to "the prospect of a world of abundance at peace."[42] E. John Russell also wrote for the series, having directed both the Rothamsted Agricultural Research Station and the Imperial Bureau of Soil Science, and having traveled extensively in India during 1936 and 1937 with the Imperial Council of

Agricultural Research.[43] After half a century of writing about soil and its fertility,[44] he offered an optimistic vision of future earthly produce. Recalling William Crookes, Russell was entirely confident that large-scale irrigation, fertilizer technologies, and changing farming practices could adequately increase global food production. His ideas were influential, picked up by high-level Rockefeller advisors as they pressed forward with their own innovations on new plant breeds.[45]

Both Huxley and Boyd Orr were serious about mass communication and education. Before the war and during it they had cooperated on various projects, engaging with the public through print, film, and radio. Having learned a good deal about the popularization of difficult ideas from his early and intense writing experience with H. G. Wells,[46] Huxley had narrated a documentary film, *Enough to Eat* (1936) that had been prompted by Boyd Orr's report *Food, Health and Income* and the League of Nations' mid-1930s nutrition reports. Together, Huxley and Boyd Orr edited a series of popular books, Targets for Tomorrow, in which Boyd Orr's own contribution was *Food and the People.*[47] In the 1940s, they both sat on Britain's Africa Research Survey Committee, transferring domestic British programs and expertise on food, population, and agriculture, to colonial ventures and thence to the ideas and infrastructure of international development.[48]

The two men shared a vision of world population growth as a problem. But they differed on the question of how fertility rates might be affected by standards of living, the core question of then-current demographic transition theories. Boyd Orr thought that raising standards of living would lead to fertility control, and that nutrition and enlarged agricultural production was what simultaneously indexed and triggered those higher standards.[49] He was consistently optimistic about agricultural revolution in the high modern era. Huxley, by contrast, was using his new position to argue more forcefully for active family-planning policies in the first instance. Through discussion of food and nutrition, population growth could be explicitly discussed by UN agencies. Through discussion on sex and reproduction, it could not.[50] Huxley chose the particular topic, "Food and People," because he thought it could be discussed intelligently at all levels; popular, semipopular, and academic. The jacket of the pamphlets featured a globe, a fork, and the UNESCO logo, and it was premised on the fact that "world population is growing at a rate of over 20,000,0000 a year."[51] Science was the answer, unsurprisingly, because UNESCO was the agency for the advocacy of science and technology.

Arguably the most influential contributor to the UNESCO Food and People series was Harvard-trained and Columbia-based sociologist Kingsley Davis, central to the development of demographic transition model launched in his important 1945 article on world population in transition.[52] A double issue in the Food and People series titled *People on the Move* included Kingsley's section, "Agriculture and Poverty," alongside economist Julius Isaac's section "Migration and Food."[53] The framework within which to comprehend the problem of food and people was still the scarcity of usable land. Davis, for example, depicted the world in a table titled "Land, Men and Food" (figure 10.1). Here, an "international unit"—a quantity of goods and services exchangeable for one dollar—measured productivity specifically in relation to density: "males in agriculture per square kilometre" was the significant international unit that had to be reckoned, seriously overriding the agricultural productivity of women's labor in many regions.[54] This kind of table began to displace the density maps of the 1920s and 1930s.

At this point, for Davis, total net growth somewhere in the economic and ecological system was perceived to be not just possible but essential; the aim was to enable an increase in the standard of living of poor countries without lowering the standard of rich countries. This was not the kind of redistribution of either wealth or food with which Boyd Orr would have been comfortable. And Davis's propositions were a conceptual and political mile from the more radical change of economic systems that Josué de Castro was then imagining in his *Geopolitics of Hunger.* Nonetheless, this study reveals the extent to which a sociologist so tightly associated with reproductive conduct and fertility transition was actively reprising the geopolitics of world population that had coalesced intellectually in the interwar years.

In 1950 when this particular small book was produced, world population policy of any significance was still speculation. The lack of a world plan troubled Davis deeply. He despaired also that national governments tended to implement public health policies that lowered death rates, and yet were reluctant to implement accompanying policy on birthrates, with the significant exception of India. The result was overpopulation that would lead to "widespread wars, famine and epidemics": a "holocaust," he wrote provocatively, that would reduce population violently. Better to avoid this and do it all another way. Like so many after the war, Davis agreed that the earth could support more people, especially with the prospects of atomic energy, solar energy, and tidal energy. But there were limits. The rate of growth that he calculated as 0.75 percent per year would double the world's population every ninety-two years. With a twist on the "standing room" idea, Davis

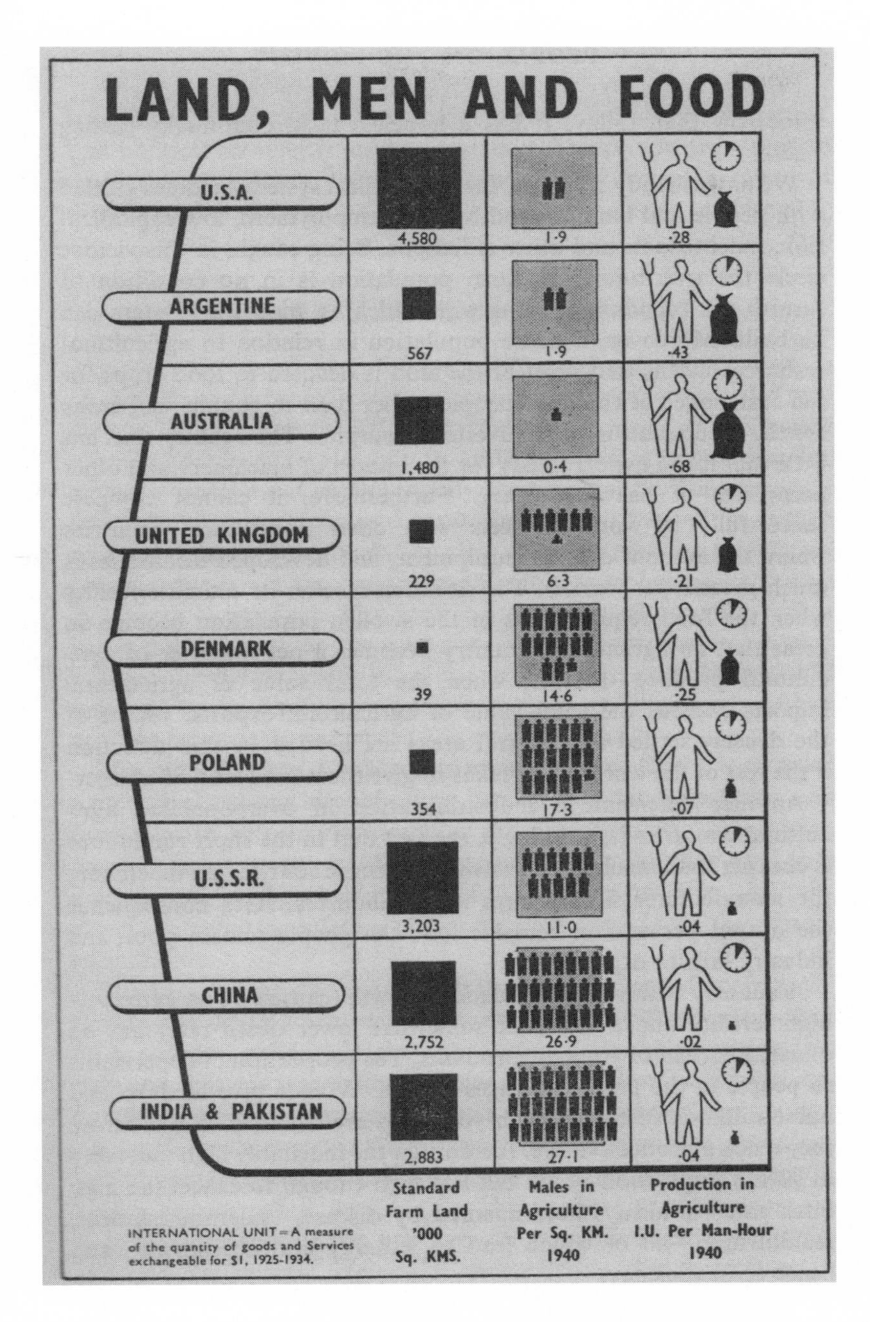

FIGURE 10.1 Density in the Cold War. The author of demographic transition was at least as interested in density as fertility rates. (From Kingsley Davis, "People and Agriculture," in Kingsley Davis and Julius Isaac, *People on the Move* [London: Bureau of Current Affairs, 1950], 13. Reproduced with kind permission, the United Nations.)

imagined all those humans having eaten the earth itself, all resources literally consumed. "Before many more centuries the number would be so large that all the substance of the earth would be incorporated in the bodies of human beings." In all, industrialization, agricultural reform, and birth control should be considered concurrently, "a campaign on all fronts."[55]

Living Earth: Global Political Ecology

In 1949, Aldous Huxley, acclaimed and well known for *Brave New World* (1932), *Eyeless in Gaza* (1936), and *Ends and Means* (1937), was living in California. He espoused vegetarianism but not yet psychedelic drugs, and had just published his latest collection of essays, *Science, Liberty and Peace* (1946).[56] His brother Julian invited him to write for the UNESCO Food and People series, and he obliged with *The Double Crisis*, a catastrophic response to Russell's optimistic vision. It was about soil erosion. "Insidiously destructive," is how Aldous Huxley described the erosion problem, even "annihilating." The younger Huxley replayed ideas put forward a generation earlier by Edward East in *Mankind at the Crossroads*:

> At the present time, our planet supports a little less than 2,250,000,000 human beings, and the area of food-producing land is in the neighborhood of 4,000,000,000 acres. It has been calculated that two and a half acres of land are needed to provide a human being with a diet which nutritionists would regard as adequate. Actually, in order to bring all the people in the world who are at a very low nutritional level up to even a modest level of adequacy within the next 25 years, the prewar food level would have to be doubled.

But what about effects on land and consequent effects on people? Soil erosion and desertification signaled lifelessness, an encroaching global sterility.[57] Indeed, Boyd Orr announced in 1948 that erosion was the greatest threat to civilization, "the wasting basis of human society." Fertile areas were daily reduced, he declared, because of population increase and greater densities.[58] Soil erosion and resulting food insecurity were the problems for which population control was increasingly an answer. This was how "family planning" came to be so closely linked to "conservation," a connection derived from the 1920s and 1930s, intensified in the postwar years and then again in the Cold War. The latter context made it all potentially apocalyptic:

atomic warfare can destroy one particular civilization, wrote Aldous Huxley, but soil erosion can put an end to the very possibility of any civilization. What's going on at the global biological and ecological levels, he asked?[59]

A suite of experts answered him. During 1948 and 1949, a spectacular series of studies raised the ecological stakes and drove population forward as part of the environmental history of the Cold War.[60] Fairfield Osborn's *Our Plundered Planet* (1948), William Vogt's *Road to Survival* (1948), and Paul Sears's *The Living Landscape* (1949) accompanied these postwar calls to action on soil, resources, and population.[61] Just as the pitch had risen in the peace after World War I, so catastrophic projections about life on Earth heightened after the next "world" war, with its new peace so tensely governed by the image and the threat of the atomic bomb. Yet by that point ecology had transformed from a fledgling science—the energizing conversation between Charles Elton and Alexander Carr-Saunders on the shores of Spitsbergen, where they put human population and nitrogen cycles together—to a complex of approaches, methods, and schools. The idea of ecology had also generated a sophisticated and increasingly institutionalized political agenda, backed by serious U.S. money and connections. This was, for the United States, *the* Malthusian moment, as Thomas Robertson has so richly demonstrated.[62]

Botanist Paul Sears's 1948 lecture to the Ecological Society of America comprehended Earth as organic and dynamic. Charles Darwin's earthworms and Daniel Hall's soil-based bacteria came together for Sears as a "living landscape." Tracing ideas from Ernst Haeckel through French microbiologist Louis Pasteur, for whom microorganisms were part of the living activity of soil, Sears claimed that one must now speak of "soil as a process."[63] It was far more than a mechanical–chemical entity. It actively integrated the organic and inorganic, the lifeless and the living, and "forces atmospheric and geological." For Sears, roots, worms, fungi, and bacteria were not just living *in* the soil, they were to be considered part of the soil.[64]

The very title "living landscape" pointed to the ways in which soil was both "bio" and "geo." One postwar writer summarized: "Fertility in human beings, as in all other forms of life, originates from and is directly related to contact with the earth."[65] Karl Haushofer would have agreed, were he not dead by his own hands in March 1946 after informal interrogation for war crimes. So would Haushofer's geopolitical predecessors, Friedrich Ratzel and Rudolf Kjellén, for whom the state was powerfully organic, embodying the vital relationship between territory, soil, and people. Put another way, if Paul Sears's landscape was living, so was *Raum;* not just area, but territory—soil—organically connected to (a particular) people. The linked intellectual genealogies between "living landscape" and "living space" are clear, though the national and global spaces and politics at issue differed: Paul Sears and Karl Haushofer shared a history of Darwinian-derived ecology, and thereby a Malthusian project of analyzing the global political "struggle for room and food."

Also in 1948, Yale zoologist Evelyn Hutchinson wrote about global resources and global waste. "Looking at man from a strictly geochemical standpoint, his most striking character is that he demands so much." And he wastes so much, added Hutchinson, reiterating Justus Liebig's nineteenth-century complaint that had made its way into Daniel Hall's Rothamsted work: "A very large quantity of a non-combustible, useful material is fated to be carried, either in solution or as sediment, into the sea." The crisis came when the population factor entered the equation. "The process is continuously increasing in intensity, as populations expand."[66] In 1948, Hutchinson was reworking an old idea for new times. An associate of Vladimir Vernadsky's son at Yale, he titled his influential essay "On Living in the Biosphere." It was via Hutchinson that the concept reached a wide international audience, culminating in UNESCO's Man and the Biosphere Programmes of the early 1970s. More locally, Hutchinson was part of the advisory group that established a bold new conservation graduate program at Yale. Funded by the Conservation Foundation, Paul Sears was appointed the program's director. The circle connected through wealthy Fairfield Osborn, cousin of eugenics leader Frederick Osborn (who was behind Princeton's Office of Population Research) and beneficiary of railway-derived New York family money. He established the Conservation Foundation in 1948.[67] "Conservation" had been part of U.S. statecraft for some time as we have seen, not least in the Soil Conservation Service, but with Osborn's backing it was turning into something else; a popularized global political ecology for the postwar world. Fairfield Osborn produced *Our Plundered Planet* (1948) followed by *The Limits of the Earth* (1953).

Historians often write about these American ecologists as precursors to, or the originary point of, the population bomb literature of the 1960s.[68] Instead, they should be seen as inheritors of the generations-old spatial history of Malthusianism and of the long-standing sciences that so closely observed the geochemical connections between soil, atmosphere, plants, and animals, including humans. In other words, this famous moment in twentieth-century environmental history is perhaps better understood by connecting it backward rather than forward in time. The premise of

Osborn's *Our Plundered Planet* is more squarely interwar geopolitics than postwar environmentalism. Human civilization had permeated every area of the earth's surface, he wrote, such that "there are no fresh lands anywhere. Never before in man's history has this been the case." The geopolitical argument that had first governed his father's generation's discussion of population and land, was still current: "When will the truth come out into the light in international affairs? When will it be openly recognized that one of the principal causes of the aggressive attitudes of individual nations and of much of the present discord among groups of nations is traceable to diminishing productive lands and to increasing population pressures?"[69] Well may he have asked "when," because identical positions had been voiced more or less continually since the end of World War I. The link between interwar "internationalism" and postwar anti-Communism becomes apparent when such alternative lineages are brought to light. Part of the reason they are often not, is the abiding division that World War II casts over the twentieth century, obscuring important continuities. Rachel Carson's *Silent Spring* (1962), for example, was certainly generated in a Cold War world, but she was also one of Raymond Pearl's assistants, working on his Baltimore-based *Drosophila* in the late 1920s.[70]

Pressure on the soil had become increasingly serious, such that human destructiveness, Osborn wrote after Hiroshima, "has turned not only upon himself but upon his own good earth—the wellspring of his life." In writing about the limits of the earth, Osborn conceptualized both the globe—"our home, the earth, is one of the smallest of the nine planets"—and the land. A revolution in human approaches to nature was required, one that understood all component parts as dependent on one another, he pressed, trying to bring ecology into popular culture, ultimately a phenomenally successful bid. And like so many before him, it began with soil, which should be considered "living." Topsoil was the most important, the most "vital" element in nature's economy, and it seemed to trump everything in the postwar years.[71] Humanity's survival depended on "the preservation of the health of the earth," because in this radical but not necessarily new ecological conception, humans were the earth: "We are, in truth, of the essence of the earth. Our bodies, as well as those of all other animals, are composed of chemical elements that are derived from the air, from water and from the soil." Humanity had turned into a geological force, he wrote, both anticipating the Anthropocene and recapitulating the ideas of George Knibbs.[72]

Osborn's *Limits of the Earth* was a culmination of these ideas, a manifesto for conservation of the "vital processes of the earth itself."[73] In a lecture titled

"Our Reproductive Potential," Osborn got specific about population. The global pressure points he nominated were India, China, and Mexico: "too many people for the land to support."[74] Picking up a demographic transition model, he wrote of these as sites where nations and people were moving into "the explosive phase of the demographic cycle." Osborn also noted the growing tendency to discuss population problems as if they were peculiar to "underdeveloped countries." But the West, he said, needed to "openly and avowedly recognize its own involvement," especially with respect to resource usage. With less than 10 percent of the "free world's" population and 8 percent of its land area, the United States consumed almost 50 percent of its produce.[75] By his own statement, Fairfield Osborn came to the population question via the consumption dimension of conservation.[76] As centrally as Osborn placed population growth as causative of soil erosion and ecological problems, at the global level this did not mean that turning to sexual and reproductive solutions was a straightforward next move. He worried deeply about this, in fact. Osborn was cautious about just what the position of the Conservation Foundation would be "if it was going to get into the birth control business."[77] His more expert and scientifically educated colleague, William Vogt, had no such qualms. For Vogt, the reduction of fertility rates through birth control was essential, the more effective and the more of it, the better.

More than others within this generation of ecologists, perhaps with the exception of Julian Huxley, it was Vogt who embodied the conceptual and institutional connection between postwar conservation and population politics. He held offices at the Population Reference Bureau that had been established in 1929 by Guy Irving Burch. This organization had early and strongly advocated birth control and later came to be funded strongly by the Ford Foundation. Later, Vogt became both director of the Planned Parenthood Federation of America in the early 1950s and secretary of the Conservation Foundation in the 1960s.[78] Vogt's ecological field work first took him to central America: "it became more and more obvious that conservation was meaningless except in terms of its relationship to populations which, there, are increasing very rapidly."[79] For Vogt, the reduction of population growth was instrumental to the ecological imperative of limiting resource use, not least managing and conserving soils. And this explains his commitment to birth control. Speaking in 1952, Vogt was firmly in favor of pursuing research that would finally produce the promised cheap and effective hormonal oral contraceptive.[80] By that point, he was clear about the mission to reduce the acceleration of population growth, and he worked closely with the aging Margaret Sanger in this regard. For her part, Sanger incorporated the ecological message into her own, multilayered arguments.[81] And yet, there was a difference. Vogt may have been president of the Family Planning Association, but for him there was a clear hierarchy of needs and benefits, in which the neo-Malthusian and feminist aspiration of local economic and health security for individual women as mothers barely registered.

Vogt was, rather, the protector of nonhuman nature. In 1949, for example, demographer Sripati Chandrasekhar sought Vogt's opinion on his plan to decrease Indian density by moving millions of Indians, for the "colonization of currently unused South Pacific areas"; the conservationist was decidedly unimpressed. The migrated people, he considered, would not "be governed by the ecological imperative, and control their birth rate as well as adapt their land-use practices. . . . [T]he result of such colonization would be destruction of these areas."[82] And in the foreword to one of Chandrasekhar's books—*Hungry People and Empty Lands*—Vogt rather ungenerously pointed to its failure to adequately analyze the "ecological vulnerability of many areas into which he would encourage emigration."[83] Of the plans for emigration to tropical zones, he said, "We simply do not know how to use most tropical soils without destroying them."[84] And he finished with the extraordinary statement: "The empty lands are not nearly so empty as some of us wish they were!" Vogt declared, by his own words and exclamation point, his final allegiance to nature without humans.[85]

World Resources and Family Planning

In August 1948, Thomas Jeeves Horder welcomed delegates to the International Congress on Population and World Resources in Relation to the Family, held in London. Lord Horder was "establishment"—more like Osborn than Vogt in this sense. But like both of them, he considered population politics to be related to food and resource politics, in the first instance. Horder had done it all. He had presided over the 1933 Birth Control in Asia Conference, was advisor to the British Ministry of Food and president of the British Family Planning Association, and served as physician to several British monarchs. On the international stage, he was now chairman of the Standing Advisory Committee on Nutrition at the FAO. With all of this credibility and authority behind him, he announced to the delegates in Cheltenham, in the cosmopolitan spirit of the times, that there were three elements that made up the "tripod of international citizenship: world food resources, standards of living, and population trends."[86] And there were two ways of acting on these to effect international security: migration and family planning. Horder was certainly concerned with land: "Many thousands of acres capable of producing food are lost to us." But this kind of territorializing of population was beginning to prove an intractable problem, and everything on the road to international citizenship was leading to a gate, he said, clearly marked "Biological Control."[87] Horder was heralding the international biological control of human reproduction.

This was, after all, a conference jointly convened by the Swedish National Association for Sexual Knowledge and the British Family Planning Association. It was proclaimed as a meeting that celebrated 150 years since the publication of Malthus's *Essay*, and twenty-one years since the Geneva Conference. World population had come of age, the delegates told each other, merrily. "150 years of birth control, from Malthus to Cheltenham," was how Sanger summed it all up, as she handed a brochure of that title to Lord Horder.[88] More substantially, though, "Malthusian" was used as a technical not a nostalgic term by the demographers present. Pascal Whelpton, Warren Thompson's colleague from Scripps and later president of Rockefeller's Population Council, divided the world into three groups: a division becoming popular, but as yet anticipating Sauvy's contribution. "The Malthusian nations" were those where the size and increase of the population was determined primarily by the amount of food available.[89] Frank Lorimer also divided the world into three, but geographically: 25 percent of the world lived comfortably in arable and temperate regions, 30 percent on the margin of food satisfaction, and the Malthusian third remained definitely "suffering from a fairly chronic degree of malnutrition." Malthus's principles were now irrelevant within the first segment of the world, but remained a "powerful force in vast regions."[90]

Old and new players gathered in Cheltenham. Some had been present in Geneva in 1927: Edinburgh's zoologist F. A. E. Crew was there, as was London's eugenics leader Carlos Paton Blacker, and New York's contraceptive expert Abraham Stone. Others present in London were second-generation demographers and population scientists who had been educated by those present at the first World Population meeting. Thirty-year-old Sripati Chandrasekhar, for example, was trained originally by New York University sociologist Henry Pratt Fairchild. The organizers were largely the British family planning establishment (Margaret Pyke, Gerda Guy, Helena Wright). But they were now organizing for a decolonizing world. If Taraknath Das had been the spokesman for anticolonialism at early neo-Malthusian meetings, in 1948, newly independent India was represented by H. Chandra and J. M. Khosla from India House, London. They brought Jawaharlal Nehru's population plans and policies to the meeting. With Partition barely implemented, Niaz Rahmat Ali and Deputy Commissioner Hassan Akhtar Khan represented Pakistan. Doctor Okli Ampofo was there from what was still the Gold Coast, British West Africa, and J. M. Mansour attended from Farouk University in Egypt. Chinese perspectives were put by Lt. Gen. Kin Cheung and Processor An-che-Li of West China University, Szechwan, and even as they spoke, the terrible civil war that was to lead to Mao Zedong's declaration of the People's Republic in October 1949, was ravaging mainland China.

Sripati Chandrasekhar did not represent his native India or his adopted United States; he represented UNESCO. In fact, he rather nervously represented Julian Huxley personally, who could not attend. Chandrasekhar's own advocacy of birth control was well known, but because he was there on behalf of UNESCO, and Huxley had recently created such a stir with his manifesto, Chandrasekhar had to deliver his talk "UNESCO and Population Problems" without mentioning family planning or birth control at all.[91] It was the connection between overpopulation and war that Chandrasekhar emphasized: "Population . . . has been a very potent factor in causing all international struggles and wars. . . . It is accordingly necessary for UNESCO to see exactly how population adjustment can help to avoid war."[92] He drew specific attention to the recent deliberate policies of Germany, as he put it, to *create* population pressure so as to expand territory, as well as the "unavoidable" pressure that Japan found itself under (rather more innocently, as he framed it, even after the war). Like so many population experts before him, Chandrasekhar directly linked population pressure and war, optimum density and peace: "Population growth—deliberate or unconscious—has been the reason behind the demand for lebensraum and declaration of wars, accession to raw materials and unrestricted freedom to migrate from one country to another." As we shall see, the use of "lebensraum" in the population context, even by anticolonials immediately after the war, was not uncommon. Chandrasekhar directly connected this to the old cosmopolitan dream mixed with a fresher discourse of economic development. "If this differential growth of peoples leading to great differences in standards of living becomes a permanent feature, the one world of our dream where wars become an ugly memory of the past can never be realized."[93] Interwar territorializing of the population problem is part of what created a new postwar geopolitics of population. The other UNESCO representative attending the conference was Joseph Needham, director of the Natural Sciences Division.[94] Chandrasekhar was criticized for omitting birth control from his talk, and Needham defended him: UNESCO was an intergovernmental body, it had to move slowly. But Needham also spoke as a biochemist, a historian of technology, and China expert: dry rice culture might be a solution to Asia's hunger problem.[95]

That was certainly what John Boyd Orr, another international statesmen addressing the 1948 Conference on Population and World Resources, wanted to hear. Entirely in accordance with Horder's notion of international citizenship, indeed himself an international citizen of the first order, Boyd Orr rather thought that biological control meant not human reproductive physiology or conduct, but plant genetics or pesticides or synthetic fertilizers that would increase yields. This is what was necessary to feed, clothe, and house both adequately and equally (Boyd Orr's signature argument) the additional 20 million people on the planet that appeared each year. In 1898, chemist William Crookes had encouraged vigilance to maximize every opportunity to use resources. John Boyd Orr could see the point. He was an agriculturalist and a human nutritionist, before he was a conservationist or an ecologist. And he really did not think much about sex at all. His prime concern was food production, consumption, and distribution, and global plans through which the mean standard of living could be raised to more equitable levels. Between physics, chemistry, biology, and engineering, plus a good dose of internationalism, solutions could be realized. He saw the problem and was effective at communicating possible solutions. With improved stock and herds, by bringing barren land back into cultivation, with irrigation in the "old Empire in the Middle East," in California, in Egypt, it could be done. The great projects of high modernity were what Boyd Orr set his hopes on, for a world population fed properly and equally: the Tennessee Valley Authority's success needed to be duplicated; the colossal plans of irrigation put in place by the Soviet Union ("even greater than the TVA"); a waterway from the Arctic to the Caspian Sea. These were the massive projects that Malthus could never have foreseen as possible and the new world could deliver.[96] Radhakamal Mukerjee agreed, writing a strongly linked story of population control and rural and regional reconstruction across multiple political and cultural sites that now appear an unlikely combination. "I have seen the achievements of the Farm Security Administration and the T.V.A. in the U.S.A., Mussolini's planning in the region of the Pontine Marshes, Hitler's agricultural reorganisation in Southern Germany and, above all, the far-reaching social experiments of the U.S.S.R."[97] Likewise, Boyd Orr's plans were for the marriage of health and agriculture through high technology: "Deserts which were once fertile can be reconditioned by re-afforestation and irrigation. It is probable that some time in the not too distant future the Sahara Desert will be irrigated by de-salted water from the Mediterranean." And indeed, UNESCO ran its Arid Zone project through this period, aiming to make the deserts bloom.[98] All this, Boyd Orr wrote, was technically possible, though politically difficult.[99]

The first question Boyd Orr received from his Cheltenham audience was from vegetarian activist Roy Walker, speaking, he made a point of clarifying, as a world citizen first and an Englishman second. What, exactly, was FAO's policy on grain fit for human consumption being used as stock feed? The second question was from Kate Frankenthal, the socialist physician who had fled Nazi Germany: Do changing labor conditions—mechanical labor, work powered by atomic energy—require a recalculation of world calorie needs? There would be need for more proteins, but fewer calories, surely?[100] Population was all about energy at macro and micro levels.

Food and Energy: The Population Council

Across the Atlantic, the same discussion on population and resources was taking place, but in surrounds different, to say the least, from the modest Cheltenham meetings rooms in war-damaged London. Deeply engaged with the population establishment on the East Coast of the United States, John D. Rockefeller III occasionally looked up from his desk and registered ideas from elsewhere. On one such occasion, he focused on Julian Huxley and asked his advisor, "Do you know Dr. Huxley and is he a fellow with ideas in our particular field which are considered sound?" Rockefeller was interested because Huxley had stated that the solution to the population problem would involve a research project that would "cost about as much as a couple of heavy bombers." Huxley, comfortable enough in material terms, but still tracking carefully the royalties and payments that came in from publications and speaking engagements, was speaking more or less in the abstract. Not Rockefeller. "Obviously when he says two heavy bombers, he is speaking terribly generally," he conceded to his advisors. "I would presume he had in mind probably something like $5,000,000 or $10,000,000." That was not outside the bounds of possibility; perhaps Huxley would be worth a letter.[101]

Rockefeller surveyed the world after the war, and like Huxley, he saw too many people, especially in a region in which he had a particular interest and some experience and for which he had a great fondness: east Asia, or the "Orient," as he tended to call Russia, Japan, China, and Korea.[102] He had recently returned from a trip to Japan with "Mr. Dulles," and his interest was much intensified. Other sectors of the Rockefeller family's philanthropy—its International Health Division and its Social Sciences Section—were already active in Japan. The trustees had approved funds for the Princeton Office of Population Research: Notestein and east Asian expert Irene Taeuber were closely involved in a reconnaissance trip.[103] By 1952, the UN Economic Commission for Asia and the Far East (ECAFE), was also seeking Rockefeller funds—$60,000—for a study of the impact of industrialization on population growth in that region.[104] The UN was coming close—through development economics—to being explicit about demographic transition and the need to promote it.

After World War II, any number of institutions organized around world population. Trying to get a grip on their range and various agendas, Rockefeller's researchers divided them into groups. First there were organizations that dealt with "control" (the Malthusian League, the Eugenics Society in England, the International Committee on Planned Parenthood, the American Eugenics Society, and the American Planned Parenthood Federation). Second, there were those that dealt with "resources" (the Conservation Foundation, the Nutrition Foundation). And finally there were those that dealt with statistics and demography (the Population Reference Bureau; the National Institute of Demographic Studies, France; the Scripps Foundation; the Office of Population Research at Princeton; the IUSIP; the American Population Association; and what the Rockefeller researchers called the London School of Economics, by which they likely meant the Population Investigation Committee that published *Population Studies* from 1947).[105] Of the third group, the most recent addition was the Population Division of the new UN.

Slightly marginalized within the Rockefeller Foundation and its various philanthropic agendas,[106] John D. Rockefeller III wanted to put his own stamp on the population issue. In 1952, he called together an advisory committee of experts to meet in Williamsburg, Virginia; this meeting was to initiate the Population Council. Rockefeller's gathering was to be one of the twentieth century's key turning points, when the geopolitics of population shifted toward a biopolitics of population. Yet what needs to be recognized is the extent to which the eventual focus on contraception was a response to geopolitics broadly understood: Cold War politics to be sure, but also a larger resource question that raised, variously, environmentalism, energy politics, and certainly food security.

Rockefeller's advisors wanted to keep the discussion open to what they framed as the "resource" side of the population question on the one hand and the "control" side on the other, bringing together experts on reproductive physiology, demography, and social sciences, as well experts on food, soil, agricultural economy, and energy:[107] Thomas Seward Lovering was an economic geologist with the U.S. Geological Survey; Theodore Shultz, Chicago professor of economics, was an expert in food and agriculture; and Sterling Hendricks from the USDA was a plant biologist. Selected specifically because of their recent controversial writings were ecologist William Vogt and conservationist Fairfield Osborn. Representing expertise in health, human reproduction, and the social and demographic sciences were Lowell Reed, the John Hopkins public health biostatistician; George W. Corner of the Carnegie Institute of Washington's Embryology Department; Marshall Balfour, the Rockefeller Foundation's face of public health in the "Far East"; Paul Henshaw, director of research with Planned Parenthood Federation of America; elder statesman of demography, Warren Thompson; and the fresher generation: Kingsley Davis, Frank Notestein, and Irene Taeuber.[108] Matthew Connelly has explained the driving anti-Communist agenda of the important leaders of the "population establishment" that Rockefeller convened, and that he so significantly enabled toward the implementation of reproductive population control. If anti-Communism was on the agenda, however, so was energy. Detlev Bronk, who presided with unsurpassed skill over the informal and loosely structured in camera conversation, was president of Johns Hopkins University, president of the National Academy of Sciences, and soon to be president of Rockefeller University. A biophysicist, he was also advisor to the U.S. Atomic Energy Commission. Lewis Strauss of that commission was also there, having prompted the whole idea of a population meeting with Rockefeller in the first place.[109]

The document that all of Rockefeller's experts read in preparation for this discussion was titled "Population and Food." At core, it was about energy, too. It was likely written by Warren Weaver, head of the Rockefeller Foundation's Natural Sciences Division between 1932 and 1955. Trained in civil engineering and mathematics, at the Rockefeller Foundation he had been responsible for encouraging and approving grants for molecular engineering and plant genetics projects in developing research into new strains of wheat and rice. He was one of the scientists at the well-catered table at the Williamsburg Inn, who thought the recent catastrophic writing about soil, population, and food was exaggerated and unnecessary. These were inaccurate warnings, because they were limited to traditional thinking about agriculture as a form of energy production, he said.[110] This was understandable, given that the discussion of world population involved the whole planet, including those with a tendency to "cling to past procedures." But the paradigm should be shifted from thinking about bread, milk, eggs, and meat—John Boyd Orr's nutritional territory—to far more creative and future-oriented scenarios. For a start, more could be derived from the sea to augment the "little things served on crackers at cocktail parties" (the kind of quip that peppered the patrician Williamsburg meeting, but that would have raised immediate objection at any of the more open world population conferences). Nonetheless Weaver's point was a familiar one: oceans needed to be utilized, land needed to be reclaimed, and the microbiological world needed to be investigated, deploying real scientific imagination.[111] On a much larger scale, Weaver was inviting a thoroughgoing reconsideration of the relationship between energy conversion and people on the planet.

Unlike Boyd Orr's vision of equitable global satiation and health, Warren Weaver's was a bare life calculation of needs. Really, he said, there are only three considerations for our metabolic processes: that they be furnished with materials from which they can derive energy, from which they can be rebuilt and repaired, and from which they can derive small amounts of the vitamins and trace minerals necessary for life. At its most foundational level, the world's food problem was about usable energy, and "every molecule and every atom of any substance whatsoever contains energy, as we now all well know." Weaver then tabled three types of energy sources available to humans. First, energy from nuclear disintegration cycles in the sun, as he put it: radiant energy direct from the sun; existing vegetation, with energy received recently from the sun; prehistoric vegetation, such as coal and oil, received from the sun in ancient geological time; waterfalls from water "raised" by the sun; wind, driven by energy from the sun; the heat of tropical waters; and atmospheric electricity (from wind, and hence from the sun). Second, there was energy from nuclear disintegration on Earth: atomic power and heat from the planet's interior. And third, there was energy of cosmic origin—tides for example.[112] Although some in this era were attempting to mobilize atomic energy—Aldous Huxley, for example, argued that it should be harnessed for the production of food for millions, in a series of new Manhattan Projects, "not of destruction, but of creation"—Warren Weaver was looking to the sun. Solar energy "can be replenished," and that is what he wanted the Rockefeller population meeting to focus upon. Humans must reckon primarily with direct solar energy, he concluded, because the sun "is some 100,000 times as great as our present

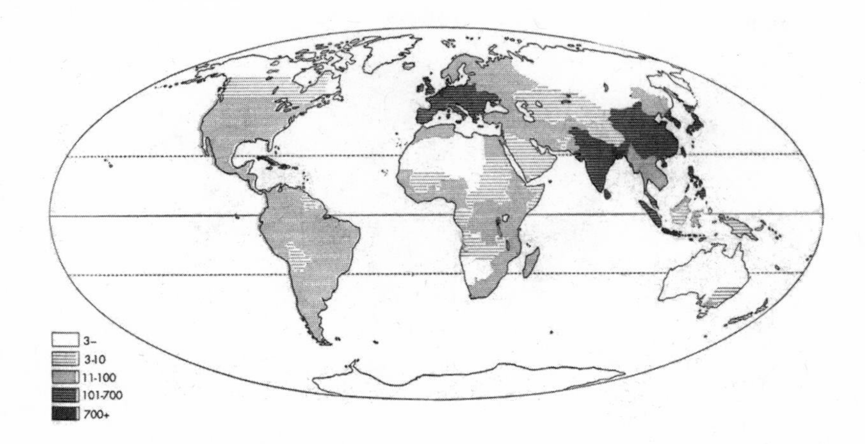

FIGURE 10.2 Global population density in 1956. The original caption for this figure read: "Density of population in various parts of the word is indicated in this map. The numbers beside the key at the left represent the number of people per square mile. In most cases the population density has been averaged within political boundaries." By the mid-1950s, comparative density and comparative calorie consumption were to be assessed together. (From Julian Huxley, "World Population," *Scientific American* 194, no. 3 [1956]: 67. Reproduced with kind permission, the estate of Bunji Tagawa.)

total energy requirements for food, fuel, and power." Indeed, if solar energy could be developed, "then the energy aspect of food requirements would completely vanish as a problem."[113]

How much energy was needed? Here, the calorie counting that had gone on since World War I continued: the aggregate of each human's energy needs per day, multiplied by the number of people in the world, provided the answer. Julian Huxley at this point was supplementing the traditional world density maps with world calorie maps, showing the number of calories per person per day (figures 10.2 and 10.3). The calorie map drew attention to east and south Asia, regions where fewer than 2,200 calories per person per day were consumed.[114] This was the geopolitics of hunger. For the United States, Weaver estimated 3,000 calories of food and 125,000 calories for heat and power were needed per capita per day. For the world as a whole, he worked from an average of 2,400 for food and 6,000 for heat and power. His own global population estimate, in 1952, was 2.1 billion.[115]

Warren Weaver was thinking creatively about the possibility of converting theoretically inexhaustible energy into forms of energy to drive human

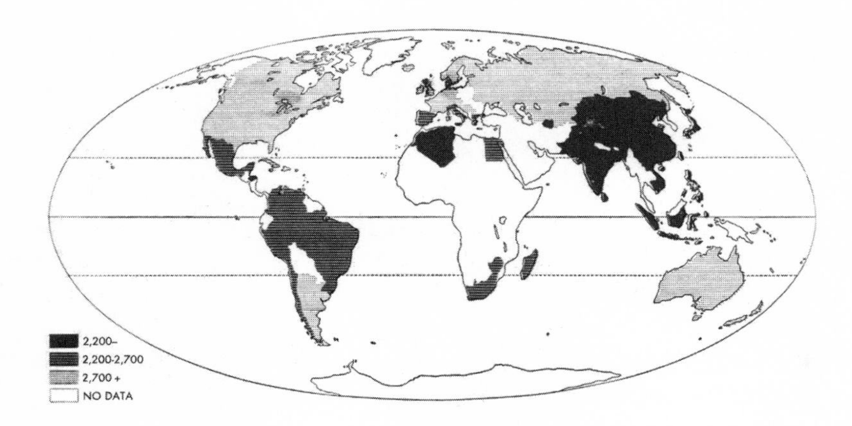

Figure 10.3 World calorie consumption. The original caption for this figure read: "Number of calories per person per day is plotted on the same projection. The minus sign in the key indicates fewer than 2,200 calories; the plus sign, more than 2,700 calories. The dotted lines are the Tropic of Cancer and the Tropic of Capricorn." (From Julian Huxley, "World Population," *Scientific American* 194, no. 3 [1956]: 68. Reproduced with kind permission, the estate of Bunji Tagawa.)

bodies directly, not just to drive communication, industry, and transport. Humankind needed to learn how to convert solar energy to a chemically stored form.[116] The same man who was famously to imagine the functional possibilities of a calculating machine was also exercised by fantastic technological and engineering solutions to the world food and population problem. He spoke to his colleagues in Williamsburg about a small device, perhaps not more than one-inch square, that might be placed on heads or on hats; a device that would receive energy from the sun, and deliver this energy, as a food substitute, directly into humans. (But how, he wondered? A pause was recorded in the transcript.) Recalling the "Standing Room Only" idea, Weaver countered that when it came down to it, humans might indeed conceivably stand closely together on the face the planet, and receive their adequate share of solar energy for life. This was an interesting new solution to an old dream: producing energy for humans—food—but without soil. In the 1820s, William Godwin had a vision of synthetic foods that bypassed agriculture, "The food that nourishes us is composed of certain elements; and wherever these elements can be found, human art will hereafter discover the power of reducing them into a state capable of affording corporeal sustenance. No good reason can be assigned, why that which provides animal nourishment must have previously passed through a process of animal or vegetable life." "Human art" might find a way to access those chemical substances directly.[117]

Any prospect of the soil-less production of food had vast implications in the twentieth century, because the population question was so tightly bound to geopolitical, territorial, and sovereignty issues. Removing soil or land from the ecological and economic equation also removed the seemingly insurmountable geopolitical element. This was why George Knibbs had bothered to collect news clippings in the 1920s about fantastic new sources of human energy: "Bread from the Breezes" for example, "Tapping the Atmosphere for Food."[118] Meat-eating Stefansson, at the same time, explored all kinds of "newer ideas . . . the farming of the ocean and the manufacture of food straight out of the air."[119] And later in the century, Boyd Orr talked about the edible fat that might be produced from coal and the possibility of "nutritious food from grass without it passing through the digestive organs of farm animals." His own FAO had determined that for every four trees cut down, only one reached the consumer, and the rest was wasted. But sawdust could turn into stock food, "and even into human food."[120] Josué de Castro also puzzled over the possibility of solving world hunger through the development of planktons, yeasts, and algae, putting science to work on synthesized foods.[121] And Aldous Huxley thought hard on various systems of soil-less farming. He discussed German wartime innovations on the conversion of waste into consumable energy: sawdust might become a sugar solution for the culturing of edible yeasts, and the synthesis of chlorophyll might also be worth pursuing, he thought, "the substance which permits the growing plant to use the sun's energy to convert air and water into carbohydrates."[122] After decades of politicizing the geopolitics of soil, territory, and population, the new food sciences might offer a way out. Remove the soil; remove the problem.

This was the tradition of interest in soil-less food production that Rockefeller's group was asked to consider in 1952. Their particular attention was drawn to the cultivation of unicellular algae, precisely because their reproduction did not depend on soil, thus offering "a far more efficient way of utilizing solar energy (which is inexhaustible for practical purposes) than the processes which take place in ordinary agriculture."[123] Another delegate spoke of the utilization of solar energy through the medium of photosynthesis, theoretically gaining 100 percent efficiency in the transformation of solar energy into organic matter.[124] It was Kingsley Davis who sounded the note of caution: "Given a simple increase of one per cent per year for the world's population, it will eventually take all of the planet's energy no matter what." He reminded his colleagues that the solution was not either to expand resources or control population, but both. "The books have to be balanced on both sides."[125]

The biological "gate" of which Horder had spoken at the Cheltenham conference on population and resources referred to human reproductive physiology and its management. Ultimately, this is where Rockefeller put his money. But for so many considering the population question, the biological options included a new microbiology, biochemistry, and a biophysics of energy that capitalized on an atomic age. Warren Weaver, for one, was putting his not inconsiderable scientific imagination to the task. New sources of energy for humans heralded a new world, as far as he was concerned: "This is the knowledge on which we must base our solution of the population and food problems. This is the understanding of life."[126] Weaver's "century of biology" was also the century of geopolitics. The whole significance of the prospect of soil-less food was that it promised to puncture the tenacious geopolitics of population, to make production of energy all but placeless.

Freedom from Hunger and Right to Land

Malthus had been confounded by food, observing in 1798, "I do not know that any writer has supposed that on this earth man will ultimately be able to live without food."[127] Yet over the twentieth century—in large part inspired by military and exploration imperatives—all kinds of condensed, distilled, dried foods and food substitutes were researched and distributed, from "pemmican," the concentrated fat and protein mixture favored by Vilhjalmur Stefansson, to the K-ration for the U.S. Army, to "Multi-Purpose Food," initially billed as "the Friendship Food for a Hungry World."[128] The latter was a high-protein, powdered, soybean-based supplement that the 1946 Meals for Millions organization created in its bid to eliminate world hunger and malnutrition, starting with the local (Californian) indigent, then the starving of Europe, and finally linking with U.S. food aid to south and east Asia. The Meals for Millions Foundation merged eventually with the American Freedom from Hunger Foundation, which had its origins in the FAO.[129]

The idea of "freedom from hunger" derived from Franklin Delano Roosevelt's Four Freedoms speech, which made its way to the 1941 Atlantic

Charter.[130] Freedom from want was Roosevelt's third freedom, repeated by William Beveridge in a 1942 report as axiomatic to a new vision of social security.[131] John Boyd Orr extended the maxim: "The first want of mankind is food."[132] The Hot Springs meeting that preceded the establishment of FAO had been called to give organizational effect to Roosevelt's Four Freedoms, and Boyd Orr fully mobilized this opening for his own brand of international politics.[133] Hunger was linked to political conduct and political possibilities. Food, in that context, was itself freedom. People on all continents, he said, were politically "awakened" and now understood hunger not as an act of God to be endured, but as part of an unequal global political economy to be challenged. The revolutions led by Sun Yat-sen, M. K. Gandhi, and Vladimir Lenin were manifestations of that "awakening." Men will not die of hunger quietly, he warned.[134] Political freedom was thus being jeopardized by hunger, a statement easy to make and hard to act upon, as Boyd Orr knew better than most, watching his own World Food Plan scuttled.[135]

If in earlier generations the links between population and food were comprehended in terms of war and peace, the discourse shifted in the postwar period to emphasize the connections between food and the very meaning of freedom. Freedom from want gave internationalists like John Boyd Orr a fresh language through which to continue to challenge the inequitable global distribution of people, land, and food. Huxley also made the connection. Population density, he pronounced, restricted human liberty and restricted other freedoms as well.[136] International Labour Organization (ILO) migration expert Imre Ferenczi first presented a similar argument about freedom from want and international population policy in 1940. By the time it was published in October 1943, the United States was part of the "total war" that he put down to Western powers' neglect of the "security factor in inter-population adjustments." Freedom from want, he argued, is what guaranteed against future wars of population. But rather than foreshadowing the United States as global liberators, Ferenczi placed responsibility for the war itself partly with the United States. Its immigration policy had closed the door to about 20 million people after 1920, "thereby aggravating the condition of impoverished Europe by millions of workless, would-be immigrants." Emigration from the Old World—what he called the safety valve of the nineteenth century—had been reversed, with fatal consequences.[137]

Roosevelt's freedom from want also morally authorized for a new generation the decades-old argument that those without food (because they lived in overpopulated regions) had a right to productive lands that lay unused elsewhere. An old version of this was lebensraum. Another version was H. G. Wells's 1939 proposed "International Declaration of the Rights of Man," in which he included a right to "move freely about the world."[138] The Indian objection to immigration restriction was another expression of the same core principle. In this way, freedom from want became another argument against the tenacious race-based immigration acts.

Sripati Chandrasekhar captured this in his 1954 title *Hungry People and Empty Lands: An Essay on Population Problems and International Tensions*. Resigning from UNESCO, he wrote this book as a Nuffield Fellow in London, and now could say what he really thought. And yet, in so many ways, it had all been said before: "These empty areas are under the control of peoples who do not desperately need them. In some cases, the people who control them are unable to fill these lands, while the people who need them and can fill them are denied access to them."[139] In the same year he broadcast for the BBC series *The Third Freedom*. The series looked at "freedom from want—the meaning and threat of world poverty and the ways it is being attacked." Initially, the BBC wanted Chandrasekhar to talk about soil and food cultivation but his contribution ended up as "The Fight Against Hunger."[140] Chandrasekhar linked the idea of freedom from hunger to the idea of freedom to move, in fact the "right" to emigrate, cultivate, and develop the potentially useful but unused arable lands of the world. His point was that freedom to move—emigration and immigration—"is yet to be recognized as one of the fundamental freedoms."[141] It was not yet a right, but was nonetheless tied intimately to the realization of another "right," freedom from hunger.

This sustained the critique of the global color line that had emerged in the 1920s and 1930s. Indeed the abandonment of the race-based acts was Chandrasekhar's first recommendation when he promoted the need for a new world population policy. The world looked like it did because of colonial "aggression that gave the possession of certain lands to certain peoples, and their consequent title to present ownership and national sovereignty." These were historical and political accidents, he insisted, and therefore entirely open to question and change. He argued that in searching for adequate nutrition, the Indian emigrant is prepared to "give up his political allegiance to India, and become a citizen of the receiving country," and should be given the right to do so freely, which meant addressing existing immigration as well as emigration restrictions. He cast this ongoing block of some people's global movement as creating a "land hunger" that would, and had, led to war. "The cause may not always be as simple and direct as it was in the case of German demand for lebensraum or Japan's desire for emigration outlets," he wrote, but the fundamental need for food and land endured. Indeed, the need had intensified, "barring conquest of the moon and other planets apart from ours." Chandrasekhar continued to argue for an international migration authority, one that would oversee peaceful population transfers. It might become a specialized agency of the new UN, he thought. The Permanent Migration Committee of the ILO and the International Refugee Organization, already in operation, went some way toward this goal, he wrote, but ignored completely Asian peoples "who are in dire need of emigration relief."[142]

Needless to say, there were inverse arguments as well, representing completely different politics but built from the same conceptual components. Some thin British arguments sought to reassert white man's entitlement to continental spaces, bolstered by postwar data on domestic rising birthrates and "overpopulation." In the United Kingdom, a group called the Migration Council held desperately to the idea of Great Britain as a world power, in the face of the United States, the UN, and the Communist bloc.[143] Mobilizing and extending the Commonwealth of British Nations would help secure world peace by feeding the hungry. But insular Britain needed to achieve its own optimum population, by emigrating at far higher rates (repeating Chandrasekhar on India): "The density of the population of the United Kingdom was 50 times that of the Commonwealth as a whole yet in the Dominions there were vast natural resources with populations far too small to exploit them." The population of the United Kingdom should be reduced to about 30 million and moved systematically to Australia, Canada, New Zealand, South Africa, and Rhodesia. Intra-Commonwealth emigration schemes were in fact implemented at a fairly large scale over the 1950s, with assisted passages, and sometimes free passages, extended to British families to enable them to settle in Australia and Canada in particular.[144] But the chair of the self-appointed Migration Council had an older idea of Greater Britain in mind, even after the fact of Greater Germany: "Why should we build satellite towns in congested Britain when vast territories lie empty? . . . The living space is there. It lies waiting, stretched over hundreds of thousands of square miles."[145] This would seem an unlikely recourse to apparently German ideas in the cause of rescuing an already-lost British global dominance. Any reference to Hitler's recent Greater Germany and living space was apparently unwitting, if telling. It was unwitting, perhaps, because the author was distracted, looking back much further in history, to more glorious British times. "Four hundred years ago, in the first Elizabethan Age, this country faced up boldly to the disaster which threatened it, and laid the foundations of a glorious future through resolute action, adventure, initiative and the development of new lands."[146] These Britons wanted to use population density to justify continued dominance in new lands, and they did so with confused nationalist referencing of a first British Empire on the one hand and "living space" on the other.

Not just the logic of "living space" but the very term lebensraum was deployed surprisingly easily after the war. Perhaps even more strangely, it occasionally served to bolster anticolonial and antiracist arguments. Mukerjee sought a Pacific charter to accompany the Atlantic Charter drawn up by Winston Churchill and Franklin D. Roosevelt in 1941, a blueprint for the postwar world. This Pacific charter should have two planks: the first, he called "Asiatic *Lebensraum*;" the second, linked, was to overturn what he called the "doctrine of the White Man's reserve," the system of racially exclusive national immigration restriction acts. Indians needed food, and they needed to be able to move. "As a matter of fact," he challenged, "the acid test of the Atlantic Charter will be the satisfactory settlement of the racial issue of the Indians in South Africa and the revision of the White Australia Policy."[147] We have seen how Mukerjee argued such aspirations through the connection between particular climates and particular human physiologies, how he repackaged an old medical geography of race into a new population ecology. Equally striking, though, is Mukerjee's borrowing of the Nazi principle of territorial expansion on food and population grounds, in this instance applied to an anticolonial end. He was able to do so because of the currency of the triad of people, land, and food, an idea that reached well beyond and well before the Nazi context.

Mukerjee's book *Races, Lands, and Food* was significantly subtitled *A Program for World Subsistence*, and was prefaced by Harold Ickes, U.S. secretary of the interior. There was a twofold global problem, Ickes suggested in his glowing preface to Mukerjee's book: "First, is that of reapportioning populations of the earth to assure the availability to them of the essentials of life, and second, the immediate institution of a program of the maximum utilization of world facilities and resources in an effort to produce sufficient food and material to equal our needs."[148] For the postwar U.S. government, however, "reapportioning populations of the earth" was less about Indians than about European Jews. Indeed it was Ickes himself who was responsible for pursuing the various American possibilities in the light of Isaiah Bowman's earlier plans in the M Project,[149] pushing Alaska at one point. On the one hand, then, Ickes was imagining the wastelands of the earth as places for resettling European Jewry, a project that was itself partly rationalized by the population pressure and war argument. On the other hand, he was writing a gushing foreword to a 1946 book that deployed lebensraum as an idea that might assist the project of world peace. What seems politically and conceptually impossible, especially for 1946, is explained by the geopolitics of population, the premise long shared by *Geopolitiker* and Anglophone economists, demographers, and geographers of population. Notwithstanding his unlikely use of lebensraum, Mukerjee's anticolonial analysis was more to the postwar point than the backward-looking British Migration Council.

It was Chandrasekhar who was to most effectively carry forward the south Asian critique of race-based immigration restriction laws, which did not unravel until the mid-1960s, and even then only in a piecemeal way. In 1965, "natural origins" as the basis of American immigration legislation was removed.[150] In Australia, the Immigration Restriction Act was repealed in 1958, replaced by a Migration Act that had no offending dictation test but retained all else besides. The "nationality" and "race" rationales for exclusion were in general being written out, but this did not introduce an era of diminishing immigration regulation per se. Indeed, just when other nations were undoing their legislative ties to nationality, race, and ethnicity, the United Kingdom was borrowing and tightening such controls and introducing new laws. In the 1968 Commonwealth Immigrants Act, immigration from the "white" dominions of the Commonwealth was facilitated—Greater Britain still—while mechanisms were retained to exclude people from other parts of the Commonwealth. Perhaps ironically, it was the newly decolonized and independent nations themselves who were now picking up the immigration restriction imperative with some considerable enthusiasm, strong statements of independent sovereignty of new nation-states. This was the beginning of the global normalization of immigration restriction.

When Sripati Chandrasekhar was in discussion with economists and demographers in the mid-1960s over the ongoing white Australia policy (notwithstanding legislative change), the new set of immigration acts in south Asia and southeast Asia were precisely to the point. Critics of white Australia, one economist explained "were often met with the comment that the non-European nations have similar policies,"[151] pointing to Malaysia, Singapore, Ceylon, and Borneo. Some decolonized nations were themselves implementing population transfers on the logic of density as well. These came to be effected rather more as internal colonization programs than the world migration policy that Chandrasekhar and Mukerjee imagined. The Indonesian transmigration programs implemented by the anti-Communist

Suharto regime, for example, were intended to equalize population density across the new Indonesian lands, and to conspicuously settle ambiguous national "frontiers" like west Papua.[152] The Soviet Virgin Lands Scheme agriculturally settled hundreds of thousands of people into northern Kazakhstan. As ever, the movement of people was tied to the production of food, and hundreds of thousands of square miles were newly plowed and sown for wheat, the Cold War incorporating a race for grain as much as a race for people, land, and even space.[153]

Chandrasekhar noted all this: he continued to think on the geopolitics of population and food, even as his responsibilities shifted further and further to its biopolitics. He still had his eyes on the lands of the British settler world that were empty and that continued to exclude would-be immigrants racially, ethnically, and nationally. How provocative should his next title be, he wondered aloud in 1965? He wanted to call it: *Asian Population Problems and White Australia*.[154] But under pressure from his publisher, and from contributing economists and demographers, including John C. Caldwell, he ended up with a "dull and prosaic" title, *Asia's Population Problems: With a Discussion of Population and Immigration in Australia*.[155]

Caldwell and Chandrasekhar had just missed each other at the Asian Population Conference in New Delhi, which was opened by Nehru.[156] Chandrasekhar was himself in Australia on a speaking tour, on the barely admissible topic of Australian emptiness, Asian density, and immigration restriction. His visit was watched carefully by a Conservative government that maintained a white Australia policy, no matter what had happened to the immigration statute itself, but that also felt under pressure from various international human rights instruments, not least the November 1963 UN Declaration on the Elimination of All Forms of Racial Discrimination.[157] Chandrasekhar was also watched by a progressive Labor prime-minister-in-waiting, who was deeply interested in Indian–Australian bilateral relations, and who once in power announced the unconditional, if belated, end of white Australia in 1972. Chandrasekhar dedicated one his last books to the prime minister, Gough Whitlam, "for guiding his countrymen and women towards a realistic and responsible view of their relations with Asia, and in particular, for his courageous and progressive act of officially abandoning the 'White Australia' policy."[158] During his very active retirement in the late twentieth century—even through the 1990s—Chandrasekhar was to undertake major historical syntheses of immigration policies, of the United States and of Australia in particular, and their relation to policy questions of assimilation and difference.

One World and the Third World: White Man's Dilemma

On Boyd Orr's departure from FAO in 1948, new Director-General Norris E. Dodd, previously U.S. under-secretary general of agriculture, maintained the message that linked population and food. Addressing the 1949 UN General Assembly he said: "Tomorrow morning there will be 55,000 more persons for breakfast than there were in the world this morning. . . . But we are not producing 55,000 more cups of milk a day for the new children . . . nor 55,000 more bowls of rice."[159] Dodd and the FAO could not implement population control, but they certainly presented the case for it. Freed from FAO and UN constraints, Boyd Orr himself went on to expand and expound his critical perspectives on development, population growth, food security, and global inequity.

In so doing, Boyd Orr caught, very dramatically, the coincidence of decolonization and ecological world crisis. He proclaimed food to be the basis of a new world unity, and hunger "the common enemy of all mankind," this time quoting Dwight D. Eisenhower, not Roosevelt.[160] "Either the white peoples give up their futile rivalries and unite to end hunger in the world—now for the first time a possibility—or the tensions between 'haves' and 'have-nots' will end in a holocaust."[161] The terminology "haves and have-nots" was strangely outdated, belonging more to the post-Versailles era than the decolonizing world of the 1950s. Julian Huxley also used it, constantly, in his postwar writings on world population.[162] It sat oddly in Boyd Orr's book, which was written for an atomic world. But the usage is also telling. Like Mukerjee, with his redeployment of lebensraum, Boyd Orr slipped fairly easily from the post-Versailles "have-nots" (Germany and Italy) to the underdeveloped "have-nots," (the new Third World, especially the nonaligned Third World). He could do so, because the issue at hand was still land, food, and population. Importantly, the new "haves" were named by Boyd Orr as white men, who wielded power in an economically unequal world. The dilemma for the West, as he saw it, was whether or not to give up this power, in order to effect equity of basic resources and deliver "freedom from want for all men." Why would white man decide to do so? Boyd Orr himself considered equity enough. There was a universal "right to life, liberty, and the pursuit of happiness," he wrote, applying U.S. not UN codifications. Indeed, Boyd Orr signaled that the American fathers were "wise to put life first, because the man who suffers premature death for lack of the necessities of life has little interest in political liberty." Life was still about food for Boyd Orr, not reproduction. He argued also that maintaining others' economic dependence (partly by maintaining their hunger), is what sustained "white man's" power: lack of food produced the submission and dependence of two-thirds of the population of the world. This is why "though it is largely subconscious," he wrote, those in power are reluctant to cooperate to abolish hunger and poverty. "The natives of Asia, Africa, and Latin America would become the equals of the white man, and as these continents became industrialized the Europeans and their descendants, the Americans, would lose the control of the world they gained in their 300 years of conquest." This, then, was what he called the white man's dilemma.[163]

Writing about the "white man" was, in some ways, as old-fashioned for the 1950s as writing about the "haves" and "have-nots." It recalled William Crookes and his white "bread-eaters of the world," who were reproducing too fast in relation to the limited land available. Even more directly, it recalled Lothrop Stoddard, who in the 1920s had looked nostalgically back at the late nineteenth century to a time when "the white man stood the indubitable master of the world."[164] The explicitly racialized "orientals" and "occidentals," rice-eaters and wheat-eaters, were critically turned around in *White Man's Dilemma* into an equally explicitly racialized Third World and First World. The threat posed by the lack of wheat for the Caucasian race, set up by Crookes, was inverted in Boyd Orr's hands, to a lack of food for the Third World caused both by overconsumption in the First World, and overpopulation in the Third. Boyd Orr certainly raised population growth as a key "biological obstacle to freedom from want," especially in the "resurgent nations of Asia."[165] Indeed, the high growth rate of population was probably the most significant long-term problem facing humankind, especially because, he wrote, there were no new continents to be discovered. But if Crookes's lecture (and later Stoddard's famous book) had been quintessentially imperial statements of white global privilege, the shift that Boyd Orr put forward was a critical one, from white man's entitlement to white man's dilemma.

This was not quite a postcolonial shift, however; for Boyd Orr, the power to decide the world's future still belonged to the white man. Yet it was certainly a position that held the West to account. In other moments, Boyd Orr indicated that this power had already diminished. "Failure to move with changing conditions . . . would mean revolution, but this time world-wide. It would be between the White man and the many times more numerous Coloured races, soon all with nuclear arms."[166] This, in fact, was a version of the arguments that Étienne Dennery, Harold Cox, and Warren Thompson had made after World War I. What had changed, though, was the power of the Asian–African bloc, soon to meet in Bandung, Indonesia, and aligned with neither Communism nor the "free" world. Boyd Orr was deeply critical of the politics of aid, thinking it petty, inadequate, and ultimately a mechanism to shore up global power with the West: the new nations wouldn't be bought. Boyd Orr wrote of the independence of Nehru, the power held by emerging nations who refused alignment. But he overstated this independence. While India sought to short-circuit the compromising and arguably recolonizing politics of food aid, Nehru was ultimately forced to buy into the system, and the India of the 1950s was strongly dependent on U.S. food aid, followed by aid linked to the implementation of population control.[167]

Boyd Orr's solution, though, was not food aid but a new deployment of Western industrial technologies that would create a "new world of plenty," wherein both the production and consumption of that plenty would be distributed equitably: one effect of that plenty would be a rising standard of living, and therefore a reduction in birthrates, locally and globally. This would all be managed as part of a new world federalism. Just like the internationalists after World War I, Boyd Orr returned to cosmopolitan visions. Citing Stringfellow Barr, who had recently established the Foundation for World Government, the white man might retain his power or else "join the human family," use industrial supremacy to develop the earth's resources, end hunger and poverty, and create worldwide prosperity: the only cost was loss of the power of superiority.[168]

Boyd Orr was a high modernist for whom science itself made world government "inevitable," because it had created intimate connection: travel and communications, satellites, planes, and wireless annihilated distance. Is there any doubt that theirs was a changing world, he asked?[169] Everything seemed newly connected, commercially, industrially, and politically, but in dystopic ways as well. The discovery of a synthetic dye in a European laboratory, he wrote, brought more or less instant ruin to millions of Indians who had made a living by growing indigo. Any economic crisis must be a world crisis; any war a world war. This situation, then, absolutely required some form of "international law and order, maintained by an international police force" or there would be continuously recurring conflict.[170] World government should begin with federation of the democracies, he had written during the war. But at the close of war, and with the new world, he saw that the difficulties were tenacious.[171] His lecture on winning the Nobel Peace Prize in 1949 was "Science, Politics, and Peace." Just as "civilization" was turning into "development," the idea of Old World and New Worlds supplanted by the idea of First World and Third World, John Boyd Orr argued for "one world or none."[172]

It is impossible to separate out the history of population control or birth control from the history and politics of food and food security and from the larger question of energy production and consumption. The connected food–people question was politically alive in an international public sphere, actively part of the process of decolonization and the reordering of global power as the world war turned into the Cold War. The massive significance and the unexpected trajectories of the very idea of "freedom from want," certainly turned into the history of human rights. But it also turned into population policy, a development that can only be understood fully by appreciating the food–people connection as clearly as did the players at the time. After all, when the U.S. Congress endorsed contraceptive "aid," as a new brand of foreign policy in the mid-1960s, it did so through none other than the Food for Peace Act.[173]

11

Life and Death

The Biopolitical Solution to a Geopolitical Problem

Death control has made birth control a moral imperative as well as a social and economic necessity.

JULIAN HUXLEY, "WHAT ARE PEOPLE FOR?" (1959)

"Freedom from want requires population limitation," Guy Irving Burch and Elmer Pendell proclaimed in 1947, authorizing their position with a long quote from Benjamin Franklin about prolific nature, crowding, and fennel.[1] Such a position was beginning to be expressed quite often in theory, if not yet in policy. Birth control offered a solution to food insecurity and therefore political insecurity, in the first instance. Statements about birth control solving a gender problem or a health problem for women certainly circulated internationally, as they had in previous decades, but they were muted. When entertained over the late 1940s and 1950s, they were often done so expediently. By the later 1960s, however, fertility regulation as a good in and of itself had risen to prominence out of the whole complex. This was the process by which world population came to be imagined and acted upon as an issue of world health, and it was in this context that the field as a whole turned increasingly seriously to think about women.

An international biopolitics was emerging. It was not just about birth, however; it was equally about death. The concept of "thanatopolitics" has been developed by scholars concerned with extreme politics of regimes that determined those who were to be killed: in Nazi concentration camps, by wartime murderers, as suicide bombers.[2] "Thanatopolitics" of the wartime labor and death camps now signals the tragedy of high modernity. Yet the population problem foregrounded a completely different kind of death:

infant mortality. It was the seemingly banal politics of those who die in the feminized perinatal sphere that was in fact the more persistent matter of state. The monitoring and management of infant mortality became increasingly normalized within modern national health structures and within international health standards and expectations. Fertility and mortality were the elements of demographic transition, and it was in this sense that the phrase "death control" came to be the constant companion to the phrase "birth control."

Kingsley Davis and World Population, 1945

Historians tend to assess the critical midcentury idea of demographic transition in terms of the Cold War.[3] However, Kingsley Davis's key intervention, "The World Demographic Transition," was written during 1944 and published in the special issue of the *Annals of the American Academy of Political and Social Science* titled "World Population in Transition" in January 1945. The outcome of the war was not yet known; Hiroshima and Nagasaki had not been bombed; the People's Republic of China did not yet exist. There had been no Berlin Blockade or Warsaw Pact, no North Atlantic Treaty Organization or Korean War. Notwithstanding the long history of U.S. anti-Communism, the Soviet Union was still an ally of the United States. In short, "demographic transition" was initially conceived and produced with the geopolitics of the two world wars in mind, not the Cold War. The twists and turns of demographic transition and modernization theories—and Kingsley Davis's owns views—came to be deeply shaped by American anti-Communism, but they did not start that way.

The need to put Cold War American anti-Communism into perspective in this politico-intellectual history of population is suggested by the famous metaphor of the population bomb. Later, in the hands of Hugh Moore (1954) and Paul Ehrlich (1968) this was a reference to "the Bomb." But the metaphor of explosion came perfectly easily to Kingsley Davis in 1944 without nuclear fission, without atomic politics, and without a nuclear arms race. He opened "The World Demographic Transition" with old-fashioned dynamite, and the real metaphorical work was done by the fuse not the charge. "Viewed in the long-run perspective, the growth of the earth's population has been like a long, thin powder fuse that burns slowly and haltingly until it finally reaches the charge and then explodes."[4] In other words, for centuries there had been no growth at all: the long

European story. The first "burst" at the end of the long fuse was population growth in Europe at the Industrial Revolution. And this, he wrote, was also the moment at which the globe became a single demographic entity. The world's entire population became linked, "as one dynamic process," because European growth was diffused around the globe, transoceanic movement en masse became feasible, and Europe began to become dependent on New World grain. The proximate context was the geopolitics of the 1920s and 1930s that was then manifesting as war. The remote context was the geopolitics of European expansion into new worlds—"Europe Overseas" as he put it, borrowing from Alexander Carr-Saunders—as well as European colonial rule in Asia and Africa.[5] This was the "fuse" of his metaphor, the long European-led world/colonial demographic history that over time would come to be left out of accounts of world population futures. The second great bang of the (prenuclear) population bomb occurred between 1900 and 1940, when the annual growth of population in the world was 0.75 percent.[6] This had not yet peaked, Davis predicted, but it would eventually. This world scenario of population growth reflected reduced mortality far more than any fertility changes, according to Davis. "All indications are that the average expectation of life at birth has practically doubled since the late seventeenth century."[7] Modernization and industrialization, famously in Davis's 1945 model, would spread over the world, effecting reduced mortality and reduced fertility globally.

The politics that Davis was writing within and against had little (yet) to do with Communism. It was far more squarely an extension of the interwar debates about land, colonialism, and migration. Unsurprisingly, then, Davis's key inspirations were that generation of population scholars. Most importantly, he drew from Carr-Saunders, especially his *World Population*. He also worked from Warren Thompson, Robert René Kuczynski, E. F. Penrose, and Alfred Lotka. Radhakamal Mukerjee's ideas and claims he knew well and cited in disagreement. Tellingly, it was Mukerjee's insistence on the right to land, the right to (agricultural) migration that Kingsley refused to concede, in a section of his influential article that dealt with "invasion by migration." Yet it would be wrong to therefore conclude a crude racism on Davis's part. At one level, for later historians to expose "racism" within demographic transition theory is only to repeat what its creator took some pains to explain. Indeed, one of Davis's own objectives in this key document was to anticipate and dispel "the implicit racialism in most Anglo-American thinking." At another level, though, Davis's own apparent antiracism itself needs historicizing and critical analysis. It was a very different 1940s antiracism than Radhakamal Mukerjee's, for example, relying on assimilation, not assertion of human difference.

In a section titled "The Racialistic Fallacy," Davis conceded that difference in infant mortality between Asia and the West would mean a changing proportion of "Asiatics in the world." But it was groundless for Westerners to therefore fear an orientalization of the world he said, since "orientals" would *become* Westerners or at least become Westernized. Modernization, a raised standard of living, would mean that "they will lose a great part of their Oriental mode of life." Davis saw his argument that oriental civilization was cultural, not "fixed in the genes of the Asiatic races," as the core of his antiracism. He distinguished between civilizational difference and biological difference and noted: "It is unlikely that future conflicts will be along racial lines unless our own prejudice fosters them."[8] As "culture," Asian-ness was open to change—improvement as he would have it—toward westernized civilization, a Western standard of living signaled and effected not least by low fertility and low mortality rates. It was a bald expression of the presumption of civilizational superiority, one that problematically underlay the entire project of modernization and development for decades.

Davis drew his demographic transition theory directly from Warren Thompson, but he was not half the anticolonial the Scripps director was. On the one hand, Davis said, Westerners fear that Asian ascendants would "demand the right to migrate to regions now held by Westerners." He cited Mukerjee's *Migrant Asia* as a key instance of this demand. But this fear was invalid, he reassured again. Not because Asians should have an inalienable right to migrate, as Mukerjee, John Boyd Orr, or Sripati Chandrasekhar would argue, but because the demographic transition was a cultural transition: Asians would essentially become Westerners and would be migrating as such. If Asians sought to migrate to "vacant lands" as civilized moderns, already with a high standard of living and an established trend toward lower mortality and lower fertility, then there should be no policy problem. If Asians sought to migrate to vacant lands as premoderns, with high fertility and mortality, however, he argued there would be good reason to exclude them, except to the extent to which they could be assimilated. What seemed progressive racial and cultural politics at the time, to some, was also one problematic end point of the idea of a cosmopolitan human species that found little room for a philosophy based on irreducible difference. This starkly reveals the limits of the idea of assimilation that persistently presumed a Western benchmark: this was not in any sense a postcolonial future. Little wonder that Davis disagreed with Radhakamal Mukerjee.

Kingsley Davis, acclaimed as the co-originator with Frank Notestein of the new midcentury theory of demographic transition, was in January 1945 holding an old line, the global color line. He did not, as yet, anticipate the great change in the ideas about human rights and nondiscrimination on the basis of race that was about to unfold. Attempting to explode "racialistic fallacies," his position was in fact nothing like Warren Thompson's or Boyd Orr's or Chandrasekhar's anticolonialism. Davis ultimately showed his hand on the geopolitics of the world population problem, in a way that explains his focus on biopolitical/reproductive solutions, at least in the medium term: "A mere extension of current Asiatic civilization to new areas does not solve any problem, and to avoid such a result the European peoples would be justified in holding the lands they have, no matter how 'vacant' these lands appear to the Asiatics."[9] Westernization of the world, but without significant population movement to the West, was Kingsley's preferred global population policy in 1945.

One of the stunning elements of Davis's thesis is the extent to which he drew on, and incorporated, the ecological model, conceptualizing human population patterns in terms of energy wasted and energy saved. The biopolitics of his aspirational system aimed to maximize not just world-level human "life," but world-level energy in a total economy of waste and use. The old demographic regime was a wasteful one, he repeated in the mode of *The Science of Life*, simply because so much energy went into creating life (birth), into maintaining human life (as infants), only to end up in death (infant mortality). What he called the new demographic balance recommended itself as "an astounding gain in human efficiency." This was not quite the recognition of reproduction as women's work—"labor"—that feminist scholars like Hera Cook have sought, because Kingsley did not fully recognize the political implications of embodied gendered difference.[10] But it did foreground ecologically wasted labor expended in fertility/mortality, an argument notable for its absence in later feminist support for birth control (this a measure of how individualized the feminist "reproductive rights" discourse became). For Kingsley Davis, the demographic balance of low fertility and low mortality recommended itself precisely because it released "a great amount of energy from the eternal chain of reproduction—energy that could be spent on other aspects of life." Thus, individual lives and aggregate human energies would be maximized in the great global change to low mortality and low fertility. Western women had begun this process, he announced, and non-Western women would follow. The efficient energy system of demographic balance will "spread throughout the world."[11]

At this point, Davis and others still considered this fertility decline to be an outcome of industrialization itself. Later, the intellectual and political orthodoxy changed toward the active inducement of fertility decline: population control.

Warren Thompson in Postwar Japan

India dominates much analysis of population in the postwar period; Kingsley Davis's own *Population of India and Pakistan*, written during his time at the Princeton Office of Population Research, contributed strongly to this focus, both at the time and since.[12] Simon Szreter has also foregrounded the significance of China's civil war and the emergence of the People's Republic on the fortunes of demographic transition theories in the United States.[13] An additional element of the Chinese question pertained to emigration: whatever prospect there had been for the international movement of Chinese people closed down with the Communist regime; the borders of the People's Republic were closed. This prompted commentators like Chandrasekhar and Warren Thompson to conclude that birth control was now the only feasible response. Their profession watched and wondered what kind of solution would be implemented by the Communists, so rigidly opposed to Malthusian theory. Chandrasekhar himself hoped that they would appreciate the pressing need for "voluntary limitation of the family."[14] And drawing on Warren Thompson's work, Chinese demographer Ta Chen urged, just before the end of the civil war, that population policies of any effective and enduring sort "should be deep-rooted in the folkways and must therefore express the 'general will' of the people."[15] Certainly the Chinese government began to look around for solutions to the economics of population in all kinds of places, not least the British Political and Economic Planning group. Carlos Paton Blacker was involved in a request from the Chinese government to translate *World Population and Resources*.[16] Eventually (from 1979) the one-child policy became its infamous solution.

For all the significance of India and China, it was Japan that was most immediately on Anglophone—especially American—demographers' radars at the war's end. Japan had featured prominently as a population "danger spot," thanks in part to Warren Thompson's work, but also to the official rationalizations of colonization and invasion emerging from imperial Japan itself. On both Allied and Japanese readings, the Pacific war was about living space. Japan also drew considerable demographic attention because its birth and death rates were known to be changing. A fall in death rates became apparent quite quickly, from an average of 17.6 per 1,000 in the 1930s to 12.0 in 1948 and 7.8 per 1,000 by 1955. The birthrate stayed high until 1948, and so the immediate demographic projection was one of rapid growth. But by 1950, birthrates had also dropped, from 34.4 to 29 per 1,000, and by 1955 to 19.4 per 1,000. As historian Deborah Oakley has noted, there had never been such a dramatic decline recorded anywhere.[17] Finally and critically, Japan drew much attention because Americans were leading the occupying forces. General Douglas MacArthur was Supreme Commander of the Allied Powers (SCAP), with the immediate job of disarming Japan and securing a food supply. Thereafter, he was to introduce liberalized and democratized governance and to work toward a new Japanese economy and polity that was aligned with U.S. interests. Population management and demographic prediction was an element in each of these objectives, necessarily part of economic and political planning. But it turned out that there was rather more consensus among the Japanese than there was among the American occupiers about programs to reduce fertility.[18] It was in the context of a declining death rate, a high birthrate, and major food scarcity immediately after the war, that population policy seemed urgent both to the U.S. occupiers and to the new Japanese Diet. This was all a unique situation in which Americans and Japanese alike were asked to think about and implement population theories; policy could be, and was, fast-tracked.

Population trends in Japan, then, were upfront and central on the postwar stage. Yet any idea of entertaining a spatial solution to Japanese population density was extremely problematic, so closely aligned was the geopolitics of population to Japanese territorial expansion that had led to war. It was one thing for Indian commentators like Chandrasekhar and Mukerjee to continue to press for "living space" into the postwar period, but quite another for Japanese commentators to argue for Japanese lebensraum, or even for emigration and immigration restrictions to be relaxed. There were, nonetheless, isolated arguments to this effect, especially from Japanese Catholics seeking to stem the growing endorsement of birth control, especially sterilization.[19] Sometimes demographers tried the idea as well. "Birth control is of the utmost importance," the director of Tokyo's Population Research Institute agreed, but he still hoped for "an ideal arrangement in which emigration and birth control could be carried out simultaneously." Sripati Chandrasekhar reported that the new Japanese prime minister hoped "that an understanding world will permit some volume of migration to the southern regions."[20] Admitting such a policy was unconscionable for the Allied powers, however, a situation compounded by the policy of repatriating diasporic Japanese who had settled in China, Korea, and the Pacific. Far from endorsing the prewar idea of transferring populations out of Japan to reduce density in an emergency moment, millions were in fact being returned to the new Japan, itself territorially straitened.

Even the great interwar proponent of global density equalization, Warren Thompson, now wanted to turn discussion away from land-based solutions. Having gripped the field in the late 1920s with his thesis about global density and the desirability of population transfers, he found himself after World War II on the front line of population thought in occupied Japan, a demographic consultant from 1948. He had just published *Population and Peace in the Pacific*, a book that was somewhere between a new edition of and a sequel to *Danger Spots in World Population*.[21] And he started to tell a quite different story: migrations, movements, and transfers were looking not just ineffective but impossible. The bloody Partition of Pakistan and India cast a shadow over the whole idea of "population transfers," and Thompson began to indicate that the opening of land, especially to the large populations of India and China, would offer only transitory relief at best. Taking the Japanese case, the sole recourse was for high fertilities to be actively reduced. Ultimately, Thompson wanted to implement a plan that would "get the Japanese to come to birth control themselves."[22]

Warren Thompson does seem to have intensified local public attention to birth control after his arrival in Japan. By some accounts, he personally urged and even persuaded the prime minister to make a statement of in-principle support for contraception in 1949. Those pressing for birth control within Japan—Baroness Kato, Koya Yoshio of the National Institute of Public Health, Kitaoka Juitsu of the Birth Control Association—would often cite Thompson's change of position. He was declaring "the almost impossible logistic factors in migration as a solution of the population problem."[23] What happens in east Asia, Thompson said, would determine whether or not there would be a third world war.

Criss-crossing with Thompson's Japanese consultancy in Japan was a Rockefeller Mission on Public Health in the Far East. Led by Marshall Balfour, it included Frank Notestein and Irene Taeuber. Balfour had confronted SCAP about the ramifications of a dramatically lowered death rate, in large part a result of programs of immunization and DDT spraying. Where were the birth control measures to counter the death control measures, he asked, in what was to become the great trope of the postwar period. But they all knew the problems in forcing, or being perceived to force, birth control too explicitly. In a later memo, Taeuber put it clearly. The problem was the "adverse propaganda that might arise from a fear that the United States was urging a decline in births . . . no-one has found as yet a formula for backing research without committing political suicide."[24] It must be indigenous, affirmed Thompson. "It must arise as the people want it. But if we can help them to get information which they can use if and when they want to undertake the control themselves, we will have contributed about all that you can hope for, from the outside."[25] Here, then, was the beginning of a Western modus operandi for population control of the East, for the emerging Third World: "international" or "Western" influence, but actual and apparent national and local implementation. The model began less in the decolonizing Third World, however, than in the idiosyncratic political circumstance of occupied (and industrializing) Japan.

Privately, MacArthur seemed to agree, and he wrote to Warren Thompson that he thought fertility decline would proceed anyway in Japan, somewhat along Davis's thesis. Public endorsement was a different matter, however. When the Catholic press asked MacArthur to clarify his position, he refused to make a statement, indicating that birth control was a Japanese problem "and that SCAP would not impose birth control on the Japanese people: However, SCAP personnel will continue to furnish the Japanese technical advice and information about birth control and SCAP will not undertake to censor statements by consultants such as Drs. Thompson and Whelpton."[26] The various U.S. players in Japan danced continuously between MacArthur's public distancing from, but private resignation to, birth control. Sometime the Americans on the ground put forward an aboveboard advocacy of birth control in postwar Japan (Thompson's position). In other cases, this was thinly disguised as public health (the Rockefeller Commission's position). Deborah Oakley concludes that the official U.S. view, while not prepared to support birth control measures openly, importantly did not prohibit them either. Matthew Connelly suggests rather differently that MacArthur's display of opposition to birth control (that included a very public denial of entry to Margaret Sanger) was orchestrated less with Japan in mind, than the domestic U.S. scene: advocacy of birth control was still risky business for anyone with public office ambitions.[27] Thus from SCAP, and from the Americans as a bloc, there were certainly prevarications and mixed messages. Yet there is no doubt that the major series of legislative changes about reproduction that ensued—the birth control advice centers that were reestablished as early as November 1945 (though outside official domains) and the new research on reproductive health—were offered and taken as U.S.

"permission to proceed." From the U.S. perspective, the more such initiatives emerged from the Japanese themselves, and appeared to the world to do so, the better.

To read this conspiratorially as neocolonial population control is to ignore pre-existing Japanese inclinations and innovations in the field and to simplify the polyvalence of fertility regulation. The swift uptake of policy, law, and practice of fertility regulation was certainly not an alien system imposed. It was, rather, a continuation of Japanese argument and activity from the 1920s, freed up from a period of pronatalist conservatism. Japanese liberals, socialists, and feminists were now in a position to put their position on birth control strongly. Critical to the new political situation was a system of governance that included women's suffrage and that saw an immediate return of a significant number of women to the new Japanese legislature after the first election in 1948.[28]

Part of what made the Japanese demographic and policy scenario unique was the legalization of both abortion and sterilization through the 1948 Eugenic Protection Law. An earlier Eugenic Law (1940) had legalized sterilization to prevent inherited mental illness. The 1948 Eugenic Protection Law that replaced it extended lawful sterilization for eugenic reasons, surrounding the procedure with new processes and regulations. The tight connections between eugenic, feminist, and economic or Malthusian rationales for various kinds of fertility control became evident. "A physician may conduct a eugenic operation on a person or spouse who has a hereditary psychopathy, bodily disease, mental deficiency, or leprosy. If the life of the mother is endangered by conception or delivery, or if she has several children and her health is threatened by delivery, a eugenic operation is legal." Here, "eugenic operation" meant sterilization, though the law allowed for abortion as well, significantly not just on medical or eugenic grounds, but when "the health of the mother might be seriously affected from the physical or economic viewpoint."[29] This "economic" indication was the stunning new development. It made the Japanese eugenic law far less like (in fact quite different to) earlier U.S., Canadian, or Nazi sterilization laws, and far more like liberal abortion laws from the later twentieth century. Both abortion and sterilization became widely and legally available in Japan, and this opportunity to prevent conception or birth was taken up by many women.

Koya Yoshio was director of the Japanese National Institute of Public Health in the postwar years, and in 1952, he summed up the developments for John D. Rockefeller III, who was funding his participation in the forthcoming Planned Parenthood conference in Bombay. Koya noted the significant extent to which the Eugenics Law had become a legal means to prevent and terminate unwanted pregnancies on health and economic grounds. He estimated around 1 million abortions in 1951, with an official number of 638,350. Although now legal, the mortality rate from abortions was problematically high: around 1 in 450 deaths. Sterilization—tubal ligation for women—he noted, was becoming more extensively used as a form of contraception, from 5,749 procedures in 1949 to 16,233 in 1951. In the early months of 1952, he projected 23,000 procedures. "These are mostly for health and economic reasons," with 717 officially recorded for eugenic indications, and 107 under the leprosy provisions of the act.[30] Irene Taeuber and Marshall Balfour agreed that the eugenic effect and use of sterilization was minimal, and 90 percent of the tubal ligations were sought and approved because pregnancy was dangerous to the mother for health or economic reasons.[31] Koya noted also that as far as birthrates were concerned, the (permanent) effect of 16,233 sterilizations far outweighed the much greater number of abortions.[32] Mixed-race children, the result of the occupation, also served as a factor.[33]

The swift uptake of abortion and sterilization to manage fertility in postwar Japan was partly because of the (initial) official sidestepping of contraception. Koya explained to Rockefeller that subsequent government-level motivation for contraceptive-based family planning in turn stemmed from this high rate of legal abortions. Contraception would minimize the number of abortions, even in a circumstance in which abortions were legal, a position explicitly articulated by the Japanese government: "Abortion has undesirable effects on maternal health. It is therefore necessary to disseminate contraception to decrease these undesirable effects."[34] This was in fact the long-standing feminist argument.

In the villages studied by Koya in the early 1950s, 92 percent of married fertile couples with fewer than three children "wanted to practice contraception enthusiastically." Most chose to use condoms over the alternatives—sponges, diaphragms, jellies, and "vaginal tablets"; or the methods of withdrawal and abstinence; or the use of the "safe period." Over his two-year study, both the birthrate and the abortion rate decreased.[35] Koya took great care to note that nothing that could even be construed as compulsion was evident in Japan: contraception was being chosen by the free will of the people.

Koya meant the free wills of individuals or couples. When Warren Thompson talked about "free will," however, he almost always meant an aggregated national will. This was linked to his anticolonialism. Policies and processes to reduce high fertility could not be imposed by a Western power but had to be an expression of the free will of a free people, in a kind of national self-determination model. In other words, for Thompson, the exercise of a national reproductive "free will" was inconsistent with colonial or even semicolonial systems of governance. "A far-sighted act of national self-discipline," is how Carlos Paton Blacker summarized good birth control policy in his positive review of Thompson's *Population and Peace in the Pacific*.[36] The ambition was to abolish the colonial system and give the nations of the Pacific full rights of self-government as independent and sovereign powers and as members of the United Nations, so they would have the freedom to implement policies to reduce fertilities. Thompson rarely individualized reproductive "freedom," then, and virtually never conceptualized it as gendered. He could only see government, even "self-government," from a national–colonial viewpoint, and never made the shift toward argument for reproductive rights of the person, even the version authorized by postwar liberal humans rights agendas that were to become so powerful.

Thompson did make other shifts on the question of world population, however: large ones. Having been the most prominent proponent of the interwar idea that "international tension" would be solved by the movement of people from global "danger spots," after the war he conceded that territorial redistribution was not going to work. This shift was starkly evident at an exchange during Rockefeller's Williamsburg meeting in 1952, where many of the U.S. actors in occupied Japan regrouped. A number of other delegates did attempt to put forward ideas about population transfers and migration, and Warren Thompson found himself arguing against people who were presenting ideas based on his own work. Dorothy Thomas, an economist and sociologist from the University of Pennsylvania who had spent a decade researching the forced mass migration of Japanese-Americans from the West Coast, was aligned with Thompson's interwar ideas. Population redistribution and economic growth was her particular research field, and she put the international population transfer angle of the problem in most unusual contexts. For her, it was part of the same broad field of inquiry as domestic forced migrations in the United States, both of Japanese-Americans during the war and earlier forced migrations of "our Indians."[37] Given this problematic history, she put to the meeting that more reasonable and careful international migration should be pursued.

Rockefeller's public health man who had led the Japanese delegation, Marshall Balfour, asked for clarification: "Does she believe international migration offers a major source of relief in population pressures in such countries as India, Indonesia, or Japan?" Most assuredly, replied Thomas, "If we had a reasonable world in regard to trade—I might say, if we had a reasonable world in regard to racial relations and movements of people." William Vogt, for one, was incredulous. Why would or should North Americans feel any responsibility at all to take surplus populations from elsewhere? Dorothy Thomas attempted a response, but her argument seemed to come from the past or else from a postcolonial future: she suggested the benefits that had accrued to the United States because of the good fortune and historical timing of the open North American continent needed to be shared with a now-closed world: "Because we have benefited from this thing you call nature, because of the fortunes of timing, which we got here, and probably not because—or, let's say, possibly not because of any superiority in the genes [*sic*]. . . . So that if you were looking at the thing from the standpoint of, let us say, something beyond national boundaries, then it seems to me that there is a moral responsibility." Besides which, she added, assimilation of Asians was entirely possible, reminding us again that in this period and context, assimilation was perceived as a progressive position. Thomas stumbled, though, unable to put her finger on the problem in any satisfactory way.

It was then that Warren Thompson—who had argued precisely that combination of positions for decades—interrupted. "If you think of the physical problems of moving out the Japanese increase, these last few years, these things would roughly figure out at 5,000 people per day that you've got to move out and settle, just to keep even, in Japan alone." The whole history of European growth and transfer to North American land was an entirely different scenario. To expect a similar movement of people now, he thought, was no longer feasible. Frank Notestein has been paying close attention and remarked that Thompson would surely be the last person to diminish migration as a solution. And Warren Thompson conceded to those who knew his lifetime's work best. "I personally have been more favorable to migration than most people, I think, but I don't think it can do the job." Dorothy Thomas ended up agreeing: "I was thinking perhaps in terms of the pattern of the past, which may have been unrealistic."[38] Her sigh is almost audible.

This frank conversation among the U.S. population experts reveals the contours of the postwar shift. The regulation of fertility was beginning to recommend itself precisely because migration and population transfers were too difficult. The spatial questions of migration and population transfers began to drop away, increasingly replaced by the conceptual axes of demographic transition: fertility and mortality began to dominate the intellectual framework.

Death Control and Birth Control: Population and World Health

As every population commentator from any discipline knew, population changes were about death as much as they were about birth. And while controversy over population control has come to focus on the gendered politics of "reproductive health" or "sexual health," which incorporated family planning, it was at least as much mortality as fertility that was significant. Probabilities of death had long been part of the management of life,[39] and the death rate that really mattered for demographers and economists, as well as feminist health lobbyists and an emerging international health network, was that of infants.

Throughout the century this was globally disparate. In the period 1950 to 1955, the lowest infant mortality rates were documented in Sweden (20 deaths per 1,000 live births); Norway (22); the Netherlands (23), Australia (24), and New Zealand (26). In the same five-year period, India's infant mortality rate was 165 deaths per 1,000 live births. But a decline in the rate was evident everywhere. Between 1950 and 1960, infant mortality (per 1,000 live births) in Japan decreased from 50 to 37; in Europe as a whole from 73 to 51; and in the United States from 30 to 27. For regions and nations with high infant mortality rates, there were decreases too: India (from 165 in 1950 to 153 in 1960); Central America (129 to 110) and in southeast Asia (165 to 142) in the same years.[40] The global decline over time in infant mortality was one important axis by which to assess world population trends. The other axis, equally significant, was the global disparity in infant mortality that nonetheless persisted. Chandrasekhar documented this: infant mortality, like the calorie consumption maps and graphs of the period, starkly represented the unacceptable difference between "undeveloped" and "developed" nations (figure 11.1).

As so often happened with respect to population, a solution became a problem. Lowered infant mortality meant, at least for the next generation, more females of reproductive age and an accelerating population growth. Successes within the health sector, then, were perceived to create short-term problems within the population sector, especially in the postwar decades, when new health interventions had sudden and sometimes one-off effects. One such instance was recorded by the *Population Bulletin*. At the end of the war, very high infant mortality rates (greater than 250) were noted in British Guiana, largely due to insect-borne diseases. This was the location for one of the earliest applications of DDT, where spraying brought "instantaneous

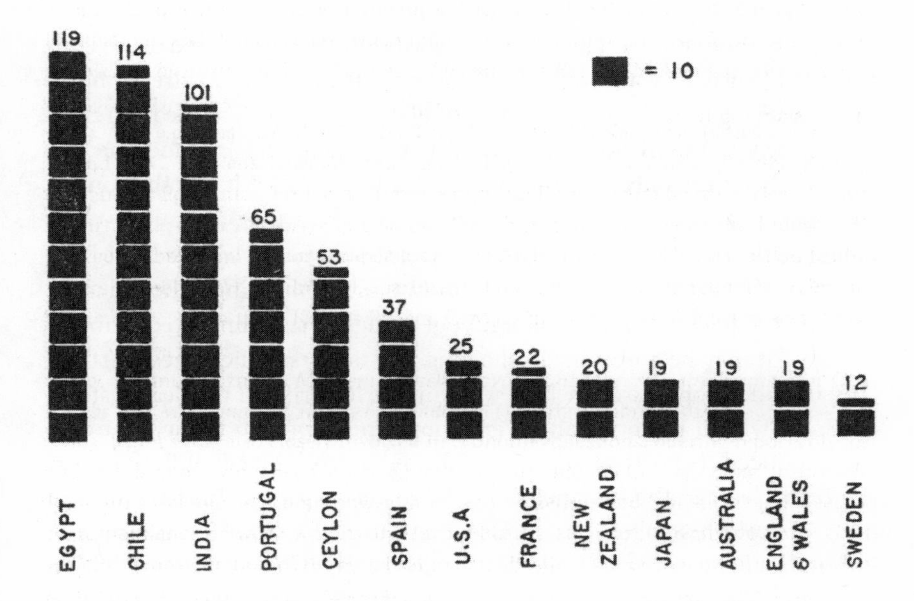

FIGURE 11.1 Infant mortality in selected countries, 1965. Chandrasekhar's table on comparative infant mortality rates was another way of "mapping" global disparity. (From Sripati Chandrasekhar, *Infant Mortality, Population Growth, and Family Planning in India* [Chapel Hill: University of North Carolina Press, 1972]. Reproduced with kind permission, University of North Carolina Press.)

results," and infant mortality had been reduced dramatically by 1948. Health writers would stop there, but the *Population Bulletin* writers indicated that the final outcome was one of the most rapid rates of population increase ever recorded.[41] Most in the population establishment would use such a case to argue for birth control, needed to balance or really offset newly effective "death control." Some began to frame it essentially as a deal. "The willing gift of mortality control to the under-developed lands" should be accompanied by "equally willing acceptance of fertility control."[42] Indeed, "Death control can get out of control," wrote Julian Huxley. "To this, there is only one desirable answer—birth control."[43] Carlos Paton Blacker constantly pushed this argument as well in the postwar years.[44] Decreasing mortality through public health measures—"death control"—was for Blacker the phenomenon that aligned family planning and eugenics.

Others turned the argument inside out, reverting to an established Malthusian position. In a prize essay honoring Margaret Sanger's visit to Japan (after the withdrawal of MacArthur), a political science student at Tokyo University thought that if birth control was neglected, unwanted death control would ensue. "Nothing can be done about the population already born. The only thing feasible is to limit the population to be born in the future and to eliminate the source of future population pressure. It is not too much to say that this is the only way of avoiding 'death control.'"[45] It was the conservationists who were most likely to take this a step further, and suggest that failing birth control, "natural" death control should be allowed to come back into play; that is to say, the active withdrawal or withholding of health measures that reduce mortality. William Vogt thought that failing a reduction of fertility, "it may be a mistake in the long run to pursue measures, particularly in undeveloped areas, which result in the lowering of the death rate."[46]

Those within public health were sometimes upfront about the problem produced by postwar success in "death control." Marshall Balfour was candid, at least in the confines of the Williamsburg Inn. "In the course of time," he told his colleagues, "I also came to realise that I and others in Public Health work were contributing to the creation of this population problem."[47] In fact, many sectors of the Rockefeller world were confronted by the increasing critique that public health itself compounded population growth, an idea that could potentially undermine the foundation's core mission to improve the lot of humankind, not least by minimizing infectious disease. Chester Barnard, president of the Rockefeller Foundation, wrote strategically to Cardinal Spellman of New York, putting the implications of its health work forward and testing the ground on population. The foundation's concern was to prevent many infectious diseases—yellow fever, typhus, malaria—he explained. It was engaged in agricultural developments aiming to increase food production, especially in Mexico. But "the ultimate utility of these efforts is rendered doubtful, at least from certain points of view, if the net effect is only an increase in population maintained at starvation level." What, he asked, is the church's view with respect to this broad problem?[48] Spellman was not going to be drawn and did not respond.

The health rationale for contraception tied infant health and maternal health together as one of three broad-based arguments: first, that spacing births was less likely to render the mother sick after confinement and for the next confinement; second, that availability of birth control would minimize dangerous abortions; and third, that spacing births through the use of contraception would significantly reduce infant mortality.[49] "The unassailable case for birth control is that it is an essential element of that part of the preventive, as well as curative medicine which cares for the health of mothers," as the Birth Control International Information Centre put it in the late 1930s, a set of arguments beginning to receive wider audience.[50] While Margaret Sanger and Edith How-Martyn were concerned with the benefits for individual women, it was the demographic effect of the reduction of maternal and infant mortality that interested most of the Malthusian men and increasingly those concerned with demographic transition. Prevention of conception became part of preventive medicine, because wherever birth-rates were high, maternal and infant death and sickness rates were also high. Following this logic, far from preventing life, birth control in fact saved lives.[51] Put another way, in preventing future lives, it also prevented untimely infant deaths, "wasteful" deaths in an ecological view.

In the middle of all this, those on the front line running birth control clinics and medical practices saw both increasing demand for and provision of new contraceptive knowledge, methods, and devices. In West Africa, medical doctor Okli Ampofo said that between the slave trade and European occupation, many African populations had diminished. What was often known as the white man's grave might very well be known as the black man's grave. But he was still constantly asked for information on birth control. Women were demanding it, he said. "They came to him, as a doctor, every day."[52] Sanger knew this, as did Blacker, and so did the medical researchers closest to the early development of the pill. Historian Elizabeth Watkins has documented the public eagerness on the part of U.S. women not to continue with multiple, close pregnancies. "I am about 30 years old have 6 children, oldest little over 7, youngest a few days. My health don't seem to make it possible to go on this way," wrote one American woman in 1957.[53]

Notwithstanding this cross-cultural demand, early attempts by the World Health Organization (WHO) to include contraception as part of health services, or even to promote research into various contraceptive methods, failed. Between about 1948 and 1952, a number of member states and observers at the World Health Assembly, as well as personnel within the WHO attempted to have family planning incorporated as part of WHO business. The Vatican and Roman Catholic-dominated countries on the one hand and Communist delegates on the other heatedly objected, and it was resolved at the 1952 World Health Assembly that WHO should not have any mandate to implement or recommend birth control or family-planning measures. There was still a strong memory of the League of Nations Health Organization's attempt to deal with contraception twenty years earlier. WHO director, the Canadian psychiatrist Brock Chisholm, privately expressed his frustration that it was still seemingly impossible to incorporate contraception into the UN agency's work. A recent study by Abraham Stone, undertaken under WHO auspices and at the invitation of the Indian government, had been confined entirely to the use of the rhythm method. Chisholm always claimed that he had been assured by the Vatican that the rhythm method was doctrinally acceptable, but even that small study became controversial.[54] The director of WHO was in a no-win situation. Personally keen to progress with health-related solutions to the population problem, his field also came under attack for creating those very problems.

William Vogt, relentless in pressing forward his conservation agenda, was on record "blaming" antimalarial insecticides and antibiotics for global growth rates.[55] He engaged with WHO during the early 1950s as president of American Planned Parenthood, arguing that health interventions caused overpopulation, and that overpopulation caused malnutrition. Vogt aggressively suggested that population increase became a health issue, not because it concerned women's health or infant mortality, but because it created malnutrition. "The pathological condition that we call malnutrition," is a result of population growth, which is the result of the fall in death rates. Therefore, he finished, the cure of malnutrition would seem to come within the purview of WHO. At a population level, contraception was cast as a cure for malnutrition. "Unless an attempt is made to reduce the demand as well to increase this supply, it would seem to me as a biologically trained layman that only half an attempt is being made to cope with the disease of malnutrition."[56] For William Vogt, and he was not alone, the improvement of health *caused* the population problem.

Gender and International Planned Parenthood

It was in 1959 that the Sixth International Conference on Planned Parenthood was held in New Delhi. Seven hunderd and fifty delegates and observers from thirty-eight countries met together over eight days. Dhanvanthi Rama Rau presided, and a frail Margaret Sanger received a standing ovation. The prime minister spoke, Jawaharlal Nehru never missing an opportunity to indicate that his was one of the very few governments in the world that included family planning as official policy. And yet, notwithstanding Rama Rau and Sanger's presence, women's health was not really the issue at hand. Rather, the family-planning movement had to be thought of and planned for in terms of "the larger movement of raising the standards of living of the people."[57] Sanger had heard that many times before and from the earliest neo-Malthusian years. She would not disagree, but nor would she limit the significance of family planning to that vision.

At the New Delhi conference, Blacker spoke on eugenics in an atomic age, offering a message entirely familiar to the conservationists and ecologists who listened, including Radhakamal Mukerjee, William Vogt, and Robert Snider of the Conservation Foundation. Julian Huxley spoke, of course. "Population Planning and the Quality of Life" was a standard for him by the late 1950s. A little thinly, he said that he wished Margaret Sanger had been listened to more carefully by earlier generations (he might have done so himself). Huxley was beginning to recast the story of birth control retrospectively as largely the feminist story of Margaret Sanger: to do so was becoming useful. Remembering the century's developments, he perceived the process as marked by her struggles and successes: the World Population Conference in 1927, the 1952 founding of the International Planned Parenthood Federation; the birth control conference in Delhi in 1959 in which, as he later remembered, she persuaded the Ford Foundation to give large sums to birth control work. "I am proud to have known her," he wrote, "and to have contributed something to the crusade she started—a crusade *against* woman's sufferings and *for* greater fulfilment—individual, national and global."[58] Sanger had long understood population as jointly individual, national and global, as early as 1922 criticizing many of the men around her for not comprehending this.

The theme of the New Delhi conference was "Family Planning: Methods and Motives," signaling the new social science and psychological focus on motivations for fertility control. The question had governed much of the work of the Princeton group under Notestein, though they, like others, never quite solved the mystery of what motivated people to limit fertility.[59] The contemporary field as a whole knew that to make any significant impact at a population level, "voluntary family limitation" had to be understood. It was the seemingly spontaneous, voluntary and individualized limitations that had reduced fertility rates in Europe in the later nineteenth century—in the face of many government policies to the contrary. This is what truly intrigued this generation, and the phenomenon somehow needed to be replicated. The invention of contraceptives itself seemed to have been a relatively minor factor, many already agreed in the 1950s. It was, in far more complex ways, about changes in expectations of family size. This, then, required research on and about women: what they wanted, what they thought, and what actions they took. Some of this research was feminist in intent, but it was not necessarily so. Nonetheless, ignoring women as objects of population inquiry in the population field was now far more difficult than it had been.

Family planning, birth control, eugenics, population control: they all impacted on women differently than men. Sometimes this was recognized, sometimes not. There were major discursive issues here, too, as different sectors engaged in public relations tussles. The links between women and maternity on the one hand, and maternity and femininity on the other, were powerful. They cut several ways, however. Various international and national Planned Parenthood chapters promoted their work through a feminized and strongly individualized maternal message. At one point the slogan "Reverence for Life" accompanied a photograph of singular maternal-infant love, suggesting that the most caring mother created life responsibly, every baby wanted and deeply cared for. This enshrined "The Right to Decide": "No one in Planned Parenthood believes that parents can or should be told how many children to have. There is no 'ideal' number of children to suit all families. . . . [E]very couple should be assured the right to determine its own family size."[60] Well might Planned Parenthood groups promote such a safe image, as *femininity* could easily become or be read as *feminism*. Sanger managed to walk the fine line between femininity and feminism with great skill. But could the men involved in the field? In 1952, Fairfield Osborn and Kingsley Davis met with Rockefeller's advisors to discuss just how conservation, family planning, and world population control could all fit together. They were candid, and Rockefeller's advisors warned about the potential problems for the fund and for the family and urged the need for backup plans. One even suggested "the possibility of studying migration as a screen . . . the importance of flanking the contraceptive project with as many others as possible."[61] Birth control could still be a public relations nightmare.

At other times, the feminist and feminine domain of birth control was seen to be too diminishing for a public sphere issue of such international significance. Rufus Miles Jr., an administrator in the Federal Security Agency, wrote a long piece in 1952 that the Rockefeller team considered closely. In casting around for how to present population control to North Americans on the one hand and to the Third World on the other, he thought the Marshall Plan technique a possible model. But he questioned the name "Planned Parenthood," because it contained no hint that the organization "is attempting to grapple with one of the most basic and important problems of humankind." People tend to think "it simply a women's organization not a world-shaking matter." Work needs to be done, he advised "to unstamp it as a women's organization." Might "Race for Peace" be a better name for the organization, he wondered. Clearly this was an unsuccessful reworking of the idea of Food for Peace, one that reveals the immediate connection in the U.S. official's mind.[62]

None of these difficulties stopped various men from deploying women's health arguments as strategically useful. It was Vogt who put on the Rockefeller table the idea that sometimes the most expedient way of talking about birth control was to frame it as a form of maternal health:

> It seems to me that perhaps we could gild the philosophic pill with two different coatings, one with maternal health, although if we are asked about that we are interested in population, overseas; but, spread birth control under the guise of maternal health. And here in the United States you can talk about "population," and what we are really getting at the same time is birth control.[63]

Elsewhere, he pinpointed the usefulness of framing family planning within a health program, in a way that might offset some of the difficulties attached to talking about "population limitation" in overseas contexts: "It is commonly said in the Orient that we want to cut their population because we are afraid of them. But the program can be sold on the basis of the mother's health and the health of the other children in the family and, as Doctor Taeuber suggested yesterday, the development of the mother. There will be no trouble getting into foreign countries on that basis."[64] Rockefeller's 1952 population meeting ended up being key to the development of new contraceptives that changed the lives of millions of women. Originally, at least, it did so less from a primary concern for reproductive health or women's health, and more through strategic use of it.

If Vogt was capitalizing on maternal health for his conservation agenda, Sanger was still capitalizing on war and global crises of one sort or another for her women's health agenda. Having theorized birth control for world statesmanship in the aftermath of World War I, Sanger was still putting birth control, war, and peace together at the end of her life.[65] It is the remarkable continuity in the arguments about birth control and peace that is most striking. When NBC organized a Chicago-based radio discussion titled "Three Billion People," the key players were largely in agreement. Dhanvanthi Rama Rau said of the low standard of living in India "there is so much hunger and starvation . . . family planning seems to be very necessary." William Vogt, introduced as chief of the Conservation Section of the Pan American Union, introduced "the total complex . . . people very generally are destroying their means of subsistence through failing to observe conservation measures." Carlos Paton Blacker, also there, introduced Margaret Sanger, noting, "She feels this to be a world problem rather than a national problem."[66] The arguments were old, but the Cold War made them seem both new and urgent.

Especially in its U.S. chapters, Planned Parenthood was strongly pressed forward from the long-standing geopolitical domain, the ecological agenda of the Conservation Foundation, and the anti-Communist foreign policy agenda. The fall 1959 issue of *Planned Parenthood News*, for example, was pumped full of the newly endorsed foreign policies. The words of the Draper Report to President Eisenhower, which recommended that birth control be linked to aid, received solid-cap treatment: family planning was not just economic aid "designed to deal with the problem of rapid population growth" but "maternal and child welfare" as well. Senator Fulbright, chair of the U.S. Senate Foreign Relations Committee was also reported in the issue, saying the population control will become imperative: "Current scientific developments indicate that a radical advance in techniques, such as a safe, effective and inexpensive oral contraceptive, could soon become available."[67] And indeed it was. This is the kind of mix that characterized family planning during the 1950s; it was in no sense solely or even mainly the preserve of feminist lobbyists. We might even say that the official foreign relations status of the issue made birth control even more of a man's domain than it had been in the 1920s.

At the 1959 annual luncheon of Planned Parenthood Federation of America, held (not a little ironically) at the Hotel Roosevelt in midtown Manhattan, all the men gathered at the head table. They represented the multiple interests and institutions that had long shaped the world population problem: Clarence Senior, chief of the Commonwealth of Puerto Rico's Department of Labor, Migration Division; F. Fraser Darling, vice president of the Conservation Foundation; B. K. Nehru, Indian Commissioner-General for Economic Affairs; Frank Notestein, now president of the (Rockefeller) Population Council; and William Vogt, president of Planned Parenthood Federation of America. The luncheon's speaker was, inevitably, Sir Julian Huxley, now pitched not as his signature generic "biologist," but as equally generic "world population authority."[68] These were years in which they could all agree that population was the problem of problems, important and immediate.[69] This was population in the Cold War, but the words derived, in effect, from John Maynard Keynes after World War I.

The solutions were looking different, however. Few were now suggesting that migration was a realistic solution to differential density. Warren

Thompson had distanced himself entirely from that idea, while some of the Indian scholars held on: they had more at stake. Chandrasekhar at this point presented an integrated world population policy that would ideally follow from five agendas: (1) colonial freedom; (2) universal birth control; (3) planned international migration; (4) industrialization; and (5) agricultural development.[70] It was the third of these—planned international migration—that was beginning to make no sense. Chandrasekhar and others slowly capitulated to the regulation of fertility as a far easier prospect than managed migration. For almost all those caught up in the Cold War frenzy around population growth, it was clear that if fertility management—whether conceptualized as birth control or family planning or population control—could be made politically acceptable, it could become a biopolitical solution to a geopolitical problem. In the context of an emerging "development" agenda, part of the usefulness of fertility regulation was that it kept people in their local place. As a primary response, it could potentially put to rest all the discussion about migration, resettlement, international population transfers, peaceful or even belligerent adjustments to density. On the face of it, world population could be shifted conceptually and politically out of the domain of "living space" entirely.

12

Universal Rights?

Population Control and the Powers of Reproductive Freedom

> Birth control knowledge must become the birth-right of every wife and mother so that she can plan and regulate the size of her family for health, economic and other reasons. Babies by choice and not by chance must be enshrined as one of mankind's fundamental freedoms.
>
> SRIPATI CHANDRASEKHAR, *HUNGRY PEOPLE AND EMPTY LANDS* (1954)

Reproductive politics in the postwar period is often assessed as involving a struggle between those seeking individualized birth control for women on the one hand, and more problematic programs of aggregated, and by implication more coerced, population control on the other.[1] Accordingly, population control with world ambitions is often understood to be a reemergence of interwar eugenics operating "under new labels."[2] Historians will thus draw a connection between compulsory sterilization—in some states of the United States, in some provinces of Canada, in National Socialist Germany—and postwar states' interest in sterilization, in particular that of India. Ian Dowbiggin's work reveals the connections between U.S. eugenics and sterilization organizations of the 1930s and the population control campaigns of the 1960s and 1970s, usefully nominating a twentieth-century "sterilization movement."[3]

For several reasons, however, it is not quite correct to understand "population control" as surreptitiously covering for an apparently underground eugenics movement after the war. Far from postwar eugenics in the international public sphere being silenced or driven underground, often enough there was an entirely explicit endorsement of it. Many on the international stage saw eugenics per se as neither controversial nor problematic, even as they understood the Nazi version of it to be so.[4] Some demographers were certainly distancing themselves from so-called quality questions,[5] and others were cautious about pursuing research about race in the aftermath of Nazism and the Holocaust. Blacker and Frederick Osborn both imagined and talked about "crypto-eugenics," as Matthew Connelly has shown.[6] But in the end neither of these eugenic leaders nor their respective associations retreated from the term in the postwar period. Far from it. They fully incorporated activity on world overpopulation *within* their eugenic activity, not as a cover for it.

The connections should be unsurprising, even if they were problematic: postwar population control was one manifestation of entirely standard existing links between eugenics and fertility regulation. Eugenic and neo-Malthusian argument had been entwined since the beginning of the twentieth century at least, and in some contexts, for example in India, had been more or less indistinguishable for several generations. And so, that many eugenic organizations morphed into family planning, birth control, or population control groups is at one level to be expected; many already functioned that way. This long historical process needs analysis and explanation, rather than exposé as a sudden postwar change.[7]

Eugenics was deeply linked to any number of compulsory and coercive measures across multiple jurisdictions. This history stands as a world historical object lesson. Yet there is another history lying within the close relationships among eugenics, fertility management, and world population control: counterintuitively, a history of reproductive "freedom." Compulsion came to be assessed as problematic and unacceptable not only by critics of eugenics but also by leading postwar advocates of eugenics. It was they who often discussed the benefits of reproductive "freedom" and how best to implement it. Especially after the war, this was clearly imperative from their point of view, if eugenics was to move forward on a world stage. Thus, eugenics and world population control were historically connected not just because of a history of coercion and compulsion, but also, strangely, because of Anglophone eugenicists' investment in the idea of modern freedom.

This approach to the history of mid-twentieth-century eugenics does not therefore suspend criticism of its practices, organizations, or personnel. Rather, it takes seriously the work performed by a late modern ethics of freedom, in Nikolas Rose's terms, the "powers of freedom." Reproductive "liberty" is plainly a public and private good. But the rise of freedom in relation to reproduction has a history and should be contextualized by a postwar liberalism, after Nazism. As Rose puts it: "Within a *Vitalpolitik* designed to create a life worth living, a new set of ethical and cultural values had to be created . . . which would accord individuals and families the power to shape their own lives."[8] The substance of this history is perhaps most intense and complex with respect to reproduction and population: the meeting point of life, death, the individual, and the state. This is the history of reproductive "freedom" that eventually, though comparatively slowly, became a "human right."

One trajectory of world population control during the twentieth century was certainly toward Indian sterilization programs and China's one-child policy. Important as these measures were and are, they were international exceptions that proved the rule for liberal democracies. The need to persuade, not force, was the key to success. World population growth did not come to be managed by a world state as H. G. Wells had early imagined. It came to be (ideally) universalized through the self-government of individual women: the exercise of a "universal" right to reproductive choice. It was "universalized" as much as "internationalized." This was far closer to Margaret Sanger's dream.

Liberal Eugenics

In Britain after World War II, the liberal strand within the Eugenics Society was extremely reluctant to dispose of the term or the title "eugenics." To do so would be to concede a connection to the racism of the Nazis. This was especially Carlos Paton Blacker's position. In 1947, he was asked to assess the scientific value of Nazi medical experiments and concluded that nothing could justify them "even had these been voluntary." And he was at pains to point out that "none of these experiments have any bearing on eugenics as the subject is understood in this country."[9] Such policies and practices were, in his view, "widely, though I think erroneously, connected with eugenics."[10] At one level, this was not disingenuous: such medical experimentation would not have been endorsed in the United Kingdom, neither by the state nor by British eugenics organizations. But on another level, Blacker had to work hard to distance his society, because the wartime medical experiments were on a continuum with earlier Nazi sterilization policies and were legitimately judged to be so. This made it the business of eugenics, whether he liked it or not.

Sterilization was clearly on eugenic agendas the world over, even if there was major disagreement about the legality of implementation. Indeed, legitimizing the procedure domestically had been Blacker's long-standing personal mission, especially given his work as a psychiatrist. But throughout decades of international discussion, jurisdictions under English common law, with the major exception of some Canadian provinces, had persistent problems with the compulsion of sterilization through legal means. Blacker well knew that eugenics, birth control, population control, indeed any management of fertility, was far more effective and genuinely more acceptable when grafted onto freedom, not force. This was Blacker's version of eugenics that he continued to promote and defend vigorously after the war and after the Nuremburg Trials.

Julian Huxley, too, continued to advocate "truly scientific eugenics," as he put it in the United Nations Educational, Scientific and Cultural Organization (UNESCO) manifesto, for decades after the war, indeed until his death in 1975.[11] How did this advocacy for eugenics align with his international work against racism? As we have seen, Huxley did not see eugenics as being *necessarily* a racist practice, because it was not necessarily a practice about race at all. For Huxley, eugenics was about intelligence and genetic physical and mental defects, as he would have it, characters that could and should be bred out of populations. Such inequality spread across all populations, he argued; it was not an inequality of "race" itself. This was a position that Huxley had put forward from the time he wrote *The Science of Life* in the late 1920s, through his bold statements with Alfred Cort Haddon in the late 1930s, and possibly with even more conviction in the postwar period. "The fact of inequality" between humans did not mean inequalities between racial or ethnic groups; it meant that within any given group, that range of inequalities existed. Expecting "eugenicists" to always be "racists," does not in the least explain how some eugenicists also led antiracism in the international public sphere. It can also obscure the problematic nature of their convictions about human inequality on other axes, and the corresponding discriminatory politics that they did hold close. In other words, just because Huxley was a leading "anti-racist" does not make him an egalitarian, still less a democrat. Thus, while Huxley might have been "progressive" in terms of the politics of race, there is no sense in which he believed that all humans were necessarily equal by virtue of being human: far from it. *This* is what made him a eugenicist and what made his eugenics so deeply dangerous. Indeed, what made both Huxley and Blacker stick with eugenics was their ongoing belief in the possibility, even the fact as they would have it, of inequality between individual humans.

In short, while Huxley and Blacker may have been uncomfortable with sciences of race difference and the kind of compulsion legislated in Nazi

Germany, this did not mean that they were opposed to eugenics. The reverse is the case: for decades to come they were English flag-bearers for eugenics, and a renewed global eugenics at that. They both thought that the new "vogue" for egalitarianism was so much biological nonsense. Julian Huxley was quite clear: "Human beings are not born equal in gifts or potentialities, and human progress stems largely from the very fact of their inequality." " 'Free but unequal' should be our motto,' he wrote quite unashamedly in a book titled *The Humanist Frame*, no less.[12]

At the same time, Huxley was acutely aware of social and economic disparity. The fact of biological inequality was a certainty for him. But difference between individuals could only be properly discerned after social and economic measures had leveled (up) everyone, a message carried forward from the Geneticists' Manifesto of 1939. Huxley transferred most of the Geneticists' Manifesto, "its purpose and its philosophy," directly into his UNESCO manifesto. Sorting out the world's quantity and quality of people was UNESCO's mission, as he saw it. There was an optimum density for the world's population that would maximize its biological and social quality. A world population policy that made birth control information, methods, and materials available would ensure this. Idiosyncratically, though, and in distinction to many of the socialist geneticists, Huxley sought "quality" but not through any endorsement of "equality." His vision for UNESCO was explicitly opposed to "The Age of the Common Man," as he put it.[13]

Huxley was an unreconstructed elitist who believed that less is more, in most spheres. "What future do we contemplate for the human race," he would often ask during the 1950s and 1960s: "Do we just have to put up with more people?" At a speech in honor of Margaret Sanger, he described current world culture as "one in which quantity is threatening quality and also, if you like, one in which the present is threatening the future." He argued that to deny millions of human beings knowledge and right to control their reproduction was wrong not because he actually believed in the right itself (let alone women's particular rights to reproductive control) but because not to have fertility control "is to give meaningless quantity the victory over meaningful quality."[14] "Quality" was a term Huxley constantly used as his public writing and speaking on world population growth accelerated, increasingly turning his internationalism into anti-Communism. Huxley's use of "quality," then, did not necessarily signal race science, or a vision of a world with less variation and more homogeneity.[15] It did, however, squarely retain an established eugenic meaning of superiority and inferiority of mental and physical capacity and fitness (and therefore, for Huxley and other eugenicists, worth). "Quality" also integrated with an older economic argument about the standard of living associated with "optimum density." "Quality of life" was a fresher discourse that was on the face of it about maximizing individual fulfillment, in which reduced population density raised the economic standard of living for all.

Postwar Race Science

Well might Huxley and the cosmopolitan left-wing geneticists insist on the elimination of spurious sciences that fostered "racial antagonism."[16] Others in the postwar period were attempting to re-invent and re-institutionalize just that. The "underground" thesis about postwar population work holds up best with respect to research specifically on race. Stefan Kühl has traced the connections between the U.S. Pioneer Fund's dreams of "race betterment" and research in the United States from the 1950s.[17] Significantly, this was the cluster of researchers who most eschewed "eugenics," the term. While the Pioneer Fund's activities in the United States have been extensively analyzed, especially the funding activity of benefactor Wickliffe Draper,[18] the latter's activity in the United Kingdom in the 1950s and 1960s has received less attention and offers telling insights into postwar eugenics and research on population and race in the Anglophone world.

Wickliffe Draper had been impressed with a broadsheet on West Indies immigration produced by the British Eugenics Society, drafted by new General Secretary Colin Bertram and toned down by an uncomfortable retired secretary, Carlos Paton Blacker. It was still anti-immigration enough to appeal to Draper, who asked anthropologist/geneticist Reginald Ruggles Gates to survey the English field for likely locations to fund research: "1) the attitude to miscegenation; 2) immigration quotas; 3) improving population." Gates, in turn, was trying to persuade Draper to fund a chair of human genetics, or possibly a chair of racial studies, in London. He was aware that Draper would not fund social anthropology, but only "the racial aspects of genetics, physical anthropology, perhaps psychology." Gates, seeking advice from the geneticist Fraser Roberts, who was then establishing early genetic counseling clinics in the United Kingdom, was aware that Draper's funds were aimed "against the propaganda for race crossing which is so prevalent in this country" (meaning the United States).[19] Fraser Roberts himself was concerned, however, with the whole question of tied funding. There could be "no thought of any curtailment of academic freedom, which would of course be unacceptable to the University."[20] The stipulations from Draper proved too strong, suggesting the extent to which biomedical race research was institutionally delicate in the 1950s. When the Eugenics Society itself was approached to accept Draper funding for race research, it also "declined all offers," because there were too many strings held by the benefactor. Yet there was nothing to prevent Bertram from personally accepting funding for a research project. This he did, and produced a study of 3,000 Liverpool schoolchildren of different racial origins and an inquiry "into the relative fertility of the immigrant West Indian, Pakistanis and others in comparison with that of the indigenous inhabitants of Birmingham."[21] Historian Gavin Schaffer has recently analyzed this series of exchanges as evidence of the Eugenics Society's willingness to have a "surreptitious relationship" with a U.S. segregationist, its connection, in other words, with racist science.[22] Yet there is equal significance in the society's refusal of the funding. Bertram met some resistance in the Eugenics Society, enough for the society not to proceed. Blacker wrote to him: "Surely colour feeling is very largely prejudice and self-interest and has hardly any biological significance? Is there good scientific justification for making it biological?"[23] Many of those who explicitly owned the term eugenics in the United Kingdom in this period would not connect themselves with U.S.-derived funding for research on race. It is this critique that is missed by crudely conflating "eugenics" with "race science" either before or after World War II.

Also missed, often enough, is the more straightforward continuity between interwar eugenics and postwar population control at the level of problematized object of inquiry: the mass poor characterized by high fertility and mortality rates. Huxley put forward a dismissive assessment of the social value of "the poor," increasingly though not exclusively figured in world terms as Asian poverty, the Asian poor. "We know in our hearts," he confided to his implicitly Western reader, "that people do not exist in order to live all their undernourished lives in the illiterate ignorance of some Asian village."[24] Huxley even ended up asking, "What are people for?" (a phrase he received from Colin Bertram).[25] Unless there was some fulfillment—"quality of life" linked to a higher standard of living—the implication was that that life was worthless. This was clearly only one conceptual step away from people, a person, being not worthy of life itself. Huxley would probably agree, certainly not about those already living—the Nazi version of unworthy lives—but with respect to those not yet born. This was, indeed, a standard Malthusian (and now Planned Parenthood) argument for birth control; that those born only to die in infancy or to live with a disability were better not to have been conceived in the first place.[26]

Negative eugenics, argued Huxley in his 1962 Galton Lecture, by which he meant minimization and even eradication of genetic defects while maintaining human variation in all other ways, would only be successful if family-planning and eugenic aims were incorporated into medicine in general and public health in particular.[27] Accordingly, Huxley lobbied for medical research into new contraceptive technologies. The ultimate solution, he announced, like everyone else, was a simple and cheap hormonal contraceptive, but until this was available, the work at hand was the preparation of public opinion, the incorporation of existing family planning into health services, and the training of personnel. There was also the method of sterilization, especially of men, of which he was a long-term proponent. This also had a curious relation to formal eugenics.

Sterilization, Eugenics, and Population Control

In the United States there were clear links. The California-based Human Betterment Foundation strongly expanded its sterilization agenda, as Dowbiggin has shown, onto a (third) world stage. Its pamphlet *Human Sterilization Today* cited population geneticist Ronald Fisher as the authority on the viability of eugenic sterilization.[28] The irony was, though, that Fisher had already distanced himself from the Eugenics Society in Britain; indeed, he resigned in 1942, partly because it was all looking too progressive for his political tastes. In the light of this U.S. pamphlet, an emergency meeting of the Eugenics Society was convened in London, which threw Alexander Carr-Saunders, Huxley, and Blacker deep into sterilization conversation, again. They were each strongly interested in promoting voluntary sterilization as a form of contraception, which would also have a population-level eugenic effect. More than their U.S. counterparts, however, they held characteristically English liberal anxieties about the compulsory powers enacted in the sterilization laws of many U.S. states and some Canadian provinces.[29]

This was also the context of the *Population Bomb* pamphlet, published in the United States by the Hugh Moore Fund, which had been established in 1944. Self-made millionaire Hugh Moore advocated sterilization in the early 1950s squarely as an anti-Communist bid—this was explicit. But like Huxley and so many others, his drive to solve a world population problem did not start that way. He was already primed as an internationalist and pacifist in the Malthusian tradition. "Can We Avoid War," he had asked in the 1930s?[30] This is why he was so involved in the League of Nations Association in the United States and why he was U.S. consultant delegate at the San Francisco conference that created the UN.[31] It was the relation between overpopulation and war versus optimum population and peace that brought him to the world population question and to sterilization in the first instance. This is the politics onto which his later Cold War anti-Communism was grafted.

As we have seen, Kingsley Davis had used the "bomb" metaphor to structure his ideas on world demographic transition. And like Davis's 1945 ideas, this *Population Bomb* pamphlet needs to be understood at least as much in terms of internationalism as anti-Communism. It signals how easily the former slid into the latter over the question of population, precisely because of the prior geopolitical framing. By this point, though, Kingsley Davis's metaphoric "fuse" had been overtaken by the H-bomb of a nuclear age: the long (European) history was gone from the population problem, replaced by instantaneous population explosion of the world, implicitly driven by the Third World (figure 12.1). "The human family is increasing at the unprecedented rate of over 500 MILLION persons IN THIS DECADE," the pamphlet announced in solid caps. "Is homo sapiens destined to consume his food supplies and perish as many other species have done in history?"[32] Like so many in this generation, Hugh Moore and his team knew the language to be used. Population control must appear to be nonracist and noncoercive, especially so after the war. And strangely, it was liberal English eugenics that could provide a model for this purported neutrality. "The problem is unrelated to color, race or geographical location. It could apply to the United States if our population was outstripping our resources—as may happen if the US population grows by 100 million in the next 22 years as the US Census predicts." This pamphlet was keen to stress the primary significance of "motivation": people *want* to limit their families to the number they can support, it stressed. "Motivation" was a code for noncoercion, and the pamphlet was carefully subtitled: *Is Voluntary Sterilization the Answer?*

Voluntary sterilization in this context was an answer to the problem of fertility rates, not an answer to the problem of the mentally or physically disabled. *The Population Bomb* pointed to activity under way in India, and Moore watched with satisfaction as sterilization became increasingly central to Indian policy, writing to Rockefeller in 1959 about the New Delhi conference on Planned Parenthood and Jawaharlal Nehru's enthusiasm.[33] In Madras and Kerala, the government had begun to encourage vasectomy, his

FIGURE 12.1 Different bombs. Kingsley Davis had focused on the metaphor of the long "fuse" in 1945, the centuries of European population growth. But Hugh Moore's metaphor was an H-bomb, population that exploded in the present, ostensibly globally, but by implication in the Third World. (From Hugh Moore Fund, *The Population Bomb* [1954].)

pamphlet declared, impressed. Men who already have at least three children and are unable to support more may have the operation free and a bonus of 40 rupees. "Thousands of men have responded."[34] This described activity within India's first and second five-year plans, predating the intensified sterilization programs over the later 1960s and early 1970s, which included both financial incentives for sterilization and disincentives for families over three children.[35] One edition of *The Population Bomb* included an endorsing letter from the World Health Organization's (WHO) Brock Chisholm. Relieved from WHO strictures in 1953, he finally found an outlet for his interest in birth control as public health. The procedure in the male is simple and painless, Brock Chisholm reassured readers, and does not interfere with sexual potency or pleasure.[36]

As late as 1970, Julian Huxley still advocated sterilization, especially for men. He was pleased that vasectomy was widely used in India, but added with regret, "though not for eugenic purposes, as I had hoped."[37] Others in the postwar period would also make this kind of distinction between sterilization as fertility control and sterilization as an instrument of eugenics. They were clearly related, but that historical actors perceived and described such a distinction is nonetheless important: they were not "hiding" eugenics as population control. Guy Irving Burch and Elmer Pendell made this kind of distinction, for example. What the new UN needed to do, they argued in 1947, was to recommend to all nations that domestic laws be adopted that would "(a) actually lead to the sterilization of all persons who are inadequate, either biologically or socially, and (b) encourage the voluntary sterilization of normal persons who have had their share of children." They imagined a new world in which the UN could insist this of "conquered countries," and that such laws would be administered under Allied supervision "for a given period as prerequisite to autonomy."[38] This was a recapitulation of the geopolitics of birth control, the 1920s arguments about nations taking responsibility for their own demographic futures. But it was also a completely disingenuous suggestion, because they knew very well that the last method of fertility regulation that would or could be endorsed by the UN would be sterilization.

While Burch and Pendell's statement has been used to denounce the eugenic population control that the West sought for the East, the First World for the new Third World,[39] a sidelined point is that precisely this recommendation was already in effect in the main "conquered country" at issue: occupied Japan. However, far from the new UN "insisting" on sterilization (nothing would be less likely in 1947), and far from the U.S.-led

Allied occupiers doing so, sterilization in occupied Japan emerged from pre-existing domestic Japanese law. And, even though legalized under a so-called Eugenic Law, sterilization was clearly being taken up as a form of increasingly normalized contraception, not as a eugenic procedure. Sterilization, plainly, could be both.

There is no doubt that vasectomy and tubal ligation as procedures stemmed from eugenic law and practice over the 1920s and 1930s, and not just in North America; indeed, they were practiced illegally in many Catholic countries.[40] But sterilization, like any method of fertility regulation, was always polyvalent. Depending on context, historian Johanna Schoen notes, the same procedure "that violated a young girl in the name of eugenics brought relief to her overburdened mother."[41] The historical challenge, then, is not to conflate individualized contraception, Malthusian, and eugenic intentions, but to show how these practices were sometimes entwined and sometimes separated. The Japanese uptake of sterilization, even though it proceeded under a Eugenic Law, should not be read as a version of eugenic compulsory sterilization of institutionalized people comparable to practices in Nazi Germany or Indiana, for example, but far more as family limitation on economic grounds. Huxley and Carr-Saunders would and did argue that such reduction of fertility rates would have a population-level eugenic effect. But this was different to arguing eugenic effect on the grounds of individual sterilization of those deemed mentally or physically disabled.

Carr-Saunders had long seen eugenics as a whole tied to "the small family system": "Whether we take long or short views, voluntary parenthood occupies the centre of the field." This was a statement he made first in his Galton Lecture of 1935, "Eugenics in the Light of Population Trends." It was republished in 1968, a very different context for such a statement, not just because of the developments in hormonal contraception but also the developments in sterilization as contraception that had stepped up quite considerably in several contexts, including in Britain itself.[42] In 1961, for example, the Eugenics Society Council resolved to encourage "public understanding of the present legal position and the potentialities of sterilization here and elsewhere in the world." This was put forward in the light of its own traditions and past efforts from the 1934 Brock committee on sterilization onward, "and fully recognizing Dr. Blacker's special interest and effort from the early 1930s until now."[43] The Eugenics Society Council commissioned a social survey, a Gallup poll, in which the first question to the curious sample groups of "scientists" and "classicists" was: "Do you think the population of the world is increasing fast, increasing slowly, or is now nearly stable?" Followed by: "In relation to the natural resources of the world, is its population too small, about right or too large?" What did people think about inheritance of intelligence, and should research on human heredity be encouraged, tolerated or stopped? Presuming world population growth, should we rely on increasing world resources or try and control population increase as well? If the latter, how? Through universal birth control? Through selective or targeted birth control? Through voluntary sterilization? Through legalizing abortion?[44] All this shows how thoroughly eugenics and world population control were one project, but not necessarily, as is often presumed, as a result of "race science."

In 1960, the editor of the Indian *Planned Parenthood* journal, lawyer Avabai B. Wadia, recorded the various local positions on sterilization. The advantage the Family Planning Association of India (FPAI) saw in sterilization was a sure and permanent means to prevent pregnancy. This put it in a category by itself, however, outside "family planning methods proper." For example, sterilization could not be recommended to newly married couples or to those with only several children. The FPAI committee considered the situation carefully and said that under the right conditions, sterilization had a "very useful part to play both from the view point of population control, and for relief from further unwanted pregnancy." This was a qualified statement, however. Certain safeguards were stipulated, and sterilization was imagined as a temporary measure at a population level, "until such time as contraceptive methods were universally in practice among married couples." Compulsory sterilization in certain cases was contemplated in eugenic terms, with respect to people with transmissible physical or mental conditions. This, though, "might lead to abuses and was not desirable." There was no public sentiment for "compulsion in the field of social action."[45] Moreover, the committee noted that there were doubts in a number of countries regarding the legality of the operation. All in all, the position of the FPAI was that if necessary legislation could and should be introduced in India, to clarify the legality of the operation, sterilization should be available free of charge at hospitals; free transport might be provided; information on sterilization, as on family planning in general, should be widely disseminated; and finally, financial incentives to sterilization were not desirable. Dhanvanthi Rama Rau was in some disagreement,[46] her dissent to some extent vindicated when such safeguards came undone in the Emergency Period of the mid-1970s.

India and the World: Sripati Chandrasekhar's Population Problem

Among others, it was Sripati Chandrasekhar who pressed early and long for sterilization in India, a process that led to the high point, or low point, in the Emergency Period (1975–1977). This involved coerced but not technically compulsory vasectomies for men and tubal ligations for women. The various government-established targets for numbers sterilized, and methods for them to be met, meant that on the ground technical noncompulsion was immaterial: people felt compelled.[47]

Freed from the constraints of UNESCO, Chandrasekhar argued actively for birth control within a health structure and continued to do so for geopolitical reasons. He was constantly irritated by the prevarications of his U.S.-, UK-, and UN-based colleagues. Partly as a younger demographer, partly in a context in which Roman Catholic opposition was not especially powerful, he thought that the moment for debating birth control had long since passed.[48] Instead, now was the time to think methods and policy and to assess the successes and failures of various interventions. It was for this reason that Chandrasekhar shifted from the international sphere back to domestic national politics, in 1964 standing successfully for India's upper house as Indian National Congress member for Madras. In the world's demographic fraternity, it was a rise compared by some to Keynes's entry to the peerage.[49] For Chandrasekhar the Indian, not the internationalist, there was ultimately a much more rewarding alliance between Indian governments and U.S. foundations, especially the Ford Foundation, than with intergovernmental world organizations.[50]

Chandrasekhar was strongly linked with eugenics through its Malthusian logic, and like Huxley or Blacker in the United Kingdom or Frederick Osborn in the United States, did not distance himself from eugenics in the least. Rather, eugenics was core business, neither to be rejected nor disavowed. He published in the *Eugenics Review* in the late 1940s. He lectured to and was elected honorary fellow by the (British) Eugenics Society in 1954. As late as 1965, he was communicating with the Japanese embassy in Washington, seeking the best English translation of the 1948 Japanese Eugenics Law.[51] He was also subject to eugenic policies himself. In 1947, for example, he was required to undertake what his New York physician called the "Marriage Wasserman," the test for syphilis.[52] For decades, Chandrasekhar sought Blacker's editorial and substantive advice both in the latter's Eugenic Society role, and his International Planned Parenthood role. Chandrasekhar's book, *Population and Planned Parenthood in India* (1955), was extensively edited by Blacker,[53] and introduced by Julian Huxley, while Jawaharlal Nehru contributed its foreword. This Indian demographer emerged in equal measures from eugenics, Malthusianism, pacifism, and anticolonial nationalism, each of which sat quite comfortably together, both before and after the war.

Chandrasekhar was at one level simply continuing the well-established conflation of eugenics and economic neo-Malthusianism that thrived in India, part of late colonialism as well as Indian anticolonial nationalism. The links between Indian economic policy, birth control methods, and overpopulation rationales were almost to be taken for granted. The Americans who eventually came financially to drive huge programs of population control, imagining their interventions as saving the Third World from itself, were talking to an Indian elite who had long before arrived at the policy connection between economic planning and family planning. This is one reason why India was such fertile ground for postwar population planning and a major site for "the birth of the third world," as Connelly has shown.[54] And as Sarah Hodges explains, the version of eugenics that operated in India was one that had long had birth control at its core, focusing far more on sex and reproductive practices, than on Mendelian or population genetics.[55]

This was all consistent enough with Radhakamal Mukerjee's eugenics as well. Influenced strongly by Carr-Saunders, he thought through Indian eugenics in "The Dysgenic Trends of Population." The "intellectual social groups," he wrote, have a small natural increase, and in some regions of India an actual decrease in numbers. By contrast, the "less literate and backward social groups are more progressive demologically, and these threaten to swamp the cultured stocks." As in the West, he concluded, "the most fertile social strata in India are inferior."[56] Mukerjee's Malthusian economics and his placement within the international network of population scholars are particularly important, because he also chaired the Population Subcommittee of the Indian National Congress's National Planning Committee, itself chaired by Nehru. Under the congress, the political economies of the limited-scale eugenic and Malthusian clinics of the 1930s gave way to large-scale strategies that were unsurprisingly central to political and economic planning—"development"—after Independence, fueled by international funding. The Planning Committee resolved, in a classic eugenic statement: "The state should follow a eugenic programme to make the race physically and mentally healthy. This would discourage marriages of unfit persons and provide for the sterilization of persons suffering from transmissible diseases of a serious nature such as insanity or epilepsy."[57] After Independence, Nehru's National Planning Commission called for free sterilization and contraception on medical, social and economic grounds, along with birth control research.[58]

Rather than there being a merging of eugenics and population control in the middle of the twentieth century, we can see in India, as elsewhere, an already-Malthusian eugenics at work. The complexity and chronology is not quite, then, as Mohan Rao has characterized it: "The parents of neo-Malthusianism were the eugenists and birth-controllers." The reverse is rather more the case; Malthusians gave birth to eugenicists, and the "world" had long been the stage. When a demographer like Chandrasekhar—trained in the United States and experienced in the new UN circles and in London-based foundations—considered world population, he did so from this strong Indian tradition of political economy of population that incorporated eugenics and from a tradition of Indian nationalism that almost automatically comprehended political and economic planning in terms of population, food, and reproduction.

Recognition of the long-standing presence of Malthusian economics surrounding, even forming, eugenics is the key to understanding how and why "population control" and even "global population control" was hardly new with the formation of the WHO contraception trials in the early 1950s; the Rockefeller, Carnegie, and Ford funding of family planning; or the linking of family planning, aid, and U.S. foreign policy in the 1960s. This is not to suggest that an apparently "global agenda" was not often a Western agenda playing out in local domains, as Sanjam Ahluwalia argues of much Western advocacy of birth control in India.[59] It does, however, explain the circumstances under which an anticolonial like Chandrasekhar aligned so neatly with eugenics and with new demographic theories. Like Mukerjee before him, he was uncompromising in pressing his anticolonial politics. And just like his interwar counterparts, indeed his pre–World War I counterparts, he was still convinced that managing population was ultimately one way to secure world peace and security.[60]

Unlike many of the men in the population sector, however, Chandrasekhar did perceive and promote the feminist history that was part of the political economic history of Malthusianism. If Mukerjee saw solutions in land and rural reconstruction, Chandrasekhar saw them in sex and gendered relations. He reconsidered abortion, and as Indira Gandhi–appointed minister of health and family planning he launched a campaign for birth control for women and for the delay of women's marriage, policies that accompanied, in his view, plans for their more central place in the workforce. All this sat alongside his advocacy of vasectomy. Chandrasekhar even made a concession to M. K. Gandhi's position, with the idea that all married couples should aim to be sex-free for one year. He later claimed that Gandhi had been supportive of sterilization for men as early as 1935.[61] Sripati Chandrasekhar called uncompromisingly for "universal birth control."

Reproductive Rights and Freedoms

In asserting the "birth-right of every wife and mother," Chandrasekhar was deploying a lexicon that had great postwar resonance. The language of "universal rights" was on everyone's lips, if not within everyone's reach. In the specific context of the intellectual history of the world population problem, it is important to recognize that the individualizing of reproductive "rights" was also their universalizing, the idea and ideal of their applicability to all humans everywhere. In other words, one element of the "internationalizing" of population over the twentieth century was this claim that all individuals, in theory, could and should determine their reproductive selves and futures.

However, neither Chandrasekhar nor anyone else would be able point to a document that encoded such a right. In what sense was birth control, technically, any kind of "right" in the third quarter of the century, when conventions and declarations on human rights abounded? The extent to which the discourse of population rights and freedoms connected with the call for freedom from want needs to be recognized. Not infrequently, "freedom from hunger" trumped reproductive freedom, but was nonetheless a common and persuasive argument for it. A motion was tabled for debate in the British House of Lords, for example, concerning "the world emergency created by the population explosion." The tabling member of Parliament pronounced: "I firmly believe that, without universal family planning, 'freedom from hunger' is an empty ideal."[62] This resonated not just because of contemporary anti-Communist politics, whereby hunger threatened capitulation to Communist forces, but also because a version of this idea had been circulating since World War I, at least. Indeed, "freedom from hunger" was arguably the most common proposition for a universalized "right" to birth control. The prospect of global "hunger" caught both John Boyd Orr–style bids for international humanitarianism and Hugh Moore–style bids for political freedom and anti-Communism.

There was another strand of thought about freedom and reproduction, however: birth control as itself a "right" and a "freedom." Chandrasekhar pressed this forward, turning inalienable rights appropriately into "birth-rights." But even this remained geopolitically driven, a means, apparently, for men to achieve global equity between West and East, North and South, rather than for women to achieve bodily autonomy. "Birth control has come to stay in the West. It is now the business of all enlightened men to spread it in Asia, Africa and Latin America."[63] Inalienable rights were the talk of the moment, and inalienable *reproductive* rights appeared as much in and around eugenics, as in and around feminism. "We should recognize that it is the inalienable right of every married couple to set a limit to the number of their offspring. . . . Women of all nationalities should be free to limit the size of the family, should they so desire." This was a typical late 1950s formulation of population planning, written by a eugenics organizer. That it constituted part of *Paths to Peace: A Study of War, its Causes and Prevention* would have made every sense at the time: population planning was here, still, a "path to peace."[64] The point to note is that this eugenic claim of a "right" to limit offspring anticipated any formal codification.

The idea of reproductive rights within the general history of human rights is not understood as well as it might be. How, precisely, was "reproduction" incorporated within the suite of conventions devolving from the 1948 Universal Declaration of Human Rights that committed member states to "the promotion of universal respect for and observance of human rights and fundamental freedoms," the preamble famously written by Jan Smuts?[65] Article 16 of the Declaration dealt not with the right *not* to reproduce, but the right to "found a family."[66] This was a response to various fascist restrictions on marriage and possibly to sterilization practices. The Convention on the Prevention and Punishment of the Crime of Genocide, signed in the same year, was also a response to state interference in individual or group capacity (and therefore right) to reproduce. Of the five acts that might constitute genocide, the fourth was defined as "imposing measures intended to prevent births within the group."[67] This right to "found a family" was asserted again in 1966 as a civil and political right: "The right of men and women of marriageable age to marry and to found a family shall be recognized."[68] All these declarations were certainly part of the emerging tradition of reproductive rights, but they were the opposite of the right *not* to reproduce, or to choose to do so less often.

This changed in the mid-1960s. Burmese UN Secretary-General U. Thant made widely publicized declarations in 1966 that "at this moment in man's history" the international community must accord to parents the right to determine the numbers of their children.[69] He was responding to a Rockefeller-led overture. A statement on population delivered to the UN was signed by twelve heads of state, a year later increasing to thirty.[70] Consequently, it was the 1968 International Conference on Human Rights in Tehran that resolved for the "basic human right to determine freely and responsibly the number and the spacing of . . . children." With this endorsement to proceed, UN agencies then came to assert an obligation for states to deliver information on contraception. This freed up the World Health Assembly to affirm family planning as a basic component of primary health care, which it did in 1968. In that same year, UNESCO adopted a resolution put forward by the Swedish delegation that the agency "should promote a better understanding of the serious responsibilities which population growth imposes on individuals, nations and the whole international community."[71] In 1969, operations commenced from the new UN Fund for Population Activities, shifting gears fairly quickly from data collection to maternal and child health, including family planning and latterly "sexual health."

During the 1970s, reproductive rights became rather more inflected by a new feminism with "liberation" firmly in sight. The 1979 Convention on the Elimination of All Forms of Discrimination Against Women addressed reproduction twice, in the context of freedom of access to information on contraception. It did so in the context of health rights: "to ensure, on a basis of equality of men and women, access to health care services, including those related to family planning." And later in the document, "rights to decide freely and responsibly on the number and spacing of their children and to have access to the information, education and means to enable them to exercise these rights."[72] The postwar history of reproduction and universal or human rights is complicated, then. On the one hand, the articulation of reproductive rights as women's rights was one culmination of the long Malthusian program. Reproductive "freedom" and "rights" over one's own body were there in Alice Vickery Drysdale's nineteenth-century Malthusianism, inherited from John Stuart Mill's ideas on liberty, as well as a French tradition of the rights of women. In the twentieth century, Margaret Sanger, Edith How-Martyn, Dhanvanthi Rama Rau, Shidzue Kato, and others each extended a feminist discourse of liberal and sometimes socialist rights and freedoms over reproduction exercised through and as contraception, and sometimes as the right not to have conjugal sex (the more standard earlier feminist position). However, the history of reproductive rights as an idea is clearly not reducible to a feminist history alone, just as the history of contraception or family planning or population politics in general are not.

The Fifth Freedom?

Often enough, the idea of reproductive rights did not come from feminists at all, but from men involved one way or another in postwar eugenics and world population. This is evident in Carlos Paton Blacker's intriguing reflections in "Eugenics in the Middle Nineteen-Sixties." Like Julian Huxley, Blacker had no time for new liberal insistence that all humans actually were equal. "Egalitarianism, personal and racial," wrote Blacker, "has become a new creed whose adherents are reluctant to admit that one individual can possess an inborn superiority over another." At the same time, and far more admissible to the feminist history of reproductive rights, he concluded strongly that "parents should be free to control the size of their families and should be enabled to avoid having unwanted children is today claimed as a new—sometimes called a fifth—freedom."[73] President Herbert Hoover occasionally referred to economic freedom as the fifth freedom. But while there was clearly a presidential genealogy being referenced, this was not what Blacker meant. He was referring to the perhaps unlikely work of Scottish obstetrician, Aberdeen's Regius Professor of Midwifery, Sir Dugald Baird.

"A Fifth Freedom?" was Baird's Sandoz Foundation Lecture delivered at University College Hospital, London, in May 1965. He began with Franklin Delano Roosevelt and his Four Freedoms speech of 1941. "You will recollect," Baird told his audience twenty years after the event, Roosevelt's vision of a secure world based on the four essential freedoms. Baird would himself add a fifth, and in most telling terms: "freedom from the tyranny of excessive fertility."[74] Baird derived freedom from tyranny of excessive fertility from freedom from hunger. Yet he did so in a different way than Guy Irving Burch, for example, for whom reproductive control was a *means* by which freedom from hunger would be secured. Instead, Baird argued "freedom from tyranny of fertility" as another freedom, an equivalent freedom. On this basis, he had established the first free family-planning clinic in Aberdeen.

The terms and scale of his lecture, combined with the tangible outcome as access to birth control for this particular local population, reveal how multifaceted population was and how "population control" and "birth control" can historically barely be distinguished. Sounding very much like the

Margaret Sanger of the 1920s, Baird insisted that such a freedom always needed to be considered from the point of view of world population and simultaneously from the point of view of "the individual man and woman." World population first, he insisted. He began, importantly, with mortality, the standard summary (consistent since Malthus's *Essay*) that in earlier and primitive times, high fertility was necessary, because death rates at all ages were also high. But the continuance of this pattern "may indeed threaten the survival of mankind," a situation worsened by recent death control, the "catastrophic effect of modern chemotherapy" (referring to DDT). In India, Nehru might authorize $10 million for population control, but he simultaneously allocated $14 million for malaria control, compounding the net increase, or so Baird claimed. The century it took for industrialized Europe to reduce birthrates is not an option, he warned. Baird then turned to industrialized societies, first documenting Japan's "effective attack" on overpopulation, a nation "short of living-space" that had reduced its birthrate from 34 to 18 per 1,000. This was all achieved, he said, under a eugenic protection law, and now the number of abortions was being replaced by proactive birth control education. Even in this national success story, the population was set to rise to 100 million by 1990, he warned.

Even the United States, "the most advanced technological society in the world," faced a doubling of its population in the next thirty years, he said, writing in a context inconceivable to Benjamin Franklin and Malthus. But the real problem with the United States, Baird wrote in his own sweep through the world, was that its standard of living was problematically high—it made disproportionate and unsustainable demands on the worlds' available resources. And Britain? Baird pointed not to the decline in birthrates as the British lesson, but the rapid increase in population since the eighteenth century, which dipped temporarily in (Enid Charles's) 1930s. His global story, then, was less the low birthrates of the West that the Third World had to catch up with, than differing rates of growth in a modern history wherein population increase was a constant, everywhere. There was, simply, "excessive fertility."[75] Ultimately the focus—the solution—was squarely biopolitical.

Baird then turned to the "problem for the individual," and offered this as a story of his local patients. In Britain, the average woman had three children, but the potential—women were still marrying generally before twenty-five—was much higher. Clearly control of conception "is the rule for almost everyone." The class difference was also very clear to Baird. In professional and managerial classes, as he structured his study, only 3 percent of women had five children or more; whereas in semi-skilled and unskilled manual classes, 14 percent did so. His use of Family Planning Association data showed that, in 1963, withdrawal was the preferred procedure in the so-called labouring classes, the cap in the managerial and professional classes. He discussed the advent of the contraceptive pill as "a great advance" if shown to be without harmful effects.

What is especially interesting is his account of the sterilization procedure he had undertaken for decades in his local Aberdeen practice. From the late 1930s, he had offered tubal ligations to women between thirty-five and forty who had born as many as eight children. There was reluctance at first, he said, but a steady increase in requests from women for the procedure. He also undertook abortions, in the interests of the mother's mental or physical health. Over a period of thirty years, as he recounted it, the birthrate had decreased, maternal mortality had decreased (because maternal death was so strongly linked to fifth and higher pregnancies), and infant mortality had decreased in Aberdeen. And he offered the feminist argument, straight up: "In Britain this means making contraceptive advice freely available under the National Health Service and offering tubal ligation to women who have had the desired number of children. This would help to free women from the tyranny of unwanted pregnancies and also make them more independent and able to choose freely how they will use their training and skills, especially after their desired number of children have been born."[76] Feminist histories would typically trace twentieth-century reproductive rights through the history of feminism itself; the progressive claiming of individual rights and freedoms over one's own reproductive self. Baird's vision of freedom and of individual and world population problems was clearly far more extensive and multifaceted. This needs to be recognized as one part of the history of twentieth-century reproductive politics; it signaled a far more ready acceptance in the Anglophone public sphere of the kind of variables that Margaret Sanger had insisted on in the 1920s.

This all raises the question of how "freedom" is understood in histories of population. The critical focus on coercive and authoritarian measures of reproductive control is as necessary as it has been forthcoming.[77] Indeed, this critique has been influential over several generations: even a hint of "population control" still makes politicians nervous in liberal democracies. In the process, for several generations now feminist scholars and commentators have sought to establish the difference between an imposed system of "population control" vis-à-vis a rights-based "birth control," an interpretation underscored especially strongly in the light of the International Conference on Population and Development in Cairo in 1994. And yet, this is an aspirational rather more than a historical distinction. Both the premise and the effect of insisting politically on the difference between "population control" and "planned parenthood" has been to separate out the history of family planning from the history of political and economic planning. Demonstrably, they were intertwined.

This critique has also had the unintended consequence of leaving "freedom" to speak for itself, unanalyzed and ahistoricized. Among other effects, the implication is that "freedom" and "voluntariness," and even "right to choose," were not the conceptual modus operandi of the twentieth-century "population establishment." They were. Coercion and compulsion is one point. More typical, and arguably therefore more important, was the growing investment on the part of states, international agencies, health organizations, philanthropic funds, and family-planning organizations to persuade, to enjoin citizens—of nations, of regions, of the world—to reproduce responsibly and to practice family planning. Individual conduct and national and international agendas were aligned, both aspirationally and to a very considerable degree actually. Further, by doing so, individuals were not just freeing themselves from the prospect of pregnancy, they were becoming subjects of freedom.

Inadmissible as it may seem to a celebratory history of birth control, the history of eugenics itself shaped one line of twentieth-century thought on the "right" to reproductive liberty and even to the nonacceptance of coerced implementation of population policies. The latter is the case, not just in reaction to the coercive eugenics of sterilization, say in the United States or in Nazi Germany, but also in a far less recognized way as a result of other eugenicists' investment in the principles and value of voluntariness. As a rule, the postwar population experts knew that they had to persuade, not force. As we have seen, they held and pursued a deep and ongoing investment in people taking responsibility for their own reproductive conduct.

In drawing a connection between interwar eugenics and postwar population control, we learn a certain amount in tracing the trajectory of sterilization from Indiana to India. We might learn at least as much from Frederick Osborn's eugenic philosophy of a "relatively freed choice." "The eugenic philosophy which we have outlined would make eugenic selection a natural and voluntary process," said Osborn, fully tuned in to the powers of freedom. "It is thus in full agreement with the concepts of individual liberty and of non-interference by government, which are so closely associated with the form of our democracy."[78] Here is the core sense in which "eugenics," "population control," and even feminist "birth control" were part of the same modern project: the creation of the modern gendered and responsible citizen, with a voluntary principle, rights, freedoms, and liberties at her core. This was less population control as an extension of authoritarian rule, and more population control as global neoliberalism. Well might we wish that such a principle had been translated into available options for women across the globe, so close and uncomfortable is the history of "freedom" shared by eugenics and family planning. But in any case, it is clear that this aspiration came to be encoded and massively endorsed through family-planning and health education structures and funds, nationally and internationally.

This agenda of persuading individuals into civic responsibility, of matching individual planning to the economic planning of nations, was evident in Indian propaganda. Family-planning programs enjoined couples to improve their families (by reducing fertility) and simultaneously improve the national and economic future of India. At the same time, the Planning Commission reminded the same citizens that pressure on inevitably limited resources must be relieved in pursuit of "civilised existence": everyone's standard of living would be raised, was the message. Nehru called for Indian citizens to practice family planning "from every point of view" (figure 12.2). "The simple sex act must be a responsible act," Chandrasekhar said at one point.[79] "A nation's, let alone the world's population problem is not the concern of population experts alone, nor even that of governments alone. It is the vital and immediate concern of every thoughtful citizen in the world."[80] This was biopolitics, relying on "freedom," the very opposite of top-down state enforcement that had characterized Nazi population control or that was to characterize changes in China, and for the Emergency Period, in Chandrasekhar's India.[81] This, as much as the deeply problematic but in the end more exceptional sterilization "camps" in India should be seen as the later twentieth-century outcome of the long world population problem.

Planetary Planning

Family planning was the modern incorporation—the literal embodiment—of national economic planning. Yet it was not just that in the 1960s: it became part of international and world planning, too. Even, one might say, planetary planning.

I have no doubt that the vast Number of People in India would welcome Family Planning and Population control from every point of view

'NEHRU'

Increasing pressure of population on natural resources (which must be inevitably limited) retards economic progress and limits seriously rate of extention of social services so essential to civilised existence. A population policy is therefore essential to planning

'PLANNING COMMISSION'

THE IDEAL ANTISEPTIC AND DEODORANT CONTRACEPTIVE TABLET

MANUFACTURED BY : COATES & COOPER LTD. ENGLAND.

KEEPS PERFECTLY IN HOT CLIMATES

The average weight of each tablet when packed in 1.2 grams and contains w/w. Sod. Bicarb, B. P. 0.146 gm. Acid Tart. B.P. 0.12 gm., Toluene-p-Sulphonso-dio-chloroamide 0.013 gm., Perfume q.s., Excipient to 1.2. gm.

MEDICAL LITERATURE AND SAMPLES GLADLY SENT ON REQUEST

DISTRIBUTORS :

V. SHARMA & Co.,
P.O. Box No. 1176,
CHANDNI CHOWK,
DELHI-6.

P. H. KHANSAHEB & CO., LTD.
P.O. Box No. 2303,
BOMBAY.

FIGURE 12.2 Family planning, economic planning, political planning. Chandrasekhar thought that "the simple sex act must be a responsible act." Contraceptive advertisement, program, International Planned Parenthood Federation Conference, Delhi, 1959. (Reproduced with kind permission, Family Planning Association of India.)

Albert Thomas might have declared himself a world citizen in 1927, but forty years later, in the new age of Aquarius, being a world citizen of reproductive age did not entail being a member of the International Labour Organization (ILO) or even knowing what it stood for. This world generation were planetary citizens: not ILO but ZPG was the acronym of choice. "Zero population growth" turned into a social movement and even a political identity that was expressed through reproductive conduct: the "thoughtful new student cause" was a crusade against too many people, *Life* announced on its April 1970 cover.

Even here, the double-edged sword of "population" revealed itself. Environmentally aware students' responsible commitment to nonreproduction to save the planet usefully doubled as freedom to have sex without pregnancy. On the other hand, though, they had perhaps not read Kingsley Davis's article "Population Policy: Will Current Programs Succeed?," which first called for "zero population growth."[82] World population growth had peaked in 1963 at 2.3 percent (currently it is 1 percent).[83] But Davis himself wondered why family-planning and population policy sectors were not setting up the explicit goal of zero population growth. To that end, he battled the "voluntarism" of both the family-planning movement and what he himself called the "population establishment," which seemed to hold reproductive freedom so close; far too close for his liking. He thought that the endless focus on contraceptive devices and technologies capitulated the population cause to the health cause, with its voluntarist injunctions that simply evaded the ultimate problem: too many people.

When the film *Z.P.G.* came out in 1972, it was a B-grade science fiction thriller, truer to Davis's dangerous impatience with "freedom" than the sexual and reproductive freedom that young Americans were taking on as planet consciousness.[84] It was a dystopia of coerced population control, in which the geopolitical context was a closed Earth, pressing not just on soil but on freedom itself. The story line's tension relied as heavily on a closed and singular globe as Wells's had done when the genre was being invented. Here, again, was "the raft to which we cling in the boundless ocean of space" that *The Malthusian* writer had so hauntingly conjured back in 1885.[85] But now it was a tarnished raft as well: "Smog covers the earth. Oxygen is depleted." And in the claustrophobic, polluted, and closed space, the freedoms of a reproductive couple were denied by an edict enforced by a World Deliberation Council: "The penalty for birth is death." Neither such dystopias in film, nor even the imminent sterilization program in India or the Chinese one-child policy—actual dystopias for many—carry the ultimate meaning of world population. The point is to read the *Z.P.G.* dystopia alongside the Indian poster for family, economic, and national planning, and each of these alongside the phenomenal successes of feminist rights to reproductive health and freedoms, and ecological warnings about food insecurity. From the Great War to the Cold War, world population growth was one social, biological, and economic phenomenon through which the meaning of freedom was shaped. The polyvalence of fertility control arose in part from the scales on which reproductive politics worked; simultaneously the intimate business of personal conduct, the business of states, and the business of a planetary future.

Conclusion

The Population Bomb in the Space Age

Narrow patriotism must go and one must become "planet conscious."

V. H. WALLACE TO C. P. BLACKER, JUNE 27, 1954

"Taking the whole earth instead of this island, emigration would of course be excluded," Malthus had written in his 1803 edition.[1] Julian Huxley was saying more or less the same thing at his Lasker Award address in 1959: "Space sets the inescapable limit to population."[2] Huxley did not, as a rule, make much mention of Malthus in his energetic world population work over the 1950s and 1960s. But there was no evading the connection between these population thinkers, even as a whole era of modern history divided them, to say nothing of the orders of magnitude by which world population growth rates had accelerated. Malthus was writing in the age of exploration, when the Pacific was the next new world. Huxley was writing in the space age, when the moon was just over the horizon. That year the first photos of Earth were beamed back from NASA's *Explorer 6* satellite.

In periodizing the point at which the planet came to be visualized in its entirety, environmental historians often isolate the era in which Earth was seen from space. "The idea that there was 'only one Earth' compelled a global imagination," as Will Steffen and Libby Robin have written in their discussion of the new geological age of the Anthropocene.[3] Yet what needs rethinking is the presumption that such planet-visualizing was a product of the space race decade. The vision of a singular planet, spaceship Earth, is better understood as a new rendition of a planetary imagination that was already many generations old. Earth, imagined as a globe in space by Franklin, by the nineteenth-century Malthusians, by George Knibbs, and H. G. Wells, now really had been orbited.

Some thought the launching of Sputnik sparked a Cold War race for rights to a whole new geopolitical living space. Yet those coming out of the Malthusian tradition were generally impatient with that idea. Aldous Huxley got straight to the point. In *Brave New World Revisited*, he warned that the epoch should not be called the space age but the age of overpopulation. When he wrote the original in 1931, he "was convinced that there was still plenty of time." But no longer. And he offered his own literary, historical, and transatlantic version of what a hundred geographers, demographers, agriculturalists, biologists, and birth control lobbyists had tried to say before, in graphs and tables, in words and pictures:

> On the first Christmas Day the population of our planet was about two hundred and fifty millions. . . . Sixteen centuries later, when the Pilgrim Fathers landed at Plymouth Rock, human numbers had climbed to a little more than five hundred millions. By the time of the signing of the Declaration of Independence, world population had passed the seven hundred million mark. In 1931, when I was writing *Brave New World*, it stood at just under two billions. Today, only twenty-seven years later, there are two thousand eight hundred million of us.

Expenditure on settling the moon, he thought, would do nothing to alleviate the poverty that such crowding brought on Earth. The new worlds in space would not lessen earthly problems, but compound them.[4] His eminent brother was also irritated with talk of emigrants to the moon and Mars that just then seemed almost possible: he was firmly earthbound.[5] Julian Huxley thought that people should see the impossibility of "shipping off to Mars or Venus, every twenty-four hours, around 150,000 people."[6] Neither population, nor hunger, nor land problems were going to be addressed by looking outward from Earth to the celestial bodies.

Still, there was something about the planet in the space age that was hard to resist for those thinking about population. Even eugenicists were enjoined to be planetary now, in their last pitch for scientific and popular credibility. Being "planet conscious" almost became eugenic orthodoxy. Proponents sought, in their own view, the welfare of "the greatest number of the earth's inhabitants, irrespective of race, religion, nationality or colour." This was a curious end point to the century's major intellectual bid to biologically authorize human inequality. That leading eugenicists even imagined foregoing "narrow patriotism" in favor of "planetary consciousness" is an unexpected turn, to say the least, given the tight connection between eugenics and nationalism. On another view, however, "planetary consciousness" was a natural extension of the cosmopolitanism that had accompanied eugenics on its long journey from its Malthusian roots. It is perhaps unsurprising, then, that aging eugenicists picked up the seemingly progressive mission of saving the species on the planet, a mission that was soon to blend into a movement to save the planet itself. For all its legacy demarcating human difference, it was Malthusian-inspired eugenics that had also long occupied the thought-zone of the "species," "the race," for better or worse. Some declared, thinly, a new nondiscriminating agenda, but when pressed generally capitulated to the idea that human difference, heterogeneity, was a social if not a biological difficulty. Eugenics advocates early in the century might have insisted on homogeneity through the mechanism of a global color line, a segregation model. Now homogeneity would be best achieved by cultural assimilation.[7] Open the doors of the empty lands to those who would come, they told each other, but expect assimilation into the West, into whiteness, into civilization, and into low birthrates and low infant mortality. Kingsley Davis's dream of modernization involved westernization without movement to the West at all. Assimilation certainly had its midcentury policy moment—it governed much of the discussion at the 1954 World Population Conference in Rome—but as the third quarter of the twentieth century progressed, something more akin to Mukerjee's recognition of human difference, less the biological and climatic determinism, prevailed.

Some were getting carried away: "World, Globe, Orb, Whole, One," Blacker wrote floridly about "the planetary problem." R. A. Fisher was perhaps vindicated in his view that British eugenics was losing its scientific edge, as Blacker indulged in purple prose. "U.N.O. Is it not the perfect word? The last of the three letters denotes our spherical planet; the first two sound the call to unite."[8] By claiming "the planet" as a space of operation, "the world" and its people could also be claimed fairly easily—too easily—with seeming political neutrality: far moreso, for example, than the efforts of the *Geopolitiker* to claim world-space by extending national "living room." It was the pacifism authorizing Blacker's version of internationalism and "planetary consciousness" that was useful. Seemingly progressive (and there is no doubt that Blacker would perceive it as progressive), this did a good deal of work to rescue population talk from connections with illiberal authority and wartime racism. It had a powerful moral claim. Postwar eugenicists like

Carlos Paton Blacker, Julian Huxley, Alexander Carr-Saunders, and Frederick Osborn each seized upon this. And it was pacifist internationalism that had long drawn in anticolonials like Radhakamal Mukerjee, Sripati Chandrasekhar, and even Warren Thompson.

Mukerjee continued to invoke one world strongly from his political and epistemological base in the new Third World. The more divided the world became, the more he pressed for a commonwealth of mankind as the only way forward, one sure basis of which was a global policy of population reduction.[9] His last book, published in 1965, was titled *The Oneness of Mankind*. By that point, his ideal model for united human organization was not economic liberalism, liberal internationalism, international socialism, or even a world federation of states. His vision now extended even beyond *Homo sapiens*; the notion of individual rights in political science would no longer do. At the end of his life, Mukerjee wrote of *Homo universalis*, a political animal freed not just from the "pathological rights of sovereignty of particular States," but also from "old" individual rights: "the abstract, unconditional and costless rights of man." Not the rights of man, but "the rights of mankind-as-a-whole." His ecological cosmopolitanism took him once step further, to propose the political vision of "mankind-and-cosmos-as-a-whole."[10]

Looking back from Mukerjee's postwar vantage point, such ideas were received in part from Malthusian internationalism. They were of a piece with the kind of integrated ecology that created Knibbs's antinationalism and planet-level visions, and H. G. Wells's cosmopolitanism. Looking forward, Mukerjee might have been—but was not—a father of 1970s environmentalism and deep ecology. Yet even as Mukerjee imagined a global population policy based on the rights of mankind, not the rights of man, it was the idea of the rights of woman—individual reproductive rights—that became far more successful in establishing the longer-term response to world population growth, even if it did not create the one world of his dreams.

The keepers of population knowledge were a long-lived set, beneficiaries of ever-increasing twentieth-century life spans. One might say that they were contributors to their own world population problem. Mukerjee died, perhaps fittingly, in the internationally resonant year, 1968. Margaret Sanger and Alexander Carr-Saunders both died in 1966, the former at eighty-six in Tucson, Arizona, the latter in Thirlmere in the United Kingdom, aged eighty. John Boyd Orr wore his Glasgow University Chancellor's robes well into his nineties, living until 1971. Warren Thompson died in 1973 in Oxford, Ohio, in the environs of both his foundation and his farm. Carlos Paton Blacker and his Oxford teacher, Julian Huxley, both died in 1975. This was the generation who had known Annie Besant personally, who had herself known Charles Darwin and John Stuart Mill. Sripati Chandrasekhar, by contrast, brought Annie Besant back to life by reprinting her work in 1981,[11] thus re-introducing neo-Malthusianism, feminism, and Besant's presidency of the Indian National Congress to a wide Anglophone audience. He saw the moon conquered and the space age come and go, lecturing until his death at eighty-three in San Diego, California, in the first year of the new millennium.

True to their intellectual bloodlines, this group could never quite give up thinking about population, the world, and Thomas Robert Malthus. If Charles Darwin had recognized something critical in the *Essay on the Principle of Population*, a century later his grandson, president of the British Eugenics Society, continued the conversation. The appropriately named Charles Galton Darwin was a physicist who looked to the vast future, much as his grandfather had looked to the deep evolutionary past. In *The Next Million Years*, he drew the limits of the earth to the public's attention, again. If Raymond Pearl had related his fruit flies in jar-sized universes to the world population problem in the 1920s, this time Boyle's law was being applied to human life on Earth: gas molecules were equivalent to people; Earth was equivalent to the "containing vessel"; and the relation between them was necessarily one of pressure. It was all related to an escalating energy crisis. Darwin worried about imminent fuel shortages, "the fifteenth generation of my descendants will get no coal at all."[12] When his ideas were discussed in British Parliament, viable life on Earth over the next million years came down to something very small: "What is needed more than anything else is the 'perfect pill'; a pill which taken orally by a woman will prevent conception for a month or perhaps a longer period."[13] The extent to which the pill's history emerges from food security, economics, war, and peace, as well as the history of feminism and women's reproductive freedom—"liberation"—sorely needs recognition.

The long history of ecological ideas drawn here link population to environmentalism. In itself no new insight, putting food production and therefore soil squarely into the equation renders legible the institutional, political, and personnel connections between "conservation" and "family planning." This was one way in which geopolitics became biopolitics. There is a more obscure history revealed as well, in bringing all these elements back into conversation. We can see how eugenics served as one key intellectual space for the emergence of "environmentalism." Indeed eugenics was not just a precursor to, but a driver of soon-to-be popular "ecology." We can also see just how strongly, explicitly, and even technically, an ecology of life and (even moreso) death shaped demographic transition theories, development, and modernization. Economies of efficiency and waste lay deep inside the very idea of "ecology." High infant mortality was the great human inefficiency that a demographic transition would turn around. We are used to thinking about infant mortality in individualized humanitarian terms or in public health terms, and Malthusians in the past did so as well. But for Kingsley Davis, what mattered most was that high infant mortality wasted energy. The global "biopolitics" of population was thus not just a matter of maximizing life for the instrumental purposes of a state or even on an international sphere. Mukerjee was right: managing "population" was about global energy in a strictly ecological model.

Understanding this ecological history of population as leading solely to a progressive, left-wing "environmentalism" (or even as leading to an out-of-favor environmentalism in which the planet is saved by enforcing fertility control), misses the sense in which "ecology" also aligned with the efficiencies sought in and of high modernity. When George Knibbs wrote a long essay on "Waste" in 1926, he praised Taylorist methods that minimized wasted physiological energy and maximized labor efficiency. War was "superwaste."[14] Planned economies, by contrast, minimized waste. Reproduction needed to be brought under control in this same logic, the scientific management of labor in the private sphere. The alignment of family planning with the economic and political planning that characterized all kinds of mid-twentieth-century states, both "developed" and "developing," thus becomes comprehensible.

The constituent parts of the world population problem—eugenics, Malthusianism, feminism, ecology, economics, environmentalism—are not often put together. Doing so makes the history all the more interesting, but also challenging, to explain. It is challenging in part because these fields are so often separated and defended individually as "good" (feminism, ecology, environmentalism) or critiqued as "bad" (Malthusianism, eugenics). One of the aims of this book has been to confound retrospective alignments of current political positions with praiseworthy or shameful histories, not for the sake of it, but in order to recognize and explain the historical complexities of the population problem. This is politically risky. For example, addressing how eugenics became ecology or how easily Malthusians were anticolonial will cause critics to worry that these problematic movements have been not just explained, but explained away. They will worry—with more justification—that such an analysis will be put to work by others to redeem that which should not be (it is naïve to think this will not unfold).[15] Nonetheless, orthodoxies of critique, especially of Malthusianism, need unsettling. This creates room to understand with rather more historical subtlety an antinationalist Knibbs or a cosmopolitan Mukerjee, for example. It enables us perhaps to expect that Margaret Sanger would pick up eugenics rather than simply be shocked by this fact. And at the same time, it should enable understanding of how she cut through it all with her sharp analysis of gendered power. Historical analysis that explains in such ways has nothing necessarily to do with whether or not we agree with what we seek to understand. This book, then, has tried to explain how and why it made sense for Chandrasekhar, to take another example, to be key to Indian sterilization programs *and* a player in the demise of the race-based immigration acts; to explain how an Indian Malthusian and eugenicist proponent of sterilization could also be an influential actor in international antiracism.

It is, if anything, a measure of the strength of long-standing critiques of eugenics, Malthusianism, and population control, and of feminism's remarkable tendency to self-critique, that such complicating factors can be added. Still, when "anti-Malthusians" seek, for all good political reasons, to explode myths that population growth is the cause of poverty and to insist on rights-based family planning, not "population control," paradoxically, something aligned with their own views lies inside the history of Malthusian thought itself, discomfiting though this might be. The recent New Internationalist cooperative's publication of the anti-Malthusian *No-Nonsense Guide to World Population* is a case in point. Not a little ironically, its global justice agenda, not to mention its "internationalism," in part derives from the work and ideas of the "Malthusian" intellectuals treated here.

In particular, this book has opened up the colonial history of world population. It has done so less as critique of colonialism or of population control as neocolonialism—this has already been done a hundred ways—and more by asking how those problematizing the world population problem themselves recognized its colonial history. It explains just how neo-Malthusianism ended up being one of the unexpected intellectual and political sites of twentieth-century anticolonialism. Typically, the historical actors here formulated anticolonialism partially and unsatisfactorily by current postcolonial benchmarks. Yet anticolonialism has a history, too, and a fascinating one. Sometimes this anticolonialism was expedient; sometimes it was honest and stark, as in John Boyd Orr's "white man's dilemma," surely an early manifesto for "whiteness studies." Often anticolonialism was not driven by indigenous politics at all. Occasionally, recognition of the world population problem's colonial history appeared in entirely unexpected places. Charles Galton Darwin, for example, registered it. Like so many, he saw that the economic and population "golden age" rested on the "accident of the nineteenth century," the agricultural opening of vast new areas of the world. Unlike most, he also named the core human cost: "It is true that this was done largely at the expense of the American Indian, and his treatment often does not make a pretty story." This was more than Julian Huxley or Margaret Sanger were ever likely to admit. But in the end, this Darwin, like the older Darwin, simultaneously recognized and problematically normalized the suffering inflicted on "a few millions" for the prospering of "many hundred millions." Done and dusted was the implication: "Now there are no frontiers or unknown parts of the world into which to expand, and so our golden age is probably near its end."[16] Perhaps the colonial past could be confronted when there seemed to be no future gain.

Charles Galton Darwin's incorporation of New World politics and New World acres into the world population story is important. It is part of a long, modern unraveling of an intertwined colonial history, population history, economic history, and global history. It has re-emerged recently in the current upsurge in "ghost acre" talk in global and international studies.[17] This, too, comes to us through Malthusian thought. It was the Swedish geographer and ecologist, Georg Borgström, professor of food science and human nutrition at Michigan State University, who invented "ghost acreage." *The Hungry Planet: The Modern World at the Edge of Famine* (1965) was Malthus's *Essay* rewritten for the Cold War. It was mainly about land and grain. But an important element was "fish acreage," continuing the conversation about who ate what that had engaged so many in Geneva fifty years earlier.[18] *The Hungry Planet* was a deeply political book, filled with the linked geopolitics and biopolitics of population, its natural history and its political economy. It detailed, for example, a "biological budget of mankind"; Wells and Huxley would be pleased. The inevitable chapter on the Australian continent was still figured as "The Safety Valve of Asia," after the work of Chandrasekhar, Mukerjee, and indeed Harold Cox. Asia itself could become either "Starvation Center or Powerhouse"; Borgström did not know. But then neither did Nehru, who had just died in office as prime minister of India, or Indira Gandhi, who was soon to take office. And neither, for that matter, did the architects of demographic transition and modernization, Kingsley Davis and Frank Notestein. The United States, Borgström explained, had expanded its tilled acreage by a massive one-fifth after World War II, through mechanization and foreign policy decisions to restrict cotton cultivation and increase wheat cultivation. It turned out that there was a good deal of cultivable land left; more than Oliver Edwin Baker reckoned in his great world soil surveys of the 1920s. It is also notable that when considering cultivation of that continent, Borgström did not have colonial politics on his radar at all, even to the limited extent that eugenics president Charles Galton Darwin did, strange though this may seem.

On the other hand, Borgström did see North America as continuing to feed the world. Its food surplus had become part of foreign policy: stored surpluses could and should be disposed of in needy foreign countries. Public Law no. 480 became "Food for Peace" under Lyndon Johnson. And critically, in the context of a linked geopolitics and biopolitics of population, it was through the Food for Peace Act that U.S. Congress allocated funds to the U.S. Agency for International Development (USAID) in 1966 to distribute medical supplies, including contraceptives. Food for peace was, for President Johnson, war on hunger. The extra step to population control was a conceptually if not politically easy one for that generation, precisely because the food–population–war triad had been a constant in the international public sphere for so long. Geopolitical problems were solved by biopolitical solutions.

Food for Peace was the Cold War's version of Margaret Sanger's 1932 Plan for Peace, when she still had World War I in mind. Sanger declared, in light of that war and in the Depression that world peace might yet be secured by implementing Woodrow Wilson's Fourteen Points and by joining an International League of Low Birth Rate Nations. The major difference between Sanger and Johnson's bid for peace through population control was Sanger's clarity and certainty that the eugenically unfit could and should be segregated and sterilized through a "stern and rigid policy": the very certainty that someone like Warren Thompson had critiqued all along.[19] It was partly this eugenic history that made Johnson choose his words very carefully indeed in the mid-1960s, as he explained food aid as contraception and population control as a war on hunger and political insecurity: "The sound population programs, encouraged in this measure, freely and voluntarily undertaken, are vital to meeting the food crisis, and to the broader efforts of the developing nations to attain higher standards of living for their people."[20] Here was Malthus's food and sex for a new age.

Borgström introduced the space age as well as the Cold War in his assessment of the hungry planet. For him, the vast sums spent on food for "space travelers" was the sticking point. The intensive research then being undertaken into human physiological requirements for space voyages was "macabre," given human physiological need on Earth. That space traffic might ease the pressure on the planet was glib talk, irresponsible nonsense he thought, just like Huxley. Not only was this earthly "spaceship in distress," but "man has already outgrown the universe."[21] This was one end point of Charles Darwin's "standing room only."

And yet, that era is so often remembered as a beginning. In 1968, conservation biologist Paul Ehrlich published *The Population Bomb: Population Control or Race to Oblivion*, an enduring historical touchstone for proponents and opponents of a dozen different political positions.[22] The text was so phenomenally successful that in one well-timed publishing move it at once incorporated and obliterated the generations of work that preceded it. So much else happened in 1968, seeming beginnings, but actually accumulations. None of it fitted neatly together, ideologically, but it was nonetheless all part of the same moment when raw human politics met life and death on Earth. This was the year the World Health Organization decided that contraceptive advice should be incorporated into primary health care. It was also the year of Pope Paul VI's encyclical on human life, *Humanae Vitae . . . On the Regulation of Birth*, proscribing contraception of all kinds. Biologist Garrett Hardin published his "Tragedy of the Commons" essay in *Science*, justifying abortion and immigration restriction on population and ultimately ecological grounds. Betty Friedan founded the National Abortion Rights Action League for completely different reasons.[23] The Eugenics Society finally gave up "eugenics," publishing its last issue of *Eugenics Review* that year. In 1968, the Club of Rome first met to discuss action on limits of the earth, and the term "green revolution" was first used. Sripati Chandrasekhar became chairman of the International Association for Voluntary Sterilization.

This was also the year the earthrise photographs were beamed back from *Apollo 8*. There it was: Edward East's little terraqueous globe, Wells and Huxley's superorganism, Raymond Pearl's universe with definite limits, Radhakamal Mukerjee's ecological whole, and John Boyd Orr's one world. This was one of high modernity's highest moments. Now masters of the planet, humans—or at least a few American men—had also turned themselves into physically insignificant denizens, just as George Knibbs predicted.

Notes

Abbreviations

APS	American Philosophical Society
BCIIC	Birth Control International Information Centre
CP	Chandrasekhar Papers
HP	Huxley Papers
IIIC	International Institute of Intellectual Cooperation
ILO	International Labour Office
IUSIP	International Union for the Scientific Investigation of Population Problems
JDR3	John D. Rockefeller III
KP	Keynes Papers
LC	Library of Congress
LNAG	League of Nations Archives, Geneva
LSE	London School of Economics
NAA	National Archives of Australia
NLA	National Library of Australia
NLS	National Library of Scotland
RAC	Rockefeller Archives Center
RFA	Rockefeller Foundation Archives
RFam.A	Rockefeller Family Archives
RG	Record Group
SPLC	Sanger Papers, Library of Congress

Introduction

1. John Maynard Keynes, Notes for a speech at Malthusian League Anniversary dinner, July 26, 1927, JMK/PS/3/115, the Papers of John Maynard Keynes, King's College, Cambridge [hereafter KP].

2. "It is hoped that those who are present will make this concession, which enables them to welcome at the dinner, one of the two defendants in the Bradlaugh–Besant trial" (printed note included in the Malthusian League Fiftieth Anniversary Dinner Menu and Program, July 26, 1927, JMK/PS/3/121, KP). See also "Notes on Guests Who Will be Present," JMK/PS/3/118, KP; Annie Besant, *The Influence of Alcohol* (1892; Madras: Theosophical Publishing House, 1930).

3. The Malthusian League Fiftieth Anniversary Dinner Menu and Program, July 26, 1927.

4. See Annie Besant, *Theosophy and the Law of Population* (London: Theosophical Publishing Society, 1891). This was publicized as "Containing Reasons for Renouncing Neo-Malthusian Teaching." John Toye, *Keynes on Population* (New York: Oxford University Press, 2000); H. G. Wells, Julian Huxley, and G. P. Wells, *The Science of Life* (London: Cassell, 1931).

5. Albert Thomas, "International Migration and Its Control," *Proceedings of the World Population Conference*, ed. Margaret Sanger (London: Edward Arnold, 1927), 266 [hereafter *1927 Proceedings*].

6. Simon Szreter, "The Idea of Demographic Transition and the Study of Fertility Change: A Critical Intellectual History," *Population and Development Review* 19, no. 4 (1993): 659–701; Susan Greenhalgh, "The Social Construction of Population Science," *Comparative Studies in Society and History* 38, no. 1 (1996): 26–66; Paul Demeny, "Population Policy and the Demographic Transition: Performance, Prospects, and Options," *Population and Development Review* 37, no. 1 (2011): 249–74.

7. "Announcement," *1927 Proceedings*, 5.

8. Ibid., 367.

9. Key historians of twentieth-century demography note the array of disciplines that made up the study of population in the early twentieth century. See Dennis Hodgson, "The Ideological Origins of the Population Association of America," *Population and Development Review* 17, no. 1 (1991): 7; Edmund Ramsden, "Carving Up Population Science: Eugenics, Demography and the Controversy over the 'Biological Law' of Population Growth," *Social Studies of Science* 32, no. 5–6 (2002): 857–99; Matthew Connelly, "Population Control is History," *Comparative Studies in Society and History* 45, no. 1 (2003): 122–47.

10. Many of these are the scholars that Elazar Barkan analyzed in *The Retreat of Scientific Racism: Changing Concepts of Race in Britain and the United States Between the World Wars* (Cambridge: Cambridge University Press, 1992).

11. With these population thinkers upfront, I reference also some of the bigger names of twentieth-century Anglophone demography (Frank Notestein, Kingsley Davis, David Glass) and explore their interaction with French and Italian scholars (e.g., Alfred Sauvy, Corrado Gini).

12. Paul R. Ehrlich, *The Population Bomb: Population Control or Race to Oblivion* (New York: Ballantine, 1968). See also Paul R. Ehrlich and Anne H. Ehrlich, "The Population Bomb Revisited," *Electronic Journal of Sustainable Development* 1, no. 3 (2009): 63–71.

13. Louis Varlez, "Suggested Plan for the Organization of an International Population Union," (1927) Mss B.P312, Folder 14, Raymond Pearl Papers, APS.

14. Robert Strausz-Hupé, *Geopolitics: The Struggle for Space and Power* (New York: Putnam, 1942). Mss B.P312, Folder 14, Raymond Pearl Papers, APS.

15. Gerry Kearns, *Geopolitics and Empire: The Legacy of Halford Mackinder* (Oxford: Oxford University Press, 2009).

16. G. H. Knibbs, "The Mathematical Theory of Population, of its Character and Fluctuations, and of the Factors which Influence Them," Appendix A, *Census of the Commonwealth of Australia, 1911, Volume 1: Statistician's Report* (Melbourne: Government Printer, 1917), 95 (italics in original).

17. See Neil Smith, *American Empire: Roosevelt's Geographer and the Prelude to Globalization* (Berkeley: University of California Press, 2003), 299-304; Warren S. Thompson, "Where Can the Indian Go?" in *Danger Spots in World Population* (New York: Knopf, 1929), 159–81.

18. Radhakamal Mukerjee, *Races, Lands, and Food: A Program for World Subsistence* (New York: Dryden, 1946), 7.

19. Marilyn Lake and Henry Reynolds, *Drawing the Global Colour Line: White Men's Countries and the International Challenge of Racial Equality* (Cambridge: Cambridge University Press, 2008).

20. Madison Grant, *The Passing of the Great Race, or, The Racial Basis of European History* (New York: Scribner, 1916); Lothrop Stoddard, *The Rising Tide of Color Against White World-Supremacy* (New York: Scribner, 1920).

21. Matthew Connelly, "To Inherit the Earth: Imagining World Population from the Yellow Peril to the Population Bomb," *Journal of Global History* 1, no. 3 (2006): 299–319.

22. Thompson, *Danger Spots*, 17.

23. Thomas, "International Migration and Its Control," *1927 Proceedings*, 256.

24. Akira Iriye, *Global Community: The Role of International Organizations in the Making of the Contemporary World* (Berkeley: University of California Press, 2002).

25. See, e.g., Sarah Hodges, "Malthus Is Forever: The Global Market for Population Control," *Global Social Policy* 10, no. 1 (2010): 120–26.

26. George Handley Knibbs, *The Shadow of the World's Future; or the Earth's Population Possibilities & the Consequences of the Present Rate of Increase of the Earth's Inhabitants* (London: Ernest Benn, 1928), 46, 72, 95.

27. Denis Cosgrove, *Apollo's Eye: A Cartographic Genealogy of the Earth in the Western Imagination* (Baltimore, Md.: Johns Hopkins University Press, 2001), 257–67.

28. Joyce E. Chaplin, "Earthsickness: Circumnavigation and the Terrestrial Human Body, 1520–1830," Special issue, *Bulletin of the History of Medicine*, 86, no. 4 (2012): 515–42; Robert Poole, *Earthrise: How Man First Saw the Earth* (New Haven, Conn.: Yale University Press, 2008); Denis Cosgrove, "Contested Global Visions: One-World, Whole-Earth, and the Apollo Space Photographs," *Annals of the Association of American Geographers*

84, no. 2 (1994): 270–94; Walter A. McDougall, "Technology and Statecraft in the Space Age," in *The Global History Reader*, ed. Bruce Mazlish and Akira Iriye (New York: Routledge, 2005), 47. See also, Sheila Jasanoff, "Image and Imagination: The Formation of Global Environmental Consciousness," in *Changing the Atmosphere: Expert Knowledge and Environmental Governance*, ed. C. A. Miller and P. N. Edwards (Cambridge, Mass.: MIT Press, 2001), 309–37; Thomas Robertson, *The Malthusian Moment: Global Population Growth and the Birth of American Environmentalism* (New Brunswick, N.J.: Rutgers University Press, 2012), 176–200.

29. Analyzed by Peder Anker, *From Bauhaus to Ecohaus: A History of Ecological Design* (Baton Rouge: Louisiana State University Press, 2010), 68–112.

30. Michel Foucault, *The Birth of Biopolitics: Lectures at the Collège de France 1978–1979*, trans. Graham Burchell (Basingstoke, U.K.: Palgrave, 2008); Andrea A. Rusnock, "Biopolitics: Political Arithmetic in the Enlightenment," in *The Sciences in Enlightened Europe*, ed. William Clark, Jan Golinksi, and Simon Schaffer (Chicago: University of Chicago Press, 1999), 49–68; Nikolas Rose, *Powers of Freedom: Reframing Political Thought* (Cambridge: Cambridge University Press, 1999); Alison Bashford, *Imperial Hygiene: A Critical History of Colonialism, Nationalism and Public Health* (Basingstoke, U.K.: Palgrave, 2004), 172–85.

31. The carrying-capacity idea was already prolific by the time demographers pursued it in the 1920s. For more recent versions, see Lester Russell Brown, *Full House: Reassessing the Earth's Population Carrying Capacity* (New York: Norton, 1994); Sandra Postel, "Carrying Capacity: The Earth's Bottom Line," in *Beyond the Numbers: A Reader on Population, Consumption, and the Environment*, ed. Laurie Ann Mazur (Washington, D.C.: Island Press, 1994), 48–70; Joan Martinez-Alier, *The Environmentalism of the Poor: A Study of Ecological Conflicts and Valuation* (Cheltenham, U.K.: Edward Elgar, 2002), 47–48.

32. Kingsley Davis, "The World Demographic Transition," *Annals of the American Academy of Political and Social Science* 237, no. 1 (1945): 1–11.

33. Pearl, "Biology of Population Growth," *1927 Proceedings*, 28.

34. Sir Bernard Mallet, "Opening Address," *1927 Proceedings*, 2; A. B. Wolfe, "The Population Problem since the World War: Part 2," *Journal of Political Economy* 36, no. 6 (1928): 679.

35. Edward M. East, "Food and Population," *1927 Proceedings*, 86.

36. For coal and land, see Kenneth Pomeranz, "Political Economy and Ecology on the Eve of Industrialization: Europe, China, and the Global Conjuncture," *American Historical Review* 107, no. 2 (2002): 443. For organic to fossil fuels, see E. A. Wrigley, *Energy and the English Industrial Revolution* (Cambridge: Cambridge University Press, 2010). For a summary of recent scholarship, see Fredrik Albritton Jonsson, "The Industrial Revolution in the Anthropocene," *Journal of Modern History* 84, no. 3 (2012): 679–96.

37. Julian Huxley, "Are We Overpopulated?" *Evening Standard*, October 25, 1927, Box 97, Folder 7, Huxley Papers, Rice University, Texas [hereafter HP].

38. E. A. Wrigley, *Population and History* (New York: McGraw-Hill, 1969), 57. See also Wrigley, *Energy and the English Industrial Revolution*, 9–25.

39. Frederick Jackson Turner, "The Significance of the Frontier in American History," *Annual Report of the American Historical Association for the Year 1893* (Washington, D.C.: American Historical Association, 1894), 197–228.

40. J. R. McNeill, "Population and the Natural Environment: Trends and Challenges," *Population and Development Review* 32, no. 1 (2006): 183–201.

41. Nick Cullather, *The Hungry World: America's Cold War Battle Against Poverty in Asia* (Cambridge Mass.: Harvard University Press, 2010), chap. 1.

42. Daniel Hall, *The Improvement of Native Agriculture in Relation to Population and Public Health* (Oxford: Oxford University Press, 1936). For earlier ideas about improvement, see Richard Drayton, *Nature's Government: Science, Imperial Britain, and the "Improvement" of the World* (New Haven, Conn.: Yale University Press, 2000).

43. One version of the long history of eco-catastrophe and its critique, although written before the current climate change discourse, is presented in Ronald Bailey, *Eco Scam: The False Prophets of Ecological Apocalypse* (New York: St. Martin's, 1993).

44. Aldous Huxley, *The Double Crisis* (Paris: UNESCO, 1949).

45. Paul Sears, *The Living Landscape* (New York: Ecological Society of America, 1949), 91.

46. Sugata Bose and Kris Manjapra, eds., *Cosmopolitan Thought Zones: South Asia and the Global Circulation of Ideas* (Basingstoke, U.K.: Palgrave, 2010). See Julian Huxley, *UNESCO: Its Purpose and its Philosophy* (Washington, D.C.: Public Affairs Press, 1947), 13; Radhakamal Mukerjee, *The Oneness of Mankind* (New Delhi: Radha, 1965); John Boyd Orr, "The Choice Ahead—One World or None," December 14, 1946, RG1.1, Series D1, Food and Agriculture Organization Archives, Rome.

47. A Doctor of Medicine, *The Question of Irish Home Rule with remarks on law and government and international anarchy, and with a proposal of the federal union of France and England as the most important step to the federation of the world* (London: Truelove, 1889).

48. Michael Geyer and Charles Bright, "World History in a Global Age," *American Historical Review* 100, no. 4 (1995): 1060.

49. For example, Pierre Desrochers and Christine Hoffbauer, "The Post War Intellectual Roots of the Population Bomb: Fairfield Osborn's Our Plundered Planet and William Vogt's Road to Survival in Retrospect," *Electronic Journal of Sustainable Development* 1, no. 3 (2009): 73–97. See Fairfield Osborn, *The Limits of the Earth* (Boston: Little, Brown, 1953); Sears, *The Living Landscape*; William Vogt, *Road to Survival* (New York: William Sloane Associates, 1948).

50. For contraceptive and heterosexual practice, see Hera Cook, *The Long Sexual Revolution: English Women, Sex, and Contraception 1800–1975* (Oxford: Oxford University Press, 2004); Simon Szreter and Kate Fisher, *Sex before the Sexual Revolution: Intimate Life in England 1918–1963* (Cambridge: Cambridge University Press, 2010).

51. F. H. Amphlett Micklewright, "The Rise and Decline of English Neo–Malthusianism," *Population Studies* 15, no. 1 (1961): 32–51; Elinor A. Accampo, "The Gendered Nature of Contraception in France: Neo-Malthusianism, 1900–1920," *Journal of Interdisciplinary History* 34, no. 2 (2003): 235–62; Henny Brandhorst, "From Neo-Malthusianism to

Sexual Reform: The Dutch Section of the World League for Sexual Reform," *Journal of the History of Sexuality* 12, no. 1 (2003): 38–67; Mausumi Manna, "Approach Towards Birth Control: Indian Women in the Early Twentieth Century," *Indian Economic and Social History Review* 35, no. 1 (1998): 35–51; Susanne Klausen, *Race, Maternity, and the Politics of Birth Control in South Africa*, 1910–39 (Basingstoke, U.K.: Palgrave, 2004); Lesley A. Hall, "Malthusian Mutations: The Changing Politics and Moral Meanings of Birth Control in Britain," in *Malthus, Medicine and Morality*, ed. Brian Dolan (Amsterdam: Rodopi, 2000), 157; Robert Jütte, *Contraception: A History* (Cambridge: Polity, 2008), 106–17.

52. For a recent summary of historical work on the size, structure, and dynamics of change in population, but not the spatial, agricultural, or geopolitical dimension, see Pat Thane, "Population," in *A Concise Companion to History*, ed. Ulinka Rublack (Oxford: Oxford University Press, 2011), 179–202. An exception is Maria Sophia Quine's work that locates population policies in European geopolitics. Maria Sophia Quine, *Population Politics in Twentieth-Century Europe* (London: Routledge, 1996).

53. Matthew Connelly, *Fatal Misconception: The Struggle to Control World Population* (Cambridge, Mass.: Belknap, 2008).

54. For this critique see Angélique Janssens, ed., *Gendering the Fertility Decline in the Western World* (Bern: Peter Lang, 2007).

55. Pearl, "Biology of Population Growth," 22.

56. Rahul Nair argues for the primacy of "health" for the construction of an Indian population problem, and recognizes, but underemphasizes the significance of both economic population arguments and the linked "geopolitical" arguments. Rahul Nair, "The Construction of a 'Population Problem' in Colonial India, 1919–1947," *Journal of Imperial and Commonwealth History* 39, no. 2 (2011): 227–47.

57. For a critique of population control, and the failure to deliver true reproductive choice, see Betsy Hartmann, *Reproductive Rights and Wrongs* (1987; Boston: South End Press, 1995). See also Sandra Lane, "From Population Control to Reproductive Health: An Emerging Policy Agenda," *Social Science and Medicine* 39, no. 9 (1994): 1303–14.

58. UNFPA, Mission Statement. http://www.unfpa.org/public/about (accessed 20 September 2011).

59. John H. Perkins, *Geopolitics and the Green Revolution: Wheat, Genes, and the Cold War* (New York: Oxford University Press, 1997).

60. For example, Lara V. Marks, *Sexual Chemistry: A History of the Contraceptive Pill* (New Haven, Conn.: Yale University Press, 2001), 15; Deborah Barrett and D. J. Frank, "Population Control for National Development: From World Discourse to National Policies," in *Constructing World Culture: International Nongovernmental Organizations since 1875*, ed. John Boli and George M. Thomas (Stanford, Calif.: Stanford University Press, 1999), 198–221; Hall, "Malthusian Mutations," 157.

61. John Sharpless, "World Population Growth, Family Planning and American Foreign Policy," *Journal of Policy History* 7, no. 1 (1995): 72–102; John Sharpless, "Population Science, Private Foundations, and Development Aid: The Transformation of Demographic Knowledge in the United States, 1945–1965," in *International Development and the Social Sciences: Essays on the History and Politics of Knowledge*, ed. Frederick Cooper and Randall Packard (Berkeley: University of California Press, 1997), 176–202; Dennis

Hodgson, "Orthodoxy and Revisionism in American Demography," *Population and Development Review* 14, no. 4 (1988): 541–69; Paul Demeny, "History of Ideas in Population since 1940," in *Demography: Analysis and Synthesis: A Treatise in Population Studies*, vol. 4, ed. Graziella Caselli, Jacques Vallin, and Guillaume Wunsch (Burlington, Mass.: Academic, 2006), 28–29; Marc Frey, "Neo-Malthusianism and Development: Shifting Interpretations of a Contested Paradigm," *Journal of Global History* 6, no. 1 (2011): 75–97. Simon Szreter's important article that unraveled the tenacity of "demographic transition" despite the evidence of empirical studies, also sought to explain 1950s orthodoxy in the light of Notestein's 1940s work, and in the context of the post–World War II significance of China, especially for the United States. Versions of demographic transition found no audience in the 1920s and 1930s, he claims. Szreter, "Idea of Demographic Transition," 664.

62. Mohan Rao, *From Population Control to Reproductive Health: Malthusian Arithmetic* (New Delhi: Sage, 2004), 112.

63. Karl Sax, *Standing Room Only: The World's Exploding Population* (Boston: Beacon, 1960); Mihajlo D. Mesarović and Eduard Pestel, *Mankind at the Turning Point: The Second Report to the Club of Rome* (New York: Dutton, 1974); J. W. Lovelock, *Gaia: A New Look at Life on Earth* (Oxford: Oxford University Press, 1979). In his excellent study, *The Malthusian Moment*, Thomas Robertson discusses the interwar uptake of Malthusianism, but still perceives "the Malthusian moment" as arriving after World War II. This chronology is unsurprising when the United States is the major object of inquiry, in this instance "American environmentalism."

64. For example, Harold Cox, *The Problem of Population* (London: Jonathan Cape, 1922); Harold Wright, *Population* (London: Nisbet, 1923); Edward M. East, *Mankind at the Crossroads* (New York: Scribner, 1923); A. M. Carr-Saunders, *Population* (Oxford: Oxford University Press, 1925); Edward Alsworth Ross, *Standing Room Only?* (New York: Century, 1927); Knibbs, *Shadow of the World's Future*; Thompson, *Danger Spots*; E. F. Penrose, *Population Theories and Their Application with Special Reference to Japan* (Stanford, Calif.: Food Research Institute, Stanford University, 1934); Frederick Sherwood Dunn, *Peaceful Change: A Study of International Procedures* (New York: Council on Foreign Relations, 1937); Radhakamal Mukerjee, *Food Planning for Four Hundred Millions* (London: Macmillan, 1938).

65. Marcel Aurousseau, Typescript review of Five Books on Population, 1925, MS 7070, Series 3, Box 16, Folder 6, Marcel Aurousseau Papers, NLA.

66. Hugh Dalton, "Review of Shadow of the World's Future," *Economica* no. 25 (1929): 83–84.

67. Emma Rothschild, "Population and Common Security," in *Beyond the Numbers*, ed. Mazur, 401–6.

68. William Petersen, "John Maynard Keynes's Theories of Population and the Concept of 'Optimum,'" *Population Studies* 8, no. 3 (1955): 228–46.

69. For Björn-Ola Linnér, the "return" of Malthus began with the post-1945 "New World Order." Björn-Ola Linnér, *The Return of Malthus: Environmentalism and Post-War Population-Resource Crises* (Isle of Harries, U.K.: White Horse, 2003), 13–31. Similarly, in the four-volume *Malthus: Critical Responses* (ed. Geoffrey Gilbert [London: Routledge, 1998]), most of the last two hundred years are covered, except for the first half of

the twentieth century. Another volume might well have been published on responses to Malthus in the 1920s alone. In Philip Appleman's Norton edition of Malthus's *Essay*, respondents are organized over time: but it is the gap left between George Bernard Shaw (1889) and Paul Ehrlich (1968) on which the present book focuses. Julian L. Simon, compiling *The Economics of Population: Classic Writings*, also omits this period. Not much happened, he claims, between Engels' and Marx's response to Malthus and the 1960s, "with the interesting exception of the double flip-flop by John Maynard Keynes in the 1920s and 1930s." Julian L. Simon, ed., *The Economics of Population: Classic Writings* (New Brunswick, N.J.: Transaction, 1998), ix.

70. Jacques Dupâquier, "Preface," in *Malthus: Past and Present*, ed. J. Dupâquier, A. Fauve-Chamoux, and E. Grebenik (London: Academic, 1983), viii–ix.

71. The term is Dipesh Chakrabarty's, *Provincializing Europe: Postcolonial Thought and Historical Difference*, 2d ed. (Princeton, N.J.: Princeton University Press, 2008).

72. Garland Allen, "From Eugenics to Population Control," *Science for the People* 12, no. 4 (1980): 22–28; Ian Dowbiggin, *The Sterilization Movement and Global Fertility in the Twentieth Century* (Oxford: Oxford University Press, 2008), 6.

73. Mohan Rao, "An Imagined Reality: Malthusianism, Neo-Malthusianism and Population Myth," *Economic and Political Weekly* (January 29, 1994): 40–52; Mohan Rao and Sarah Sexton, eds, *Markets and Malthus: Population, Gender and Health in Neo-Liberal Times* (New Delhi: Sage, 2010); Steven W. Mosher, *Population Control: Real Costs, Illusory Benefits* (New Brunswick, N.J.: Transaction, 2008); Connelly, *Fatal Misconception*; Hodges, "Malthus Is Forever," 120–26. See also Ines Smyth, "Gender Analysis of Family Planning: Beyond the Feminist vs. Population Control Debate," in *Feminist Visions of Development: Gender Analysis and Policy*, ed. Cecile Jackson and Ruth Pearson (London: Routledge, 1998).

74. Jacqueline Kasun, *The War Against Population: The Economics and Ideology of World Population Control* (San Francisco: Ignatius, 1988). For the antifeminism of this argument, see 165.

75. Connelly, *Fatal Misconception.*

76. Radhakamal Mukerjee (1931), cited in Sanjam Ahluwalia, *Reproductive Restraints: Birth Control in India, 1877–1947* (Urbana: University of Illinois Press, 2008), 41.

77. John Maynard Keynes, "Population," Typescript [1913–1914] JMK/SS/1/27, KP; see also Toye, *Keynes on Population*, 66. The date of this typescript is not clear, as John Toye explains. Throughout I have indicated its likely date as either 1913 or 1914, or possibly both, in that it is a heavily corrected and annotated set of folios.

78. East, "Food and Population," 85.

1. Confined in Room

1. Thomas Robert Malthus, *An Essay on the Principle of Population*, 2d ed. (London: J. Johnson, 1803), 5. For islands in the period, see Richard H. Grove, *Green Imperialism: Colonial Expansion, Tropical Island Edens and the Origins of Environmentalism, 1600–1860* (Cambridge: Cambridge University Press, 1995).

2. Thomas Robert Malthus, *An Essay on the Principle of Population,* 1st ed. (London: J. Johnson, 1798), 21–22.

3. Malthus, *Essay* (1798), 25, 15.

4. Malthus, *Essay* (1803), 46.

5. See A. M. C. Waterman, "Reappraisal of 'Malthus the Economist' 1933–97," *History of Political Economy* 30, no. 2 (1998): 304–8.

6. Malthus, *Essay* (1798), 27, 47–48. Keynes noted Malthus's original use of the "Darwinian" phrase; see J. M. Keynes, "President's Address," *Report of the Fifth International Neo-Malthusian and Birth Control Conference*, ed. Raymond Pierpoint (London: Heinemann, 1922), 60.

7. Charles Darwin, Notebook D, September 28, 1838, DAR208.39, Darwin Papers, Cambridge University Library. Cited also in Janet Browne, *Charles Darwin: Voyaging* (Princeton, N.J.: Princeton University Press, 1995), 388.

8. Robert H. MacArthur, *Geographical Ecology: Patterns in the Distribution of Species* (New York: Harper and Row, 1972); G. N. von Tunzelmann, "Malthus's 'Total Population System': A Dynamic Reinterpretation," in *The State of Population Theory: Forward from Malthus*, ed. David Coleman and Roger Schofield (Oxford: Blackwell, 1986).

9. E. A. Wrigley, *Poverty, Progress, and Population* (Cambridge: Cambridge University Press, 2004), 431–32; E. A. Wrigley and R. S. Schofield, *The Population History of England, 1541–1871: A Reconstruction* (Cambridge: Cambridge University Press, 1981). For Malthus's lack of awareness of early manufacturing and industrialization, see E. A. Wrigley, "Elegance and Experience: Malthus at the Bar of History," in *The State of Population Theory*, ed. Coleman and Schofield, 46–64.

10. Malthus, *Essay* (1798), 122–25.

11. Wrigley and Schofield, *The Population History of England*, 533–34; Jacques Vallin, "Europe's Demographic Transition, 1740–1940," in *Demography: Analysis and Synthesis: A Treatise in Population Studies*, 4 vols., ed. Graziella Caselli, Jacques Vallin, and Guillaume Wunsch (Amsterdam: Elsevier, 2006), 3:42.

12. Malthus, *Essay* (1798), 11.

13. Malthus, *Essay* (1803), 483. See book IV, chap. 1 of this edition, "Of moral restraint, and the foundations of our obligation to practise this virtue."

14. See Warren S. Thompson, *Population: A Study in Malthusianism* (New York: Columbia University, 1915), 16–17.

15. Malthus, *Essay* (1798), 101, 104, 105–6; in the 1803 edition, 4. See also Dennis Hodgson, "The Ideological Origins of the Population Association of America," *Population and Development Review* 17, no. 1 (1991): 3.

16. Joyce E. Chaplin, *Benjamin Franklin's Political Arithmetic: A Materialist View of Humanity* (Washington, D.C.: Smithsonian Institution Libraries, 2008), 22.

17. Malthus, *Essay* (1803), 2. The version Malthus read and cited was the republication of "Observations on the Increase of mankind, peopling of countries &c" in Benjamin Franklin, *Political, Miscellaneous, and Philosophical Pieces* (London: J. Johnson, 1779), 9. See Chaplin, *Benjamin Franklin's Political Arithmetic*, 40–41.

18. For Keynes's use of doubling American populations see "Population," Typescript [1913–1914], JMK/SS/1/28, KP.

19. Frederick Jackson Turner, "The Significance of the Frontier in American History," *Annual Report of the American Historical Association for the Year 1893* (Washington, D.C.: American Historical Association, 1894), 197–228.

20. Malthus, *Essay* (1798), 188. In the 1803 edition, this was repeated in the context of hypothetical increases of English land. "Where is the fresh land to turn up? Where is the dressing necessary to improve that which is already in cultivation?" (Malthus, *Essay* [1803], 371).

21. Malthus, *Essay* (1798), 342.

22. William Petty, *Several Essays in Political Arithmetick* (London, 1699). In the early chapters of the *Essay*, Malthus indicated that Petty "supposes a doubling [of population] possible in so short a time as ten years." Malthus *Essay* (1803), 5. See also Ted McCormick, *William Petty and the Ambitions of Political Arithmetic* (Oxford: Oxford University Press, 2009).

23. Peter Buck, "People Who Counted: Political Arithmetic in the Eighteenth Century," *Isis* 73, no. 1 (1982): 28–45; Terence Hutchison, *Before Adam Smith: The Emergence of Political Economy, 1662–1776* (Oxford: Blackwell, 1988); Andrea A. Rusnock, *Vital Accounts: Quantifying Health and Population in Eighteenth-Century England and France* (Cambridge: Cambridge University Press, 2002).

24. Michel Foucault, *The Birth of Biopolitics: Lectures at the Collège de France 1978–1979*, trans. Graham Burchell (Basingstoke, U.K.: Palgrave, 2008).

25. Jacques Vallin and Graziella Caselli, "The United Nations' World Population Projections," in *Demography*, ed. Caselli, Vallin and Wunsch, 3:197. See also John A. Taylor, *British Empiricism and Early Political Economy: Gregory King's 1696 Estimates of National Wealth and Population* (London: Greenwood Press, 2005).

26. David Hume, "Of the Populousness of Ancient Nations," Discourse X, *Political Discourses*, 4 vols. (Edinburgh: Kinkaid and Donaldson, 1752), 4:155–261.

27. Adam Smith, *An Inquiry into the Nature and Causes of the Wealth of Nations*, 5th ed. (London: Strahan and Cadell, 1789), book 3, chap. 1.

28. E. A. Wrigley, "The Limits to Growth: Malthus and the Classical Economists," in *Population and Resources in Western Intellectual Traditions*, ed. Michael S. Teitelbaum and Jay M. Winter (Cambridge: Cambridge University Press, 1989), 40. In pre-industrial societies, not just food but all the raw materials of production were organic: even the processing of mineral raw materials was dependent on the availability of organic raw materials, especially wood. E. A. Wrigley, *Population and History* (New York: McGraw Hill, 1969), 56.

29. Malthus, *Essay* (1798), 315–16.

30. Malthus, *Essay* (1803), 147, 149.

31. Malthus, *Essay* (1798), 56. For China, Britain, and diverging economies and ecologies, see Kenneth Pomeranz, *The Great Divergence: China, Europe, and the Making of the Modern World Economy* (Princeton, N.J.: Princeton University Press, 2000).

32. Edward East, "Food and Population," *Proceedings of the World Population Conference*, ed. Margaret Sanger (London: Edward Arnold, 1927), 89.

33. William Godwin, *Of Population: An enquiry concerning the power of increase in the numbers of mankind, being an answer to Mr Malthus's essay on that subject* (London:

Longman, 1820), 448–49. See also Emma Rothschild, "Echoes of the Malthusian Debate at the Population Summit," *Population and Development Review* 21, no. 2 (1995): 351–59.

34. William Godwin, *Political Justice*, cited in Malthus, *Essay* (1803), 368–69.

35. Nassau William Senior, *Two Lectures on Population* (London: Saunders and Otley, 1828), 32.

36. Friedrich Engels, "Outline of a Critique of Political Economy" (1844), in *An Essay on the Principle of Population*, ed. Philip Appleman (New York: Norton, 2004), 147–48. For Karl Marx, Friedrich Engels, and Malthus, see William Petersen, "Marxism and the Population Question: Theory and Practice," in *Population and Resources in Western Intellectual Traditions*, ed. Teitelbaum and Winter, 77–101.

37. James Belich, *Replenishing the Earth: The Settler Revolution and the Rise of the Anglo-World, 1783–1939* (Oxford: Oxford University Press, 2009).

38. John Stuart Mill, *Principles of Political Economy*, 2 vols. (Boston: Little and Brown, 1848), 1:231.

39. William Thomas Thornton, *Over-Population and Its Remedy* (London: Longman, Brown, Green, and Longmans, 1846). See "occupation of waste lands," 429–35. See also David P. Nally, *Human Encumbrances: Political Violence and the Great Irish Famine* (Notre Dame: University of Notre Dame Press, 2011).

40. Darwin read the final 1826 edition of Malthus's *Essay*, and it was this edition he cited in *Descent of Man*. Charles Darwin to Ernst Haeckel, August 1864 [after August 10], Darwin Correspondence Project, Letter 4631. This has been extensively discussed. See Robert M. Young, "Malthus and the Evolutionists: The Common Context of Biological and Social Theory," *Past and Present* 43, no. 1 (1969): 109–45; Silvan S. Schweber, "The Origin of the *Origin* revisited," *Journal of the History of Biology* 10, no. 2 (1977): 229–316; Peter J. Bowler, "Malthus, Darwin and the Concept of Struggle," *Journal of the History of Ideas* 37, no. 4 (1976): 631–50; Sharon Kingsland, "Evolution and the Debates over Human Progress from Darwin to Sociobiology," in *Population and Resources in Western Intellectual Traditions*, ed. Teitelbaum and Winter, 167–98; Robert M. Young, "Malthus on Man: In Animals No Moral Restraint," in *Malthus, Medicine and Morality*, ed. Brian Dolan (Amsterdam: Rodopi, 2000), 73–91.

41. Malthus, *Essay* (1798), 47–48.

42. Charles Darwin, *On the Origin of Species* (London: John Murray, 1859), 63.

43. Darwin, *On the Origin of Species*, quoted in Annie Besant, *The Law of Population: Its consequences and its bearing upon human conduct and morals* (Chicago: Wilson, 1880), 8. This was first published in 1877 by Annie Besant and Charles Bradlaugh's own London press, Freethought Publishing Company. A first American edition of *Law of Population* was published in 1878.

44. Mill, *Principles of Political Economy*, book 1, chap. 10.

45. Cited in George Drysdale, *Elements of Social Science, Physical, Sexual and Natural Religion: An exposition of the true cause and only cure of the three primary social evils, poverty, prostitution and celibacy*, 4th ed. (London: Truelove, 1861), 457.

46. Darwin, *On the Origin of Species*, 65, 64.

47. Besant, *Law of Population*, 14.

48. Drysdale, *Elements of Social Science*, 463.

49. Malthus, *Essay* (1798), 14–15.

50. The first edition was published anonymously by a Student of Medicine, *Physical, Sexual, and Natural Religions* (London: Truelove, 1855).

51. This statement introduces his chapter "Woman the Physician." For the book's place in histories of birth control, see Peter Fryer, *The Birth Controllers* (London: Secker & Warburg, 1965); Vern L. Bullough, ed., *Encyclopedia of Birth Control* (Santa Barbara, Calif.: ABC–Clio, 2001), 100–1. The most thorough research on the Drysdale family remains J. Miriam Benn, *Predicaments of Love* (London: Pluto Press, 1992). *Elements of Social Science* was translated into German (1871), French (1872), Dutch (1873), Italian and Portuguese (1876), Russian (1877), and Danish and Hungarian (1879). In later editions, Drysdale indicated that portions were first published in *The Political Economist and Journal of Social Science*, a journal he edited as "A Student of Medicine" for sixteen numbers from January 1856 to April 1857.

52. Drysdale, *Elements of Social Science*, 456.

53. Ibid., 323 (emphasis in original); 461, 457–64.

54. Theodore Besterman, *Mrs Annie Besant: A Modern Prophet* (London: Paul, Trench, Trubner, 1934).

55. Besant, *The Law of Population.*

56. Ibid., 5. This was drawn straight from Malthus's second edition (1803), by which time Franklin had gained the prominence of paragraph three of the greatly enlarged *Essay*.

57. Arthur Hobart Nethercot, *The First Five Lives of Annie Besant* (Chicago: University of Chicago Press, 1960); Arthur Hobart Nethercot, *The Last Four Lives of Annie Besant* (Chicago: University of Chicago Press, 1963).

58. In the few instances in which ecology and Besant are connected, it is typically through her theosophy and spiritualism, linked to a later twentieth-century nonexpert usage of "ecology" as holism, including a spiritual holism. See John Algeo and Shirley Nicholson, *The Power of Thought: A Twentieth-First Century Adaptation of the Annie Besant's Classic Work, Thought Power* (Wheaton, Ill.: Theosophical Society of America, 2001), 5.

59. Donald Worster, *Nature's Economy: A History of Ecological Ideas* (Cambridge: Cambridge University Press, 1994), 149.

60. Besant, *The Law of Population*, 15–16, 11–12.

61. A. Elley Finch, *Malthusiana: Illustrations of the Influence of Nature's Law of the Increase of Human Life, Discovered and Verified by Malthus* (London: Standring, 1904), 17–18.

62. Benn, *Predicaments of Love*; Rosanna Ledbetter, *A History of the Malthusian League, 1877–1927* (Columbus: Ohio State University Press, 1976).

63. Besant, *The Law of Population.*

64. John Stuart Mill, *On Liberty* (London: Parker, 1859).

65. For Darwin, on sexual selection and the process of writing *The Descent of Man*, see Adrian Desmond and James Moore, *Darwin's Sacred Cause: Race, Slavery and the Quest for Human Origins* (London: Lane, 2009), 362–76.

66. "Notes [to H. G. Wells] on Guests Who Will be Present," Typescript, 1927, JMK/PS/3/118, KP.

67. See, e.g., Alice Drysdale Vickery, *A Women's Malthusian League* (London: Standring, n.d.), explained as "A Women's League for the Extinction of Poverty and Prostitution Through the Rational Regulation of the Birth–Rate." See also Alice Drysdale-Vickery, *Ligue Malthusienne de Femmes: Branche de la Fédération universelle de la Régeneration Humaine* (Paris: Ligue de la Régeneration Humaine, n.d.).

68. See Barbara Caine, *English Feminism, 1780–1980* (Oxford: Oxford University Press, 1997).

69. For example, Lesley A. Hall, "Malthusian Mutations: The Changing Politics and Moral Meanings of Birth Control in Britain," in *Malthus, Medicine and Morality*, ed. Dolan, 141–63; Hera Cook, *The Long Sexual Revolution: English Women, Sex, and Contraception, 1800–1975* (Oxford: Oxford University Press, 2004).

70. Charles Darwin to Charles Bradlaugh, June 6, 1877, Darwin Correspondence Project, Letter 10988.

71. Annie Besant dealt with Darwin's argument "against scientific checks," in *Law of Population*, 71–72.

72. T. H. Huxley, "The Struggle for Existence," *Nineteenth Century* (1888), quoted in C. R. Drysdale, "The Malthusian Question at Home and Abroad," *The Malthusian* 19, no. 8 (1895): 58.

73. Akira Iriye, *Global Community: The Role of International Organizations in the Making of the Contemporary World* (Berkeley: University of California Press, 2002). See also Mark Mazower, *Governing the World: The History of an Idea* (New York: Penguin, 2012).

74. Neo-Malthusian organizations were also established in Italy before the war. See Maria Sophia Quine, *Population Politics in Twentieth-Century Europe* (London: Routledge, 1996), 224–25.

75. International Neo-Malthusian Bureau of Correspondence and Defence, *Report and Accounts for 1916* (London: International Neo-Malthusian Bureau of Correspondence and Defence, 1916). Pamphlet in Widener Library, Harvard University, Soc 5362.791.

76. C. V. Drysdale, "The Future of the Neo-Malthusian Movement," *The Malthusian* 30, no. 1 (1906): 3.

77. C. V. Drysdale, "Neo-Malthusian Philosophy," in *International Aspects of Birth Control*, ed. Margaret Sanger (New York: American Birth Control League, 1925), 26.

78. See Christine Bolt, *Sisterhood Questioned? Race, Class and Internationalism in the American and British Women's Movements, c. 1880s–1970s* (London: Routledge, 2004).

79. Jay Winter, *Dreams of Peace and Freedom: Utopian Moments in the Twentieth Century* (New Haven, Conn.: Yale University Press, 2006), 13.

80. Casper Sylvest, *British Liberal Internationalism, 1880–1930: Making Progress?* (Manchester: Manchester University Press, 2009).

81. Immanuel Kant, *Perpetual Peace: A Philosophical Essay*, trans. M. Campbell Smith (1795, London: Allen & Unwin, 1917).

82. Besant cited in A Doctor of Medicine [George Drysdale], *The Question of Irish Home Rule, with remarks on law and government and international anarchy, and with a proposal for the federal union of France and England as the most important step to the federation of the world* (London: Truelove, 1889), 11–12. This was Annie Besant's speech at a Home Rule meeting in St. James's Hall.

83. John S. Partington, *Building Cosmopolis: The Political Thought of H. G. Wells* (Aldershot, U.K.: Ashgate, 2003); Keynes read *A Modern Utopia* in 1905; see Donald Markwell, *John Maynard Keynes and International Relations: Economic Paths to War and Peace* (Oxford: Oxford University Press, 2006), 27.

84. A Doctor of Medicine, *The Question of Irish Home Rule*, 11.

85. Gregg Mitman, *The State of Nature: Ecology, Community, and American Social Thought, 1900–1950* (Chicago: University of Chicago Press, 1992), 62–88.

86. V. I. Lenin, "The Working Class and Neo Malthusanism," *Pravda* 137 (June 16, 1913). See also William Petersen, "Marxism and the Population Question: Theory and Practice," in *Population and Resources in Western Intellectual Traditions*, ed. Teitelbaum and Winter, 77–101.

87. See for example, C. V. Drysdale, *Diagrams of International Vital Statistics with Description in English and Esperanto* (London: William Bell, 1912), PP/EPR/G.20, Eileen Palmer Collection, Wellcome Library.

88. Gabriel Giroud, *Population et Subsistances* (Paris: Schleicher, 1904); Malthusian League, *The Semi-Starvation of the Human Race* (London: Author, 1906).

89. "Mr. Symes on the Cause of Poverty," *The Malthusian* 72 (1885): 582.

90. E. G. Ravenstein, "Lands of the Globe Still Available for European Settlement," *Proceedings of the Royal Geographical Society* 13 (1891): 27–35. See also Paul Demeny, "Limits of Growth," in *Population and Resources in Western Intellectual Traditions*, ed. Teitelbaum and Winter, 213–44.

91. Ravenstein, "Lands of the Globe," 32.

92. For Britain, see Richard Soloway, *Demography and Degeneration: Eugenics and the Declining Birthrate in Twentieth-Century Britain* (Chapel Hill: University of North Carolina Press, 1990). For France, see Alisa Klaus, "Depopulation and Race Suicide: Pronatalist Ideologies in France and the United States," in *Mothers of a New World: Maternalist Politics and the Origins of the Welfare State*, ed. Seth Koven and Sonya Michel (London and New York: Routledge, 1993); Elisa Camiscioli, "Producing Citizens, Reproducing the 'French Race': Immigration, Demography, and Pronatalism in Early Twentieth–Century France," *Gender and History* 13, no. 3 (2001): 593–621. For the United States, see Laura L. Lovett, *Conceiving the Future: Pronatalism, Reproduction, and the Family in the United States, 1890–1938* (Chapel Hill: University of North Carolina Press, 2007); Miriam King and Steven Ruggles, "American Immigration, Fertility, and Race Suicide at the Turn of the Century," *Journal of Interdisciplinary History* 20, no. 3 (1990): 347–69. For Australia, see Neville Hicks, *This Sin and Scandal: Australia's Population Debate, 1891–1911* (Canberra: Australian National University Press, 1978). For Italy, see Carl Ipsen, *Dictating Demography: The Problem of Population in Fascist Italy* (Cambridge: Cambridge University Press, 1996).

93. Malthusian League, *The Semi-Starvation of the Human Race*.

94. C. V. Drysdale, "The Future of the Neo-Malthusian Movement," *The Malthusian* 30, no. 1 (1906): 3.

95. John M. Robertson, "Malthusianism in America," *The Malthusian* 29, no. 4 (1905): 29.

96. "Protest," *The Malthusian* 27, no. 7 (1903): 49.

97. Waterman minimizes this period in the long history of the reappraisal of Malthus, focusing on Keynes's work in the 1930s. But it is the turn-of-the-century generation

of economists who educated Keynes, not least through their readings of Malthus. See Waterman, "Reappraisal of 'Malthus the Economist' 1933–97," 296–98.

98. Richard T. Ely, *Outlines of Economics* (New York: Macmillan, 1908); F. W. Taussig, *Principles of Economics*, 2 vols. (New York: Macmillan, 1911).

99. Taussig, *Principles of Economics*, 2:212, 221.

100. Alfred Marshall, *Principles of Economics*, 6th ed. (London: Macmillan, 1910), 179–80.

101. Alfred Marshall, "Social Possibilities of Economic Chivalry" (1907), in *Memorials of Alfred Marshall*, ed. A. C. Pigou (London: MacMillan, 1925), 323–46. See also John Toye, *Keynes on Population* (Oxford: Oxford University Press, 2000), 28.

102. It was Keynes's own first edition that was used to produce the 1925 facsimile reprint of the 1798 *Essay*, published by the Royal Economic Society and edited by James Bonar. See James Bonar to John Maynard Keynes, December 18, 1925, JMK/SS/1/95.

103. Toye, *Keynes on Population*, 25–35.

104. John Maynard Keynes [1912], Notes bound as "Principles of Economics, volume 2," JMK/UA/6/9/27, KP [hereafter 1912 Notes on Population].

105. Keynes, "Population," Typescript [1913–1914], JMK/SS/1/17, KP.

106. Keynes, 1912 Notes on Population, JMK/UA/6/9/26, KP.

107. Keynes, pencil notes for speech, 26 July 1927, JMK/PS/3/110–11, KP.

108. Keynes, 1912 Notes on Population, JMK/UA/6/9/24, KP. See also Markwell, *John Maynard Keynes and International Relations*, 25–27

109. Keynes, "Population," Typescript [1913–1914], JMK/SS/1/28, KP.

110. Darwin, *On the Origin of Species*, 64.

111. George Bernard Shaw, "The Basis of Socialism: Economic," in *Fabian Essays in Socialism*, ed. G. Bernard Shaw (London: Fabian Society, 1889), 11.

112. G. H. Knibbs, *The Mathematical Theory of Population, of Its Character and Fluctuations, and of the Factors which Influence Them* (Melbourne: Government Printer, 1917), 30.

113. A. M. Carr-Saunders, *Standing Room Only: A Study in Population* (London: British Broadcasting Corporation, 1930).

114. Keynes, "Population," Typescript [1913–1914], JMK/SS/1/ 24, KP.

115. Thompson, *Population*, 162, 113.

116. Ibid., 163, 165.

117. Ibid., 130.

2. War and Peace

1. Mark Bassin, "Imperialism and the Nation State in Friedrich Ratzel's Political Geography," *Progress in Human Geography* 11, no. 4 (1987): 477. For a recent account of German war aims, see William Mulligan, *The Origins of the First World War* (Cambridge: Cambridge University Press, 2010).

2. Robert Strausz-Hupé, *Geopolitics: The Struggle for Space and Power* (New York: Putnam, 1942), 157.

3. See Sebastian Conrad, *Globalisation and the Nation in Imperial Germany* (Cambridge: Cambridge University Press, 2010); Paul Weindling, "Fascism and Population in Comparative European Perspective," *Population and Development Review* 14 (1988): 102–21; Holger H. Herwig, "*Geopolitik*: Haushofer, Hitler and Lebensraum," *Journal of Strategic Studies* 22, no. 2 (1999): 218–41.

4. Strausz-Hupé, *Geopolitics*, vii–x. See Gerry Kearns, *Geopolitics and Empire: The Legacy of Halford Mackinder* (Oxford: Oxford University Press, 2009), 131–37.

5. For Friedrich Ratzel's indebtedness to the German tradition of natural geography, see Franco Farinelli, "Friedrich Ratzel and the Nature of (Political) Geography," *Political Geography* 19, no. 8 (2000): 943–55. See also Gertjan Dijkink, *National Identity and Geopolitical Visions: Maps of Pride and Pain* (London: Routledge, 1996), 17–35.

6. Friedrich Ratzel and Ellen C. Semple, "Studies in Political Areas. III: The Small Political Area," *American Journal of Sociology* 4, no. 3 (1898): 368–69.

7. Ibid., 378.

8. Friedrich Ratzel, *Der Staat und sein Boden* (Leipzig: S. Hirzel, 1896). Translation by Ellen C. Semple in review of *Der Staat und sein Boden, Annals of the American Academy of Political and Social Science* 9 (1897): 274. See also Gearóid Ó Tuathail, *Critical Geopolitics* (Minneapolis: University of Minnesota Press, 1996), 36.

9. Rudolf Kjellén, quoted in Strausz-Hupé, *Geopolitics*, 42. See Ola Tunander, "Swedish-German Geopolitics for a New Century: Rudolf Kjellén's 'The State as a Living Organism,'" *Review of International Studies* 27, no. 3 (2001): 451–63.

10. Strausz-Hupé, *Geopolitics*, vii–x. For the intellectual history of geopolitics, see David Thomas Murphy, *The Heroic Earth: Geopolitical Thought in Weimar Germany, 1918–1933* (Kent, Ohio: Kent State University Press, 1997).

11. David T. Murphy, "'A Sum of the Most Wonderful Things': *Raum*, Geopolitics and the German Tradition of Environmental Determinism, 1900–1933," *History of European Ideas* 25, no. 3 (1999): 125–6; Woodruff Smith, "Friedrich Ratzel and the Origins of Lebensraum," *German Studies Review* 3, no. 1 (1980): 51–68; J. H. Paterson, "German Geopolitics Reassessed," *Political Geography Quarterly* 6, no. 2 (1987): 107–114.

12. Hans W. Weigert, *Generals and Geographers: The Twilight of Geopolitics* (New York: Oxford University Press, 1942).

13. See Strausz-Hupé, *Geopolitics*, 225.

14. Friedrich Ratzel, "Studies in Political Areas: The Political Territory in Relation to Earth and Continent," trans. Ellen C. Semple, *American Journal of Sociology* 3, no. 3 (1897): 297.

15. Ibid., 299.

16. Ó Tuathail, *Critical Geopolitics*, 27–30; Klaus Dodds, *Geopolitics* (Oxford: Oxford University Press, 2007), 28.

17. George Steinmetz, *The Devil's Handwriting: Precoloniality and the German Colonial State in Qingdao, Samoa, and Southwest Africa* (Chicago: University of Chicago Press, 2007).

18. See Kearns, *Geopolitics and Empire*, 146–47.

19. Ellen Churchill Semple, *Influences of Geographic Environment: On the Basis of Ratzel's System of Anthropo-Geography* (New York: Holt, 1911).

20. Isaiah Bowman, *The New World: Problems in Political Geography* (New York: World Book, 1921).

21. Neil Smith, *American Empire: Roosevelt's Geographer and the Prelude to Globalization* (Berkeley: University of California Press, 2003).

22. The jurist Carl Schmitt made the most of the idea of a German "Monroe Doctrine." See William Hooker, *Carl Schmitt's International Thought* (Cambridge: Cambridge University Press, 2009). For a discussion of the place of the Monroe Doctrine in the establishment and presumptions of the peace settlement in 1919, see Jay Winter, *Dreams of Peace and Freedom: Utopian Moments in the Twentieth Century* (New Haven, Conn.: Yale University Press, 2006), 56–60.

23. Frederick Jackson Turner, "The Significance of the Frontier in American History," *Annual Report of the American Historical Association for the Year 1893* (Washington, D.C.: American Historical Association, 1894), 197–228.

24. Walter Nugent, *Habits of Empire: A History of American Expansionism* (New York: Random House, 2009).

25. Fergus Chalmers Wright, *Population and Peace: A Survey of International Opinion on Claims for Relief from Population Pressure* (Paris: IIIC, 1939), 129–30.

26. Friedrich von Bernhardi, *Germany and the Next War* (New York: Longmans, Green, 1914), 24. The 1912 translation was by Allen Powles.

27. Cited in Wallace Notestein and Elmer E. Stoll, *Conquest and Kultur: Aims of the Germans in their Own Words* (Washington, D.C.: Government Printing Office, 1917), 50–51. For Arthur Dix, see Murphy, *The Heroic Earth*, 91–98.

28. "The increase of population in Europe is having one of Malthus' 'positive checks' administered with very great sharpness." Edwin Canna, "Review of Warren Thompson, *Population*," *Economic Journal* 26, no. 102 (1916): 219. See also Matthew Connelly, *Fatal Misconception: The Struggle to Control World Population* (Cambridge, Mass.: Belknap, 2008), 46–47.

29. John Maynard Keynes, *The Economic Consequences of the Peace* (New York: Harcourt, Brace, and Howe, 1920), 10. For Keynes and the Paris Peace Conference, see Donald Markwell, *John Maynard Keynes and International Relations: Economic Paths to War and Peace* (Oxford: Oxford University Press 2006), 54–138.

30. Keynes, *The Economic Consequences of the Peace*, 4. See also John Toye, *Keynes on Population* (Oxford: Oxford University Press, 2000), 161–62.

31. Keynes, *The Economic Consequences of the Peace*, 25, 26.

32. John Maynard Keynes [1912], Notes bound as "Principles of Economics, volume 2," JMK/UA/6/9/17, KP.

33. J. M. Keynes, "A Reply to Sir William Beveridge," *Economic Journal* 33, no. 132 (1923): 476; Bowman, *The New World*.

34. W. H. Beveridge "Population and Unemployment," *Economic Journal* 33, no. 132 (1923): 447–75. See also discussion in Toye, *Keynes on Population*, 172–80.

35. A. B. Wolfe, "The Population Problem since the World War: A Survey of Literature and Research. Part 1," *Journal of Political Economy* 36, no. 5 (1928): 534.

36. Harold Cox, *The Problem of Population* (London: Jonathan Cape, 1922), 13. The book was published in New York by Putnam in 1923.

37. Ibid., 71–72.

38. Harold Wright, *Population* (London: Nisbet, 1923), 121; For Keynes's preface, see typescript, April 14, 1922, JMK/CEB/1/1, KP.

39. A. M. Carr-Saunders, *The Population Problem: A Study in Human Evolution* (Oxford: Clarendon Press, 1922); A. M. Carr-Saunders, *Population* (London: Oxford University Press, 1925), 72–94; A. M. Carr-Saunders, *World Population: Past Growth and Present Trends* (Oxford: Clarendon Press, 1936).

40. Carr-Saunders, *The Population Problem*, 5.

41. Edward M. East, *Mankind at the Crossroads* (1923: New York: Scribner, 1924), 3.

42. O. E. Baker, "Land Utilization in the United States: Geographical Aspects of the Problem," *Geographical Review* 13, no. 1 (1923): 1–26.

43. Edward Alsworth Ross, *Standing Room Only?* (London: Chapman and Hall, 1928), 170.

44. Warren S. Thompson, *Danger Spots in World Population* (New York: Knopf, 1929), 131.

45. Cox, *The Problem of Population*, 80.

46. Percy Dearmer, "Population, Eugenics and Birth Control," Sermon in Westminster Abbey, January 10, 1932, PP/EBR/F.1/1, Eileen Palmer Collection, Wellcome Library.

47. Carr-Saunders, *Population*, 84–85.

48. Carr-Saunders, *The Population Problem*, 304–05.

49. Carr-Saunders, *Population*, 92, 84.

50. Radhakamal Mukerjee, *The Political Economy of Population* (London: Longmans, Green, 1942), 21.

51. Erez Manela, *The Wilsonian Moment: Self-Determination and the International Origins of Anticolonial Nationalism* (Oxford: Oxford University Press, 2007), 23–24.

52. Jo-Anne Pemberton, *Global Metaphors: Modernity and the Quest for One World* (London: Pluto, 2001).

53. George Handley Knibbs, *The Shadow of the World's Future* (London: Ernest Benn, 1928), 128.

54. George H. Knibbs, "The Fundamental Elements of the Problems of Population and Migration," *Eugenics Review* 19, no. 4 (1928): 289.

55. George Knibbs to Bernard Mallet, June 20, 1927, Box 191, SPLC.

56. Knibbs, *The Shadow of the World's Future*, 97. Knibbs's book is discussed in Michael Kile, *No Room at Nature's Mighty Feast: Reflections on the Growth of Humankind* (Perth, Australia: Demos, 1995), 175–84.

57. Knibbs, *The Shadow of the World's Future*, 5.

58. Ibid., 101, 116.

59. Arne Fisher, review of G. H. Knibbs, "The Mathematical Theory of Population, of Its Character and Fluctuations, and of the Factors which Influence Them," *Publications of the American Statistical Association* 16, no. 123 (1918): 156–58. In 1906, Knibbs was appointed the (first) chief statistician of the new Commonwealth of Australia. He had entered public service from relatively humble conditions in 1877 and was to become the polymath president of the Royal Society of New South Wales, editor of its *Journal and Proceedings, and* fellow of the Royal Astronomical Society and the Royal Statistical Society. From 1889, Knibbs lectured at the University of Sydney in astronomy and hydraulics, and later, physics.

60. G. H. Knibbs, "The Mathematical Theory of Population, of Its Character and Fluctuations, and of the Factors which Influence Them," Appendix A, *Census of the Commonwealth of Australia, 1911, Volume 1: Statistician's Report* (Melbourne: Government Printer, 1917), 455–56.

61. See, e.g., East, *Mankind at the Crossroads*, 67; M. Kacprzak (chief statistician, Poland) to C. Wickens, November 17, 1924; G. H. Knibbs, "The World's Possible Population and Its Population Problem," Typescript, 1924; Extract from South African Census Report, 1924; all in A8510/175/4, Commonwealth Institute of Science and Industry, Sir George Knibbs, semi-personal papers, National Archives of Australia. [Hereafter Knibbs Papers]; Knibbs to D. F. Talman, Department of Meteorology, Washington D.C., September 29, 1932, A8510 175/1, Knibbs Papers; C. Wickens to Knibbs, August 7, 1922, A8510 175/2, Knibbs Papers.

62. Emanuel Czuber, *Mathematicische Bevölkerungstheorie, auf Grund von G.H. Knibbs' The Mathematical Theory of Population* (Leipzig: Teubner, 1923).

63. Knibbs was also closely involved in economic questions, offering reports on cost and standard of living. See George Knibbs, *Inquiry into the Cost of Living in Australia 1910–1911* (Melbourne: Commonwealth Bureau of Census and Statistics, 1911). See also correspondence between Knibbs and Irving Fisher, A8510 175/1, Knibbs Papers; G. H. Knibbs, "Consideration of the Proposal to Stabilize the Unit of Money," *American Economic Review* 9, no. 2 (1919): 244–55.

64. Knibbs, *The Shadow of the World's Future*, 84.

65. Charles Drysdale to George Knibbs, May 2, 1922, and Program leaflet, 1922, A8510 175/2, 175/6, and 175/3, Knibbs Papers.

66. Knibbs, *The Shadow of the World's Future*, 6.

67. C. V. Drysdale, "Neo-Malthusian Philosophy," in *International Aspects of Birth Control*, ed. Margaret Sanger (New York: American Birth Control League, 1925), 12, 14.

68. C. P. Blacker, *Birth Control and the State: A Plea and a Forecast* (London: Kegan Paul, Trench, Trubner, 1926), 74–75.

69. "Riscattare le terra, con la terra gli uomi e con glie uomi la razza." Cited (approvingly) in Benoy Kumar Sarkar, *The Sociology of Population with Special Reference to Optimum, Standard of Living and Progress: A Study in Societal Relativities* (Calcutta: Ray-Chowdhury, 1936), 36.

70. Carl Ipsen, *Dictating Demography: The Problem of Population in Fascist Italy* (Cambridge: Cambridge University Press, 1996), 90–144; See also Maria Sophie Quine, *Italy's Social Revolution: Charity and Welfare from Liberalism to Fascism* (Basingstoke, U.K.: Palgrave, 2002).

71. Ipsen, *Dictating Demography*, 34, 119–35.

72. R. R. Kuczynski, *"Living Space" and Population Problems* (Oxford: Clarendon, 1939), 10–11.

73. George Knibbs to Bernard Mallet, June 20, 1927, Box 191, SPLC.

74. Isaiah Bowman, "Population Outlets in Overseas Territories," reprinted for private circulation, MS 58, Bowman Papers, Milton Eisenhower Library Special Collections, Johns Hopkins University, 16.

75. Mark Mazower, *Hitler's Empire: Nazi Rule in Occupied Europe* (London: Lane, 2008).

76. Cited in Wright, *Population and Peace*, 49.

77. Adolf Hitler, *Mein Kampf*, trans. Ralph Manheim 1925, 1926; repr. (London: Pimlico, 2005), 587.

78. Ellen C. Semple, Review of Friedrich Ratzel, *Der Staat und sein Boden*, *Annals of the American Academy of Political and Social Science* 9 (1897): 274.

79. Hitler, *Mein Kampf*, 590.

80. David Blackbourn, *The Conquest of Nature: Water, Landscape, and the Making of Modern Germany* (New York: Norton, 2006), 251–309.

81. Hans Grimm, *Volk ohne Raum* (München: A. Langen, 1926).

82. Karl Haushofer, *Geopolitik des Pazifischen Ozeans* (Heidelberg: Vowinckel, 1938). Translated typescript, Karl Haushofer, "Geopolitics of the Pacific Ocean," DF17398, Fisher Library, University of Sydney, 309 [translator unknown].

83. Haushofer, "Geopolitics of the Pacific Ocean," 2.

84. Henning Heske, "Karl Haushofer: His Role in German Geopolitics and in Nazi Politics," *Political Geography Quarterly* 6, no. 2 (1987): 135–144; Holger H. Herwig, "*Geopolitik:* Haushofer, Hitler and Lebensraum," in *Geopolitics, Geography and Strategy*, ed. Colin S. Gray and Geoffrey Sloan (Oxford: Cass, 1999), 218–41; Christian W. Spang, "Karl Haushofer Re-Examined: Geopolitics as a Factor of Japanese-German Rapprochement in the Interwar Years?," in *Japanese-German Relations, 1895–1945: War, Diplomacy, and Public Opinion*, ed. Christian W. Spang and Rolf-Harald Wippich (New York: Routledge, 2006), 140; Lewis A. Tambs, "Preface," in Karl Haushofer, *An English Translation and Analysis of Major General Karl Ernst Haushofer's "Geopolitics of the Pacific Ocean,"* ed. Lewis A. Tambs, trans. Ernst J. Brehm (Lewiston: Mellen, 2002), v–ix.

85. Adolf Hitler, Statement in the Reichstag, 28 April 1939, quoted in Kuczynski, *"Living Space"*, 5.

86. Corrado Gini, "Considerations of the Optimum Density of a Population," *Proceedings of the World Population Conference*, ed. Margaret Sanger (London: Edward Arnold, 1927), 120. [Hereafter *1927 Proceedings*].

87. Corrado Gini, "The Scientific Basis of Fascism," *Political Science Quarterly* 42, no. 1 (1927): 105.

88. Sarkar, *The Sociology of Population*, 12, 13–14.

89. For Gini, see Ipsen, *Dictating Demography*, 45–47, 80 ff.; Maria Sophia Quine, *Population Politics in Twentieth Century Europe* (New York: Routledge, 1996), 29–34.

90. Henry Pratt Fairchild, "Optimum Population," *1927 Proceedings*, 84.

91. Mark R. Peattie, "The Japanese Colonial Empire, 1895–1945," in *The Cambridge History of Japan: The Twentieth Century*, ed. Peter Duus (Cambridge: Cambridge University Press, 1988), 6: 226.

92. A good contemporary summary of the Manchurian–Japanese relationship, critical of Japan, is Royama Masamichi, "Japan's Position in Manchuria," in *Problems of the Pacific, 1929: Proceedings of the Third Conference of the Institute of Pacific Relations,* ed. J. B. Condliffe (Chicago: University of Chicago Press, 1930), 524–93. For context and a study of Royama Masamichi, see William Miles Fletcher, *The Search for a New Order:*

Intellectuals and Fascism in Prewar Japan (Chapel Hill: University of North Carolina Press, 2011).

93. Tsurumi Yusuke, "The Ideals and Aspirations of Japan," *News Bulletin (Institute of Pacific Relations)* (September 1927), 29.

94. The publications mentioned are: *Geographische Grundlagen der japanischen Wehrkraft* (1910); *Dai Nihon* (1913); *Der deutscher Anteil an der geographischen Erschliessung Japans* (1914); *Die politischen Parteien in Japan* (1914), in Tambs, "Preface," Haushofer, *An English Translation and Analysis of Major General Karl Ernst Haushofer's "Geopolitics of the Pacific Ocean,"* viii.

95. For Japanese use of geopolitics, see Li Narangoa, "Japanese Geopolitics and the Mongal Lands, 1915–45," *East Asian Studies* 3 (2004): 45–67. For Haushofer and Japan, see also Kearns, *Geopolitics and Empire*, 20–21.

96. Peattie, "The Japanese Colonial Empire," 238; Robyn Lim, *The Geopolitics of East Asia: The Search for Equilibrium* (London: Routledge, 2003), 55–56.

97. See, e.g., Masamichi Royama, "Japan's Position in Manchuria," 528.

98. Keiichi Takeuchi, "Japanese Geopolitics in the 1930s and 40s," in *Geopolitical Traditions: A Century of Geopolitical Thought*, ed. Klaus Dodds and David Atkinson (New York: Routledge, 2000), 72. See also Goro Ishibashi, "Political Geography and Geopolitics," *Chigaku Zasshi* [Journal of Physical Geography], 42, no. 10 (1930): 611–14.

99. Emil Lederer, "Fascist Tendencies in Japan" *Pacific Affairs* 7, no. 4 (1934): 378.

100. "Manchuria, as a solution for the problem of overpopulation in Japan has not materialized." Royama, "Japan's Position in Manchuria," 573.

101. But see A. M. Carr-Saunders, "Fallacies about Overpopulation," *Foreign Affairs* 9, no. 4 (1931): 654–55.

102. Cited by *New York Times*, August 27, 1932, 2. See E. F. Penrose, *Population Theories and their Application with Special Reference to Japan* (Stanford, Calif.: Stanford University, 1934), 98.

103. See also John E. Orchard, "The Pressure of Population in Japan," *Geographical Review* 18, no. 3 (1928): 374–401.

104. Francesco Coppola, "Land for Italy" (1926), cited in Thompson, *Danger Spots*, 226.

105. Penrose, *Population Theories*, 99.

106. Cox, *The Problem of Population*, 77–78.

107. East, *Mankind at the Crossroads*, 92–93.

108. Editor, "Japan's Outbreak: Its Cause, Its Cure," *New Generation* 17, no. 1 (1938): 1.

109. Grzegorz Frumkin, "Japan's Demographic Expansion in the Light of Statistical Analysis," *Sociological Review* 30, no. 1 (1938): 1–28.

110. Thompson, *Danger Spots*, 278.

111. For explication and critique, see C. Walter Young, *Japan's Special Position in Manchuria* (Baltimore, Md.: Johns Hopkins University Press, 1931). Albert Thomas wrote of "the right to exist of overpopulated communities." See also Thomas, "International Migration and its Control," *1927 Proceedings*, 261.

112. J. Merle Davis, "Preface," in *Problems of the Pacific, 1929*, ed. Condliffe, v.

113. Tomoko Akami, *Internationalizing the Pacific: The United States,, Japan and the Institute of Pacific Relations in War and Peace 1919–45* (London: Routledge, 2001).

114. "Food and Population in the Pacific: Questions for Discussion," in *Problems of the Pacific, 1929*, ed. Condliffe, 37.

115. Ibid., 55, 63.

116. See Institute of Pacific Relations [Office Files] 1927–1962, Columbia University Rare Book and Manuscript Library. American delegates included Carl Alsberg of Stanford's Food Research Institute; Frederick Keppel, president of the Carnegie Corporation; James Weldon Johnson, secretary of the National Association for the Advancement of Colored People, New York; and Chicago sociologist Robert E. Park.

117. Cited in "Food and Population in the Pacific," in *Problems of the Pacific, 1929*, ed. Condliffe, 45.

118. Haushofer, "Geopolitics of the Pacific Ocean," 78–79.

119. Akami, *Internationalizing the Pacific*, 82–85, 193–94.

120. Nitobe Inazō, Handwritten note, December 24, 1920, Box R 642, International Eugenics, Social Section 12 9789/7260 (1920), League of Nations Archive, Geneva. [Hereafter LNAG].

121. Malthusian League, *Fifth International Neo-Malthusian and Birth Control Conference* (London: Author, 1922), 2. Promotional leaflet. Ishimoto was billed as "daughter-in-law of the late Japanese Minister of War. She is a Liberal in politics and passionately in favour of this great birth-control knowledge being spread among the people of her country. . . a pioneer of this cause in the East."

122. American Birth Control League, in Margaret Sanger, *The Pivot of Civilization* (New York: Brentano's, 1922), 283.

123. František Jaroslav Netusil, Comment, *1927 Proceedings*, 48–49.

124. League of Nations, Report of the Economic Committee to Council, 1938, quoted in Richard Symonds and Michael Carder, *The United Nations and the Population Question, 1945–1970* (London: Chatto & Windus, 1973), 17–19.

125. Note by the Secretariat, March 3, 1939, Box R 4440, Economic Section, Demographic and Migration Problems, LNAG.

126. Wright, *Population and Peace*, 5, xiii.

127. Ibid., 296, 121.

128. League of Nations Economic Committee, "Preliminary Observations on the Drawing-up of a Scheme of Work for the Study of Demographic Problems," June 30, 1938, Box R 4440, Economic Section, Demographic and Migration Problems, LNAG.

129. Dr Yoshisaka, comment in *Peaceful Change: Procedures, Population, Raw Materials, Colonies. Proceedings of the Tenth International Studies Conference 28 June–3 July 1937* (Paris: IIIC, 1937), 375.

130. Adolphe Landry, *La notion de surpeuplement*, cited in Wright, *Population and Peace*, 117. See also Adolphe Landry, *La revolution démographique: études et essais sur les problèmes de la population* (Paris: Librairie du Recueil Sirey, 1934).

131. André Touzet in *Le Problème colonial et la Paix du monde* (1937), cited in Wright, *Population and Peace*, 35.

132. "It was recognized generally by the experts that these demographic maladjustments do constitute a source of trouble and friction in international relations." Leonard Cromie, in *Peaceful Change*, 487.

133. See Carr-Saunders, *World Population*, v.

134. H. F. Angus, *The Problem of Peaceful Change in the Pacific Area: A Study of the Work of the Institute of Pacific Relations and Its Bearing on the Problem of Peaceful Change* (Oxford: Oxford University Press, 1937).

135. Imre Ferenczi, in *Peaceful Change*, 365

3. Density

1. Albert Thomas, "International Migration and Its Control," in *Proceedings of the World Population Conference*, ed. Margaret Sanger (London: Edward Arnold, 1927), 264 [hereafter 1927 Proceedings].

2. Raymond Pearl, "The Biology of Population Growth," *1927 Proceedings*, 22–38.

3. Gearóid Ó Tuathail, *Critical Geopolitics* (Minneapolis: University of Minnesota Press, 1996), 36.

4. In chapter 6, the connections with an emerging ecology and the establishment of a political ecology are discussed.

5. Margaret Sanger, *Margaret Sanger: An Autobiography* (New York: Norton, 1938); Madeline Gray, *Margaret Sanger: A Biography of the Champion of Birth Control* (New York: Marek, 1979); Ellen Chesler, *Woman of Valor: Margaret Sanger and the Birth Control Movement in America* (New York: Doubleday, 1992); Patricia Coates, *Margaret Sanger and the Origin of the Birth Control Movement, 1910–1930* (Lewiston: Mellen, 2008).

6. Emma Goldman, *Living My Life* (1931; New York: Da Capo, 1970), 2. See also Alice Wexler, *Emma Goldman: An Intimate Life* (London: Virago Press, 1985), 209; F. Ronsin, "Between Malthus and the Social Revolution: The French Neo-Malthusian Movement," in *Malthus: Past and Present*, ed. J. Dupâquier, A. Fauve–Chamoux, and E. Grebenik (London: Academic, 1983), 329–39.

7. Goldman quoted in Wexler, *Emma Goldman*, 210.

8. Raymond Pierpoint, ed., *Report of the Fifth International Neo-Malthusian and Birth Control Conference* (London: Heinemann, 1922), 3.

9. See news clipping report on the private session for medical men and women, Neo-Malthusian and Birth Control Conference, *Lancet*, July 22, 1922, 195–96, JMK/SS/3, KP.

10. Havelock Ellis, "The Evolutionary Meaning of Birth Control," in *Medical and Eugenic Aspects of Birth Control*, ed. Margaret Sanger (New York: American Birth Control League, 1926), 2.

11. Margaret Sanger to Clarence Little, November 30, 1925, Box 191, SPLC. Minutes of the international committee at Mrs. Sanger's Residence, March 31, 1925, Box 190, SPLC.

12. East had publicly supported Sanger, but questioned the statistical validity of neo-Malthusian population research in his important 1923 book. Edward M. East, *Mankind at the Crossroads* (New York: Scribner, 1923), 10.

13. Clarence Little to Margaret Sanger, March 25, 1926; Margaret Sanger to Raymond Pearl, February 7, 1927, Box 191, SPLC.

14. Raymond Pearl to Margaret Sanger, April 19, 1926, Box 191, SPLC. Pearl published in Sanger's journals, e.g., "The Menace of Population Growth," *Birth Control Review* 7, no. 3 (1923): 65–67.

15. "Announcement," *1927 Proceedings*, i.

16. Bernard Mallet to Margaret Sanger, August 3, 1927, Box 191, SPLC. The conference was funded by Sanger personally, as well as by a grant from the Laura Spelman Rockefeller Memorial Fund. Raymond Pearl to Margaret Sanger, April 13, 1927; Margaret Sanger to Raymond Pearl, October 17, 1927, Box 191, SPLC. The Bureau of Social Hygiene (funded by John D. Rockefeller Jr) contributed, and the U.S. National Research Council funded a further $10,000 for the formation of a permanent union of scientists. Sanger's husband Noah Slee financed the journal. World Population Conference Budget, November 11, 1927; Margaret Sanger to Clinton Chance, January 16, 1928, Box 191, SPLC.

17. Corrado Gini to Bernard Mallet, August 22, 1927; Bernard Mallet to Corrado Gini, August 26, 1927; Box 191, SPLC. For Mussolini, see D. V. Glass, "Italian Attempts to Encourage Population Growth," *Review of Economic Studies* 3, no. 2 (1936): 106–19. See also Maria Sophia Quine, *Population Politics in Twentieth Century Europe* (London: Routledge, 1996), 33–35.

18. Bernard Mallet to Margaret Sanger, August 3, 1927, Box 191, SPLC.

19. Otto L. Mohr to Margaret Sanger, March 12, 1927; Otto L. Mohr to Julian Huxley, March 12, 1927, Box 191, SPLC.

20. Sir Eric Drummond to Under-Secretaries General of the League of Nations, August 22, 1927, Dossier Concerning World Population Conference, 1927, Box 1602, LNAG.

21. Dame Rachel Crowdy to Sir Eric Drummond, May 16, 1927, Dossier Concerning World Population Conference, 1927, Box 1602, LNAG. See also, Richard Symonds and Michael Carder, *The United Nations and the Population Question, 1945–1970* (London: Chatto & Windus, 1973), 13–14.

22. Dame Rachel Crowdy to Sir Eric Drummond, June 7, 1927, Dossier Concerning World Population Conference, 1927, Box 1602, LNAG.

23. Frank Ricardreau to Margaret Sanger, November 9, 1927; Mrs. Alfred Zimmern to Margaret Sanger, February 1, 1926, Box 191, SPLC; Norman White to Deputy Secretary-General, June 17, 1927, Dossier Concerning World Population Conference, 1927, Box 1602, LNAG.

24. A. E. Johnson to Margaret Sanger, October 19, 1927, Box 191, SPLC.

25. That biologists started talking with economists is something Julian Huxley made a point of noting; see Huxley, "Are We Overpopulated?," October 25, 1927, News clipping, Box 97, Folder 7, HP.

26. "Announcement," *1927 Proceedings*, i.

27. Thomas, "International Migration and Its Control," 264, 266.

28. P. F. Verhulst was a student of early nineteenth-century Belgian mathematician Adolphe Quetelet. See Enid Charles, *The Twilight of Parenthood: A Biological Study of the Decline of Population Growth* (London: Watts, 1934), 150–55. On Quetelet, see Ian Hacking, *The Taming of Chance* (Cambridge: Cambridge University Press, 1990); Libby Schweber, *Disciplining Statistics: Demography and Vital Statistics in France and England, 1830–1885* (Durham, N.C.: Duke University Press, 2006).

29. Raymond Pearl, *The Biology of Population Growth* (New York: Knopf, 1925), 22.

30. Ibid., 45, 208–9, 3.

31. Ibid., 28–29, 38.

32. Radhakamal Mukerjee, *The Political Economy of Population* (London: Longmans, Green, 1942), 165.

33. H. S. Jennings, *Biographical Memoir of Raymond Pearl* (New York: National Academy of Science, 1942), 297.

34. Raymond Pearl, *The Biology of Poultry Keeping* (Orono: Maine Agricultural Experiment Station, 1913); Raymond Pearl, *Triplet Calves* (Orono: Maine Agricultural Experiment Station, 1912). See also Thomas Robertson, *The Malthusian Moment: Global Population and the Birth of American Environmentalism* (New Brunswick, N.J.: Rutgers University Press, 2012), 16–19; Edmund Ramsden, "Carving up Population Science: Eugenics, Demography, and the Controversy over the 'Biological Law' of Population Growth," *Social Studies of Science* 32, no. 5/6 (2002): 862.

35. Raymond Pearl, *Studies in Human Biology* (Baltimore, Md: Williams and Wilkins, 1924); Raymond Pearl, *The Biology of Death* (Philadelphia: Lippincott, 1922); Raymond Pearl, *The Nation's Food: A Statistical Study of a Physiological and Social Problem* (Philadelphia: Saunders, 1920).

36. G. H. Knibbs, "The Mathematical Theory of Population, of Its Character and Fluctuations, and of the Factors which Influence Them," Appendix A, *Census of the Commonwealth of Australia, 1911, Volume 1: Statistician's Report* (Melbourne: Government Printer, 1917), 95. George Knibbs argued strongly against Pearl's logistical curve slightly earlier in "The New Malthusianism in the Light of Actual World Problems of Population," *Scientia* 40 (1926): 379–88.

37. George Knibbs to Bernard Mallet, June 20, 1927, Box 191, SPLC. Knibbs insisted on writing, pointedly, of "Verhulst's logistic curve." See George H. Knibbs "The Laws of Growth of a Population. Part 1," *Journal of the American Statistical Association* 21, no. 156 (1926): 381–98. "What Verhulst has called the logistic curve does *not* very accurately represent rationally the facts for human populations, although the curve may be empirically fitted to the actual facts, approximately" (382). For Knibbs on Pearl, see also Sharon Kingsland, *Modeling Nature: Episodes in the History of Population Ecology* (Chicago: University of Chicago Press, 1985), 78–79.

38. Knibbs, "The Laws of Growth of a Population. Part 1," 381–83.

39. William Rappard, Comment, *1927 Proceedings*, 354.

40. J. B. S. Haldane, Comment, *1927 Proceedings*, 39–40.

41. Ramsden, "Carving up Population Science"; Ana Millan Gasca, "Mathematical Theories versus Biological Facts: A Debate on Mathematical Population Dynamics in the 1930s," *Historical Studies in the Physical and Biological Sciences* 26, no. 2 (1996): 347–403; Kingsland, *Modeling Nature*, chap. 3–4.

42. Gregg Mitman discusses Pearl's theory and density in the context of Chicago ecologist Warder Clyde Allee's work on aggregation. For funding purposes, more than intellectual purposes, Mitman argues, Allee "encountered the need to reclothe his aggregation research in population garb" (Gregg Mitman, *The State of Nature: Ecology, Community, and American Social Thought, 1900–1950* [Chicago: University of Chicago Press, 1992], 90).

43. Pearl, "The Biology of Population Growth," 28.

44. Edward M. East, "Food and Population," *1927 Proceedings*, 85.

45. Donald F. Jones, *Biographical Memoir of Edward Murray East* (Washington, D.C.: National Academy of Sciences, 1945), 229. See also Robertson, *The Malthusian Moment*, 19–23.

46. E. M. East, "The Agricultural Limit of Our Population," *Scientific Monthly* 12, no. 6 (1921): 551–57; E. M. East, "Population," *Scientific Monthly* 10, no. 6 (1920): 603–24.

47. Sanger–East correspondence in American Birth Control League Papers, MS Am 2063, Houghton Library, Harvard University.

48. Margaret Sanger, "The World We Live In," *Birth Control Review* 7, no. 12 (1923): 315.

49. East, "Food and Population," 85.

50. Harold Wright, *Population* (London: Nisbet, 1923), 147.

51. F. J. Netusil, "Comment," *1927 Proceedings*, 48.

52. E. M. East, "Civilization at the Crossways," *Birth Control Review* 7, no. 12 (1923): 328–29.

53. East, "The Agricultural Limit of Our Population," 551–52.

54. Ibid., 556; East, "Food and Population," 89.

55. East, *Mankind at the Crossroads*, 69.

56. East, "Food and Population," 85.

57. A. M. Carr-Saunders, *World Population: Past Growth and Present Trends* (Oxford: Clarendon, 1936), 17, 29–45; W.F. Willcox and Imre Ferenczi, *International Migrations*, 2 vols. (New York: National Bureau of Economic Research, 1929–1931). The first volume compiled statistics on behalf of the ILO. The second volume compiled interpretations by a number of scholars.

58. E. G. Ravenstein, "Lands of the Globe Still Available for European Settlement," *Proceedings of the Royal Geographical Society* 13, no. 1 (1891): 31.

59. Knibbs, "The World's Possible Population and Its Population Problem," n.d., A8510 175/4, Knibbs Papers. See Knibbs cited in A. M. Carr-Saunders, *Population* (London: Oxford University Press, 1925), 68–69.

60. Henry Pratt Fairchild, "Optimum Population," *1927 Proceedings*, 81.

61. Fergus Chalmers Wright, *Population and Peace: A Survey of International Opinion on Claims for Relief from Population Pressure* (Paris: IIIC, 1939), 72.

62. Ravenstein, "Lands of the Globe," 28.

63. C. B. Fawcett, "Some Factors in Population Density," in *Problems of Population: Proceedings of the Second General Assembly of the International Union for the Scientific Investigation of Population Problems*, ed. G. H. L. F. Pitt-Rivers (London: Allen & Unwin, 1932), 191.

64. Typescript, "Population Problems," July 4, 1931, "Nature" in Bureau of Social Hygiene, 1931, Series 3, Box 9, Folder 191, RAC.

65. George Knibbs, "Agriculture and Population—Questions of Agriculture in Connection with World's Possible Population," 1922, A8510 7/2, Knibbs Papers.

66. Knibbs, "The New Malthusianism in the Light of Actual World Problems of Population," 384.

67. George Handley Knibbs, *The Shadow of the World's Future* (London: Ernest Benn, 1928), 110.

68. Hugo Grothe, Comment, *1927 Proceedings*, 297. See also Warren D. Smith, "World Population," *Scientific Monthly* 40, no. 1 (1935): 41.

69. Radhakamal Mukerjee addressed "the Biological Factors." "These are analysable and not immutable forces of nature, as assumed in Pearl's somewhat mystical hypothesis" ("First Indian Population Conference, Lucknow," January 27, 1936, *Madras Bulletin Birth Control Bulletin,* Coll. Misc. 639/1/25, Eileen Palmer Collection, LSE). For uptake of the economics of optimum population by biologists, see Mitman, *The State of Nature*, 94.

70. Huxley, "Are We Overpopulated?" For a useful summary of the theory of optimum population, see Manuel Gottlieb, "The Theory of Optimum Population for a Closed Economy," *Journal of Political Economy* 53, no. 4 (1945): 289–316.

71. "Birth Control in Asia," *British Medical Journal* (December 2, 1933).

72. C. V. Drysdale, Comment, *1927 Proceedings*, 104.

73. Sir Charles Close, Comment, *1927 Proceedings*, 95. President of the International Geographical Congress, Close also worked with Raymond Pearl on the IUSIP, especially in its difficult years during the 1930s, when the apparently apolitical body found itself deep inside a politicized population question, centered on German involvement, a planned Berlin meeting, and the relations with director of the Kaiser Wilhelm Institute of Anthropology, Human Heredity and Eugenics, Eugen Fischer. Correspondence between Sir Charles Close and Raymond Pearl, B: P312, Folders 9–12, Pearl Papers, APS.

74. "The material basis of National Happiness is that a country should not have too many inhabitants for comfort nor too few for safety." Binnie Dunlop, *National Happiness under Individualism, an Explanation and Solution of the Poverty and Riches Problem*, January 1912, 1. Bound in Pamphlets, Soc 5362/791, Widener Library, Harvard University.

75. Alexander Carr-Saunders, Comment, *Birth Control in Asia,* ed. Michael Fielding (London: BCIIC, 1935), 13. Understanding Malthus to have predicted "uncontrolled population growth" is as common now as it was then. See, e.g., Robertson, *The Malthusian Moment*, 5.

76. Alexander Carr-Saunders, Typescript paper for the Birth Control in Asia Conference, 1933, Box 194, SPLC.

77. Carr-Saunders, *Population*, 67.

78. A. M. Carr-Saunders, *Standing Room Only: A Study in Population* (London: British Broadcasting Corporation, 1930), 13–14. This lecture series was broadcast between September 30 and November 4 1930.

79. Fairchild, "Optimum Population," *1927 Proceedings*, 79.

80. Ibid., 76.

81. Corrado Gini, "Considerations on the Optimum Density of a Population," *1927 Proceedings*, 118.

82. Corrado Gini, "The Scientific Basis of Fascism," *Political Science Quarterly* 42, no. 1 (1927): 99–115.

83. G. H. Knibbs, "The Fundamental Elements of the Problems of Population and Migration," *Eugenics Review* 19, no. 4 (1928): 276.

84. Knibbs, "The World's Possible Population."

85. Adolphe Landry quoted in *Peaceful Change: Procedures, Population, Raw Materials, Colonies. Proceedings of the Tenth International Studies Conference 28 June–3 July 1937* (Paris: IIIC, 1938), 122.

86. George H. T. Kimble, *The World's Open Spaces* (London: Thomas Nelson, 1939), 24.

87. Karl Thalheim, Comment, *1927 Proceedings*, 291.

88. Ravenstein, "Lands of the Globe," 27.

89. Knibbs, *The Shadow of the World's Future*, 20–22.

90. Carr-Saunders, *Population*, 97.

91. "Population Problems," July 4, 1931, RAC.

92. Huxley, "Are We Overpopulated?"

93. "If I were to tackle the problem from a strictly nationalistic point of view," she qualified. Liu Chieh, Transcript of Birth Control in Asia Conference, London, 1933, Box 194, SPLC.

94. Professor H.H. Chen, Transcript of Birth Control in Asia Conference, 1933, Box 194, SPLC.

95. For trends in global mapping, see Denis Cosgrove, *Geography and Vision: Seeing, Imagining and Representing the World* (London: Tauris, 2008).

96. Marcel Aurousseau to Margaret Sanger, May 16, 1927, Box 191, SPLC.

97. John Scott Keltie, *Applied Geography: A Preliminary Sketch* (London: Philip, 1890). Reviewed in "New Maps," *Proceedings of the Royal Geographical Society* 13, no. 1 (1891): 60.

98. Radhakamal Mukerjee, *Migrant Asia* (Rome: Failli, 1936).

99. V. C. Finch and O. E. Baker, "Geography and World Agriculture [map]," in Warren S. Thompson, *Danger Spots in World Population* (New York: Knopf, 1929), 5.

100. Clyde V. Kiser, "Warren S. Thompson," *Population Index* 40, no. 1 (1974): 21–23.

101. Thompson, "Scripps Foundation for Research in Population Problems," *1927 Proceedings*, 350.

102. See, e.g., Warren Thompson, review of Karl Julius Beloch, *Bevölkerungsgeschichte Italiens* (Berlin: De Gruyter, 1937); Marcel Dutheil, *La Population Allemande* (Paris: Payot, 1937); Michel Huber, Henri Bunle, and Fernard Boverat, *La Population de la France* (Paris: Hachette, 1937); Statistique Generale de la France, *Statistique du Mouvement de la Population en 1934* (Paris: Imprimerie Nationale) all reviewed in *American Sociological Review* 4, no. 2 (1939): 288–90. Thompson published in the French journal *Population*, e.g., "La paix et l'accroissement de la population dans le Pacifique," *Population* 2, no. 4 (1947): 647–62.

103. See Dennis Hodgson, "Thompson, Warren S.," in *Encyclopedia of Population*, ed. Paul Demeny and Geoffrey McNicoll (New York: Macmillan, 2003), 939–40; Warren Thompson, *Population Problems* (New York: McGraw–Hill, 1930). This ran to a 5th edition in 1965.

104. Thompson, *Danger Spots*, v.

105. Ibid., 13.

106. Warren S. Thompson, "Population," *American Journal of Sociology* 34, no. 6 (1929): 975.

107. Thompson, *Danger Spots*, 183.

108. Ibid., 329–30.

4. Migration

1. Charles Davenport, Comment, *Proceedings of the World Population Conference* ed. Margaret Sanger (London: Edward Arnold, 1927), 242 [hereafter *1927 Proceedings*].

2. Davenport's deputy at the Eugenics Record Office, Harry Laughlin, was instrumental to development of immigration restriction law and policy in Washington D.C. Daniel J. Kevles, *In the Name of Eugenics : Genetics and the Uses of Human Heredity* (Cambridge, Mass.: Harvard University Press, 1995), 103–4.

3. Immigration Act of 1924 (U.S.), H.R. 7995, 68th Cong. (1924).

4. Edward Alsworth Ross, *Standing Room Only?* (London: Chapman and Hall, 1928); Marilyn Lake and Henry Reynolds, *Drawing the Global Colour Line: White Men's Countries and the International Challenge of Racial Equality* (Cambridge: Cambridge University Press, 2008); Paul A. Kramer, "Empire Against Exclusion in Early 20th Century Trans-Pacific History," *Nanzan Review of American Studies* 33 (2011): 13–32. This phenomenon was not limited to the Anglophone world: Adam McKeown details the racially exclusive laws of many of the Central and South American republics. Adam M. McKeown, *Melancholy Order: Asian Migration and the Globalization of Borders* (New York: Columbia University Press, 2008), 47–51, 320–21.

5. A. B. Wolfe, "The Population Problem since the World War: A Survey of Literature and Research, Part 1," *Journal of Political Economy* 36, no. 5 (1928): 533.

6. Warren Thompson, "The Demographic and Economic Implications of Larger Immigration," in *Postwar Problems of Migration*, ed. Lowell Jacob Reed (New York: Milbank Memorial Fund, 1947), 100–1. For a retrospective view on population and international migration, see Kingsley Davis, "Social Science Approaches to International Migration," in *Population and Resources in Western Intellectual Traditions*, ed. Michael S. Teitelbaum and Jay M. Winter (Cambridge: Cambridge University Press, 1989), 245–61.

7. See, e.g., Andrea Geiger, *Subverting Exclusion: Transpacific Encounters with Race, Caste, and Borders, 1885–1928* (New Haven, Conn: Yale University Press, 2011). For the implications of Chinese regional movement within and beyond east Asia, see McKeown, *Melancholy Order.* Indian emigration was tied to the abolition of slavery in the British Empire in 1834, a system of indenture that lasted until its suspension in 1917 and abolition in 1920. See Sunil Amrith, "Indians Overseas? Governing Tamil Migration to Malaya, 1870–1941," *Past and Present* 208, no. 1 (2010): 231–61; Sunil Amrith, *Migration and Diaspora in Modern Asia* (Cambridge: Cambridge University Press, 2011), 32–37.

8. Albert Thomas, "International Migration and Its Control," *1927 Proceedings*, 256–65. See also ILO catalog of immigration and emigration acts, Box 191, SPLC.

9. G. de Lapouge, "Contribution to the Fundamentals of a Policy on Population," *Eugenics Review* 19, no. 3 (1927): 194.

10. Warren S. Thompson, "Race Suicide in the United States," *The Scientific Monthly* 5, no. 1 (1917): 22. For Theodore Roosevelt and Edward Alsworth Ross, see Laura L. Lovett, *Conceiving the Future: Pronatalism, Reproduction, and the Family in the United States, 1890–1938* (Chapel Hill: University of North Carolina Press, 2007), 91–102.

11. Matthew Connelly, "To Inherit the Earth: Imagining World Population, from the Yellow Peril to the Population Bomb," *Journal of Global History* 1, no. 3 (2006): 299–319.

12. Theodore Roosevelt, "National Life and Character," in *American Ideals and Other Essays, Social and Political* (1897; New York: Putnam, 1898), 287.

13. Lothrop Stoddard, *The Rising Tide of Color Against White World-Supremacy* (London: Chapman and Hall, 1923). First published in New York by Scribner in 1920; Madison Grant, *The Passing of the Great Race, or, The Racial Basis of European History* (New York: Scribner, 1916).

14. Stoddard, *The Rising Tide of Color,* 6–7, 154, 8.

15. Madison Grant, "Introduction," in Stoddard, *The Rising Tide of Color*, xxx. For Benjamin Franklin, see Joyce E. Chaplin, *Benjamin Franklin's Political Arithmetic: A Materialist View of Humanity* (Washington, D.C.: Smithsonian Institution Libraries, 2008).

16. Stoddard, *The Rising Tide of Color*, 200, 198–99, 201, 179.

17. "Food and Population in the Pacific," in *Problems of the Pacific, 1929*, ed. J. B. Condliffe (Chicago: University of Chicago Press, 1930), 49.

18. Warren S. Thompson, *Population: A Study in Malthusianism* (New York: Columbia University, 1915), 91.

19. Enid Charles, *The Twilight of Parenthood: A Biological Study of the Decline of Population Growth* (London: Watts, 1934), 94.

20. A. M. Carr-Saunders, "An Outline of Population History," *Population* 1, no. 1 (1933): 28.

21. A. M. Carr-Saunders, *World Population: Past Growth and Present Trends* (Oxford: Clarendon, 1936), 261.

22. R. K. Das, Comment, *1927 Proceedings*, 107.

23. Benoy Kumar Sarkar, *The Sociology of Population: with Special Reference to Optimum, Standard of Living and Progress* (Calcutta: Ray-Chowdhury, 1936), 18.

24. Teijiro Uyeda in *Pacific Affairs* 6, no. 6 (1933): 297–304; See also W. R. Crocker, *The Japanese Population Problem* (London: Allen & Unwin, 1931).

25. E. F. Penrose, *Population Theories and Their Application with Special Reference to Japan* (Stanford, Calif.: Stanford University Press, 1934), ix–xi, 95.

26. Ibid., v, 99.

27. John Maynard Keynes [1912], Notes bound as "Principles of Economics, volume 2," JMK/UA/6/9/22–24, KP.

28. Carr-Saunders, *World Population*, 261.

29. Edward M. East, *Mankind at the Crossroads* (1923: New York: Scribner, 1924), 111, 12.

30. Ibid., 115–16, 13.

31. Ibid., 125.

32. Warren S. Thompson, "Review of *Mankind at the Crossroads*," *Science, New Series* 58, no. 1509 (1923): 445–46.

33. Warren S. Thompson, *Danger Spots in World Population* (New York: Knopf, 1929), 6.

34. Sean Brawley, *The White Peril: Foreign Relations and Asian Immigration to Australasia and North America, 1919–1978* (Kensington: University of New South Wales Press, 1995); Naoko Shimazu, *Japan, Race, and Equality: The Racial Equality Proposal of 1919* (London: Routledge, 1998).

35. "Food and Population in the Pacific," in *Problems of the Pacific, 1929*, ed. Condliffe, 44.

36. G.W. Silverstople, "Discussion on Food and Population," *1927 Proceedings*, 100.

37. C. P. Blacker, *Birth Control and the State: A Plea and a Forecast* (London: Kegan Paul, Trench, Trubner, 1926), 27.

38. Thompson, *Danger Spots*, 212–13, 222–23, 294. "Hence Russia and Siberia—the soviets—may be expected to do their part in caring for the population growth of the world during the next half-century" (230–31). See also Warren S. Thompson, "The Population Barrier," in *The Underdeveloped Lands: A Dilemma of the International Economy*, ed. DeVere E. Pentony (San Francisco: Chandler, 1960), 113–30.

39. Ernest Mahaim, Comment, *1927 Proceedings*, 265.

40. Étienne Dennery, *Asia's Teeming Millions: And Its Problems for the West* (1931; New York: Kennikat, 1970), 232, 83, 198. "The protests are made not only in the name of the Indian nation, but more generally, in that of all peoples of colour. From India are heard the voices that echo in China and Japan, claiming equal rights of emigration. A solidarity of the coloured races is aroused, far more intense than mere nationalist feeling" (198).

41. See Rajani Kanta Das, *Hindustani Workers on the Pacific Coast* (Berlin: Gruyter, 1923); Harold Cox, *The Problem of Population* (London: Jonathan Cape, 1922).

42. Dennery, *Asia's Teeming Millions*, 235–36; See Sun Yat-sen, *The International Development of China* (New York: Putnam 1922).

43. Friedrich Ratzel, *Die Chinesische Auswanderung* (Breslau: Kern, 1876) and Ernst Grünfeld, *Die Japanische Auswanderung* (Tokyo: Druck der Hobunsha, 1913).

44. Fergus Chalmers Wright, *Population and Peace: A Survey of International Opinion on Claims for Relief from Population Pressure* (Paris: IIIC, 1939), vii.

45. For example, Alastair Pennycook, *English and the Discourses of Colonialism* (London: Routledge, 1998), 169.

46. Dennery, *Asia's Teeming Millions*, 229.

47. Taraknath Das, "The Population Problem in India," in *Religious and Ethical Aspects of Birth Control*, ed. Margaret Sanger (New York: American Birth Control League, 1926), 196, 224, 195.

48. Matthew Connelly, "Population Control Is History," *Comparative Studies in Society and History* 45, no. 1 (2003), 122–47; Matthew Connelly, *Fatal Misconception: The Struggle to Control World Population* (Cambridge, Mass.: Belknap, 2008), 65.

49. Mohan Rao, *From Population Control to Reproductive Health: Malthusian Arithmetic* (New Delhi: Sage, 2004), 91–92.

50. Das, "The Population Problem in India," 195.

51. Tapan K. Mukherjee, *Taraknath Das: Life and Letters of a Revolutionary in Exile* (Kolkata: National Council of Education, 1998); "Das, Taraknath (Dr.)," in *Dictionary of*

National Biography, 4 vols., ed. S. P. Sen (Calcutta: Institute of Historical Studies, 1972), 1:363–64.

52. See, e.g., P. K. Wattal, *Population Problem in India: A Census Study* (Bombay: Bennett, Coleman, 1916). For the significance of the British census and other Indian population studies, see David Arnold, "Official Attitudes to Population, Birth Control and Reproductive Health in India, 1921–1946," in *Reproductive Health in India: History, Politics, Controversies*, ed. Sarah Hodges (New Delhi: Orient Longman, 2006), 22–50.

53. Radhakamal Mukerjee, *Migrant Asia* (Rome: Failli, 1936), 156.

54. Ibid., 179; J. W. Gregory, *The Menace of Colour* (Philadelphia: Lippincott, 1925).

55. For Mukerjee's relation to liberalism in late colonial and early Independent India, see C. A. Bayly, *Recovering Liberties: Indian Thought in the Age of Liberalism and Empire* (Cambridge: Cambridge University Press, 2011), chap. 10.

56. Mukerjee, *Migrant Asia*, 12, 22.

57. Ibid., 18–19.

58. Ibid., 156, 132.

59. See also, Radhakamal Mukerjee, *The Political Economy of Population* (London: Longmans, Green, 1942), 403–28.

60. Thompson, *Danger Spots*, 327.

61. Harold Wright, *Population* (London: Nisbet, 1923), 2.

62. Harold Cox, "Foreword," in Dennery, *Asia's Teeming Millions*.

63. Frederick Sherwood Dunn cited in Wright, *Population and Peace*, 121 (emphasis in original).

64. J. B. Condliffe, "The Pressure of Population in the Far East," *Economic Journal* 42, no. 166 (1932): 206.

65. Wright, *Population and Peace*, 121.

66. "That was a period of liberty, and this liberty had far-reaching results" (Thomas, Comment, *1927 Proceedings*, 267). For the significance of the free migrant, see McKeown, *Melancholy Order*, 66–89.

67. A. Koulisher, Comment, *1927 Proceedings*, 102.

68. T. Eguchi in *Birth Control in Asia*, ed. Michael Fielding (London: BCIIC, 1935), 22. In the typescript version of the original paper, Eguchi wrote: "Controls and restrictions became the Mistress of the World, we are controlling birth, exchange, currencies, imports, and restricting production of sugar, rubber, tin, and even burnt coffee and cotton to protect the market prices." Report of the Conference Held at the London School of Hygiene and Tropical Medicine, First Session, November 24, 1933, Box 194, SPLC.

69. See Richard Symonds and Michael Carder, *The United Nations and the Population Question, 1945–1970* (London: Chatto & Windus, 1973), 17.

70. Cox, "Foreword," in Dennery, *Asia's Teeming Millions*, 14.

71. Das, "The Population Problem in India," 216–17.

72. Lovett, *Conceiving the Future*, 91–102.

73. Sir William Crookes, "Address," *Science, New Series* 8, no. 200 (1898): 564.

74. O. E. Baker, "The Potential Supply of Wheat," *Economic Geography* 1, no. 1 (1925): 16.

75. Ross, *Standing Room Only?*, v-vi. Ross was also author of *The Old World in the New: The Significance of Past and Present Immigration to the American People* (New York: Century, 1914) and *Italians in America* (New York: Century, 1914).

76. Ross, *Standing Room Only?*, 341–42, 344.

77. Ibid., 208, 355.

78. Harold Cox, "League of Low Birth-rate Nations," in *Problems of Overpopulation*, ed. Margaret Sanger (New York: American Birth Control League, 1926), 145–56.

79. Charles V. Drysdale, "Neo-Malthusian Philosophy," in *International Aspects of Birth Control*, ed. Margaret Sanger (New York: American Birth Control League, 1925), 25.

80. Cox, *The Problem of Population*, 85.

81. H. P. Fairchild, "Migration," Harris Foundation Round Table, Chicago 1929, Typescript, Box 192, SPLC.

82. "First Indian Population Conference, Lucknow," January 27, 1936 in *Madras Bulletin Birth Control Bulletin*, Coll. Misc. 639/1/25, Eileen Palmer Collection, LSE.

83. Mukerjee, *Migrant Asia*, 201.

84. "Mrs. Sanger's Address," *The Hindu*, January 25, 1936, 24.

85. Margaret Sanger, "Address," *Proceedings of the International Congress on Population and World Resources in Relation to the Family* (Family Planning Association: London, 1948), 91–92.

86. George Knibbs to Bernard Mallet, June 20, 1927, Box 191, SPLC.

87. G. H. Knibbs, "The New Malthusianism in the Light of Actual World Problems of Population," *Scientia* 40 (1926): 384.

88. Thompson, *Danger Spots*, 332.

89. Irene Taeuber, "Migration and the Population Potential of Monsoon Asia," in *Postwar Problems of Migration*, ed. Reed, 7–29.

90. Hugh Gibson, "Director of Intergovernmental Committee for European Migration," *International Catholic Migration Congress* (Geneva: Catholic Institute for Socio-Ecclesiastical Research, 1954), 16.

91. J. Belehradek to Margaret Sanger, October 31, 1927, Box 191, SPLC.

92. O. E. Baker, "Review of Danger Spots in World Population," *American Journal of Sociology* 36, no. 2 (1930): 302.

93. John Coatman, "Migration in the Twentieth Century," *Population* 1, no. 2 (1934): 57.

94. Jean Bourdon, "Is the Increase in the Population a Real Danger for the Food Supply of the World?" *1927 Proceedings*, 113.

95. Wilhelm Keilhau, Comment, *1927 Proceedings*, 273.

96. Dr. Carl L. Alsberg, in International Institute of Intellectual Cooperation, *Peaceful Change: Procedures, Population, Raw Materials, Colonies. Proceedings of the Tenth International Studies Conference 28 June–3 July 1937* (Paris: IIIC, 1938), 490.

97. Thomas, "International Migration and Its Control," 262, 269, 256.

98. Ibid., 260, 262, 269.

99. "Population and Migration," 1939, League of Nations Committee of Experts on Demography, Economics and Finance Section, Box R4467, LNAG.

100. W. F. Willcox and Imre Ferenczi, *International Migrations*, 2 vols. (New York: National Bureau of Economic Research, 1929–1931).

101. *Le Chomage et les migrations internationales des travailleurs* (Jena: Fischer, 1913) cited in Imre Ferenczi, "Freedom from Want and International Population Policy," *American Sociological Review* 8, no. 5 (1943): 537.

102. For example, Carter Goodrich, "Possibilities and Limits of International Control of Migration," in *Postwar Problems of Migration*, ed. Reed, 79–81. Goodrich sat with Carr-Saunders on the League of Nations Committee of Experts for the Study of Demographic Problems,1939.

103. Permanent Migration Committee, *Report* (Geneva: ILO, 1946). 24.

104. Lawrence D. Levien, "A Structural Model for a World Environmental Organization: The ILO Experience," *George Washington Law Review* 40, no. 3 (1972): 464–95.

105. Ferenczi, "Freedom from Want and International Population Policy," 538.

106. Thompson, *Danger Spots*, 324–25.

107. Mukerjee, *Migrant Asia*, 22, 241.

108. Mukerjee, *The Political Economy of Population*, 449–50.

109. *Peaceful Change*, 211.

110. Notes by the Secretariat, Committee of Experts on Demographic Problems, April 17, 1939, Economics and Finance Section, Box R4467, LNAG.

111. "Population and Migration," 1939, League of Nations Committee of Experts on Demography, Economics and Finance Section, Box R4467, LNAG.

112. Thomas, "International Migration and Its Control," 299. See also commentary at the 1927 Geneva conference by Dr. Henriette Furthe, *1927 Proceedings*, 286; M. Jean Bourdon, *1927 Proceedings*, 296.

113. Carr-Saunders, *World Population*, 152–53. Carr-Saunders would not have understood this to be about crude "overpopulation" of Poland, but others received the scheme in this way.

114. See, e.g., Frederick Sherwood Dunn, *Peaceful Change: A Study of International Procedures* (New York: Council on Foreign Relations, 1937), 74.

115. See, e.g., H. L. Wilkinson, *The World's Population Problems and White Australia* (London: King, 1930), 14–16.

116. Sir Arthur Salter, "The Economic Causes of War," in *The Causes of War: Economic, Industrial, Racial Religious, Scientific, and Political*, ed. Arthur Porritt (New York: Macmillan, 1932), 17.

117. Dunn, *Peaceful Change: A Study of International Relations*, 140–41.

118. Mark Mazower, *No Enchanted Palace: The End of Empire and the Ideological Origins of the United Nations* (Princeton, N.J.: Princeton University Press, 2009), 110. See, e.g., Committee of Experts on the Problem of Refugees and Over-population, *Refugees and Surplus Elements of Population* (Strasbourg: Council of Europe, 1953).

119. Franklin D. Roosevelt to Isaiah Bowman, November 2, 1938, cited in Neil Smith, *American Empire: Roosevelt's Geographer and the Prelude to Globalization* (Berkeley: University of California Press, 2003), 295.

120. Cited in Mazower, *No Enchanted Palace*, 111–12.

121. Bowman, "Population, Migration, Settlement" (1942), cited in Smith, *American Empire*, 301.

122. Joseph B. Schechtman, *European Population Transfers, 1939–1945* (New York: Oxford University Press, 1946); Joseph B. Schechtman, *Population Transfers in Asia* (New York: Hallsby, 1949). See Mazower, *No Enchanted Palace*, 133–142.

123. Mazower, *No Enchanted Palace*, 113.

124. Sandra M. Sufian, *Healing the Land and the Nation: Malaria and the Zionist Project in Palestine, 1920–1947* (Chicago: University of Chicago Press, 2007).

125. Isaiah Bowman, "Population Outlets in Overseas Territories," Typescript, n.d., MS.58, 28, Bowman Papers, Milton S. Eisenhower Library Special Collections, Johns Hopkins University.

5. Waste Lands

1. For example, Fairfield Osborn, "Our Reproductive Potential," n.d., GC/142/22, Wellcome Library, London.

2. John Stuart Mill, *Principles of Political Economy*, 2 vols. (Boston: Little and Brown, 1848), 2:550.

3. John C. Weaver, *The Great Land Rush and the Making of the Modern World, 1650–1900* (Montreal: McGill-Queens University Press, 2003); James Belich, *Replenishing the Earth: The Settler Revolution and the Rise of the Anglo-world, 1783–1939* (Oxford: Oxford University Press, 2009).

4. G. de Lapouge, "Contributions to the Fundamentals of a Policy of Population," *Eugenics Review* 19, no. 3 (1927): 192.

5. Henry Pratt Fairchild, "A New Aspect of Population Theory," in *Problems of Population: Proceedings of the Second General Assembly of the International Union for the Scientific Investigation of Population Problems*, ed. G. H. L. F. Pitt-Rivers (London: Allen & Unwin, 1932), 322.

6. Friedrich Ratzel, "Studies in Political Areas: The Political Territory in Relation to Earth and Continent," trans. Ellen C. Semple, *American Journal of Sociology* 3, no. 3 (1897): 297–98.

7. W. H. Beveridge, "Population and Unemployment," *Economic Journal* 33, no. 132 (1923): 460.

8. Henry Pratt Fairchild, "Optimum Population," *Proceedings of the World Population Conference*, ed. Margaret Sanger (London: Edward Arnold, 1927), 83 [hereafter *1927 Proceedings*].

9. O. E. Baker, "The Potential Supply of Wheat," *Economic Geography* 1, no. 1 (1925): 16.

10. Isaiah Bowman, ed., *Limits of Land Settlement: A Report on Present Day Possibilities to the 10th International Studies Conference* (New York: Council on Foreign Relations, 1937); Isaiah Bowman, *The Pioneer Fringe* (New York: American Geographical Society, 1931).

11. J. W. Gregory, *The Foundation of British East Africa* (London: Marshall, 1901); Maps of the Angolan expedition, 1912, GB 0248, John Walter Gregory Papers, University of Glasgow.

12. J. W. Gregory, *Report on the Work of the Commission Sent out by the Jewish Territorial Organization, under the Auspices of the Governor-General of Tripoli to Examine the Territory Proposed for the Purpose of a Jewish Settlement in Cyrenaica* (London: Jewish Territorial Organization, 1909).

13. Ibid., 4.

14. Murdoch Smith, cited in Israel Zangwill, "Preface, Historical and Political," in Gregory, *Report on the Work of the Commission*, v.

15. Gregory, *Report on the Work of the Commission*, 13. "Redjeb Pasha told us that the *word* autonomy must not be used in connection with any concession, but that fiscal and religious liberty would be granted to the colony" (14). Emphasis in original.

16. See Robert G. Weisbord, "Israel Zangwill's Jewish Territorial Organization and the East African Zion," *Jewish Social Studies* 30, no. 2 (1968): 89–108.

17. Vilhjalmur Stefansson, *The Friendly Arctic: The Story of Five Years in Polar Regions* (New York: Macmillan, 1921).

18. Vilhjalmur Stefansson, "The Population Question and the World's Potential Food Supply," in *Economic Problems: A Book of Selected Readings*, ed. Fred Rogers Fairchild and Ralph Theodore Compton (New York: Macmillan, 1928), 418–35. See Gîsli Palson, "Arcticality: Gender, Race, and Geography in the Writings of Vilhjalmur Stefansson," in *Narrating the Arctic: A Cultural History of Nordic Scientific Practices*, ed. Michael Bravo and Sverker Sörlin (Canton, Mass.: Science History, 2002), 275–310.

19. Tom Griffiths, *Slicing the Silence: Voyaging to Antarctica* (Cambridge, Mass.: Harvard University Press, 2010).

20. Andrew Fitzmaurice, "The Genealogy of Terra Nullius," *Australian Historical Studies* 38, no. 129 (2007): 1–15; Peder Anker, *Imperial Ecology: Environmental Order in the British Empire, 1895–1945* (Cambridge, Mass.: Harvard University Press, 2001), 87–89.

21. R. N. Rudmose-Brown, "Spitsbergen, *Terra Nullius*," *Geographical Review* 7, no. 5 (1919): 311.

22. Martin Conway, *No Man's Land: A History of Spitsbergen from its Discovery in 1596 to the Beginning of the Scientific Exploration of the Country* (Cambridge: The University Press, 1906).

23. William Martin Conway et al., *The First Crossing of Spitsbergen* (London: Scribner, 1897).

24. "Steel Mills May Rise in Spitzbergen," *New York Times*, October 6, 1919: 14.

25. G. H. Knibbs, "The Fundamental Elements of the Problem of Population and Migration," *Eugenics Review* 19, no. 4 (1928): 279.

26. J. W. Gregory, *The Dead Heart of Australia: A Journey Around Lake Eyre in the Summer of 1901–1902* (London: John Murray, 1906).

27. Vilhjalmur Stefansson to Griffith Taylor, May 20, 1924; Diary, 1924, Box 12, Folder 8, Stefansson Papers, Dartmouth College; Vilhjalmur Stefansson, *Central Australia: A Report* (Melbourne: Government Printer, 1924).

28. J. W. Gregory, "The Principles of Migration Restriction," *1927 Proceedings*, 303.

29. J. W. Gregory, *Human Migration & the Future: A Study of the Causes, Effects & Control of Emigration* (London: Seeley, Service, 1928), 197.

30. Albert Thomas, "International Migration and Its Control," *1927 Proceedings*, 261–62.

31. George Handley Knibbs, *The Shadow of the World's Future* (London: Ernest Benn, 1928), 76.

32. A. M. Carr-Saunders, "An Outline of Population History," *Population* 1, no. 1 (1933): 26–27.

33. A. M. Carr-Saunders, *World Population: Past Growth and Present Trends* (Oxford: Clarendon, 1936), 170; cited also in International Institute of Intellectual Cooperation, *Peaceful Change: Procedures, Population, Raw Materials, Colonies. Proceedings of the Tenth International Studies Conference 28 June–3 July 1937* (Paris: IIIC, 1938), 139–40.

34. Allen Buchanan and Margaret Moore, "The Making and Unmaking of Boundaries" in *States, Nations and Borders: The Ethics of Making Boundaries*, ed. Allen E. Buchanan and Margaret Moore (Cambridge: Cambridge University Press, 2003), 12.

35. Martti Koskenniemi, *The Gentle Civilizer of Nations: The Rise and Fall of International Law 1870–1960* (Cambridge: Cambridge University Press, 2002).

36. Benedict Kingsbury, "People and Boundaries," in *States, Nations and Borders*, ed. Buchanan and Moore, 305.

37. Harold Cox, *The Problem of Population* (London: Jonathan Cape, 1922), 78.

38. Radhakamal Mukerjee, *Migrant Asia* (Rome: Failli, 1936), 57.

39. "Population Problems," Typescript, Bureau of Social Hygiene, July 4, 1931, Series 3, Box 9, Folder 191, RAC.

40. Harold Wright, *Population* (London: Nisbet, 1923), 123.

41. J. B. S. Haldane, *Human Biology and Politics* (London: Norman Lockyer Lecture, 1934), 8; cited also in Radhakamal Mukerjee, *The Political Economy of Population* (London: Longman, Green, 1942), 366.

42. H. L. Wilkinson, *The World's Population Problem and a White Australia* (London: King, 1930), 199.

43. Michael Fielding, ed., *Birth Control in Asia* (London: BCIIC, 1935), 22–23.

44. "The Distribution of Population: Claims of National Expansion," Speech by Sir Archibald Parkhill at the Empire Parliamentary Association, May 4, 1937, A5954/69, NAA. This file continues with correspondence from the prime minister, Joseph Lyons, marked "Confidential."

45. Warren S. Thompson, *Danger Spots in World Population* (New York: Knopf, 1929).

46. See also Warren Thompson, Review of Norman Angell, *Raw Materials, Population Pressure and War* (Boston: World Peace Foundation, 1936); League of Nations, "The Economic Interdependence of States," *American Sociological Review* 1, no. 3 (1936): 518.

47. "It is one of the peculiar and interesting characteristics of our modern civilization that the urge to economic expansion and control persists after the urge to biological expansion ceases. Hence we have the spectacle of nations holding on to vast economic resources (lands) after there ceases to be any reasonable hope of their being able to hold them through actual settlement on the land" (Thompson, *Danger Spots*, 7).

48. Ibid., 42–45, 119, 124. See also Walter R. Crocker, *The Japanese Population Problem* (London: Allen & Unwin, 1931).

49. Isaiah Bowman, "Population Outlets in Overseas Territories," Bowman Papers, MS.58, p. 16, Milton Eisenhower Library Special Collections, Johns Hopkins University.

50. Neil Smith, *American Empire: Roosevelt's Geographer and the Prelude to Globalization* (Berkeley: University of California Press, 2003).

51. This is detailed by each of the authors in Sebastian Conrad and Dominic Sachsenmaier, eds., *Competing Visions of World Order: Global Moments and Movements, 1880–1930s* (Basingstoke, U.K.: Palgrave, 2007); Erez Manela, *The Wilsonian Moment: Self-Determination and the International Origins of Anticolonial Nationalism* (Oxford: Oxford University Press, 2007).

52. Petition of the Annual International Convention of Negro People of the World, New York, August 1924, Section 1, 37672/21159, Racial Equality Files, LNAG.

53. Fairchild, "Optimum Population," 84.

54. Fergus Chalmers Wright, *Population and Peace: A Survey of International Opinion on Claims for Relief from Population Pressure* (Paris: IIIC, 1939), 264.

55. T. N. Carver, "Some Needed Refinements of the Theory of Population," *1927 Proceedings*, 124.

56. Mabel Buer, Comment, *1927 Proceedings*, 57.

57. Gregory, "The Principles of Migration Restriction," 303.

58. See Carolyn Strange and Alison Bashford, *Griffith Taylor: Visionary, Environmentalist, Explorer* (Toronto: University of Toronto Press, 2009).

59. "To regard the exploitation of all unused, or poorly used, tropical land and all backward peoples as his [the white man's] special prerogative is wholly unjustified" (Thompson, *Danger Spots*, 101).

60. Ibid., 164–65, 177.

61. Ibid., 163.

62. Koskenniemi, *Gentle Civilizer of Nations*, 128.

63. "Since almost no settlement took place after the acquisition. These are being held today as areas of pure exploitation" (Thompson, *Danger Spots*, 7).

64. Ibid., 128–29.

65. Sripati Chandrasekhar, *Hungry People and Empty Lands: An Essay on Population Problems and International Tensions* (London: Allen & Unwin, 1954), 74.

66. Mukerjee, *Migrant Asia*, 78, 84, 79.

67. Thompson, *Danger Spots*, 163–4.

68. David N. Livingstone, "Human Acclimatization: Perspectives on a Contested Field of Inquiry in Science, Medicine and Geography," *History of Science* 25 (1987): 359–94; Mark Harrison, *Climates and Constitutions: Health, Race, Environment and British Imperialism in India, 1600–1850* (Oxford: Oxford University Press, 1999).

69. Montesquieu, "The Spirit of the Laws, 1748," quoted in Koskenniemi, *The Gentle Civilizer of Nations*, 100.

70. Program, International Congress for Studies Regarding Population, Rome, September 7–10, 1931, Box R2872, Economic Section, 10d Series, 16131, File 15768, LNAG.

71. Andrew Balfour, "Problems of Acclimatisation," *Lancet* 202, no. 5214 (1923): 245; cited also in Mukerjee, *The Political Economy of Population*, 396–97.

72. George H. Knibbs, "The Fundamental Elements of the Problems of Population and Migration," *Eugenics Review* 19, no. 4 (1928): 289.

73. G. H. Knibbs, *Mathematical Analysis of Some Experiments in Climatological Physiology* (London: Taylor and Francis, 1912).

74. Sir Arthur Salter, "The Economic Causes of War," in *The Causes of War*, ed. Arthur Porritt (New York: Macmillan, 1932), 17.

75. Thompson, *Danger Spots*, 89.

76. See Alison Bashford, "'Is White Australia Possible?' Race, Colonialism and Tropical Medicine," *Ethnic and Racial Studies* 23, no. 2 (2000): 248–71.

77. Thompson, *Danger Spots*, 127, 293, 94.

78. See Bashford, "Is White Australia Possible?," 248–71; Warwick Anderson, *The Cultivation of Whiteness* (Melbourne: Melbourne University Press, 2002).

79. This is developed in Alison Bashford, *Imperial Hygiene: A Critical History of Colonialism, Nationalism, and Public Health* (Basingstoke, U.K.: Palgrave, 2004).

80. For Australia as Asia's "safety valve," see Georg Borgström, *The Hungry Planet: The Modern World at the Edge of Famine* (New York: Macmillan, 1965).

81. See Sripati Chandrasekhar, ed., *Asia's Population Problems: With a Discussion of Population and Immigration in Australia* (London: Allen & Unwin, 1967).

82. Mukerjee, *The Political Economy of Population*, 360–401.

83. Mukerjee, *Migrant Asia*, 9.

84. Mukerjee, *The Political Economy of Population*, 396.

85. Mukerjee, *Migrant Asia*, 28–30.

86. Ibid., 61–69, 71–72.

87. Ibid., 264, 266.

88. F. A. E. Crew, "The Biological Aspects of Migration," *Population* 1, no. 1 (1933): 38.

89. Ibid., 38–39.

90. Ibid., 35–38.

91. Knibbs, *The Shadow of the World's Future*, 76.

6. Life on Earth

1. John Maynard Keynes to Julian Huxley, July 28, 1927, Box 9, Folder 5, HP.

2. H. G. Wells, Julian Huxley, and G. P. Wells, *The Science of Life* (London: Cassell, 1931), 578.

3. Gregg Mitman, *The State of Nature: Ecology, Community, and American Social Thought, 1900–1950* (Chicago: University of Chicago Press, 1992); Sharon E. Kingsland, *Modeling Nature: Episodes in the History of Population Ecology* (Chicago: University of Chicago Press, 1985).

4. Kingsland, *Modeling Nature*, 23.

5. A. B. Wolfe, "The Population Problem since the World War: A Survey of Literature and Research, Part 2," *Journal of Political Economy* 36, no. 6 (1928): 663.

6. Christopher Lawrence and George Weisz, eds., *Greater Than the Parts: Holism in Biomedicine, 1920–1950* (Oxford: Oxford University Press, 1998). See also, on biology and holism, Anna Bramwell, *Ecology in the Twentieth Century: A History* (New Haven, Conn.: Yale University Press, 1989), 39–61.

7. Phen Cheah and Bruce Robbins, eds., *Cosmopolitics: Thinking and Feeling Beyond the Nation* (Minneapolis: University of Minnesota Press, 1998).

8. Julian Huxley, *UNESCO: Its Purpose and its Philosophy* (Washington D.C.: Public Affairs, 1947). For a larger history of world government, see Mark Mazower, *Governing the World: The History of an Idea* (New York: Penguin, 2012).

9. Charles Elton, "The Oxford Expedition to Spitsbergen, 1921." Typescript, NPOLAR, 91(08):(*32) 1921, pp. 1, 1a, Norwegian Polar Institute, Tromsø.

10. A. M. Carr-Saunders, *The Population Problem: A Study in Human Evolution* (Oxford: Clarendon Press, 1922), 34–35, 5. This book influenced Chicago-based ecologists as well. See Mitman, *The State of Nature*, 94–96.

11. Peder Anker, *Imperial Ecology: Environmental Order in the British Empire, 1895–1945* (Cambridge, Mass.: Harvard University Press, 2001), 94.

12. Victor S. Summerhayes and Charles S. Elton, "Contributions to the Ecology of Spitsbergen and Bear Island," *Journal of Ecology* 11, no. 2 (1923): 214–86. See Anker, *Imperial Ecology*, 91.

13. Frank Benjamin Golley, *A History of the Ecosystem Concept in Ecology: More Than the Sum of the Parts* (New Haven, Conn.: Yale University Press, 1993), 44. See also Richard Weigert, *Ecological Energetics* (Stroudsburg: Dowde, Hutchinson and Ross, 1976).

14. Julian Huxley, *Memories*, 2 vols. (London: Allen & Unwin, 1970), 1:128; Julian Huxley, "Science and Spitzbergen," *Cornhill Magazine*, 1921, Box 97, Folder 2, HP.

15. *Spitsbergen Papers: Scientific Results of the First Oxford University Expedition to Spitsbergen (1921)*, vol 1. (London: Oxford University Press, 1925). For the process of writing up this account, see their letters in Box 8, Folder 1, HP; and Alexander Carr-Saunders to Julian Huxley, February 9, 1925, Box 8, Folder 4, HP.

16. Huxley, *Memories*, 1:149–50.

17. Julian Huxley, Comment on "The Biology of Population Growth," *Proceedings of the World Population Conference*, ed. Margaret Sanger (London: Edward Arnold, 1927), 53.

18. H. Eliot Howard to Julian Huxley, November 2, 1920, Box 6, Folder 2, HP; H. Eliot Howard, *Territory in Bird Life* (New York: Dutton, 1920). Huxley was to reissue the book in 1948.

19. Huxley considered Wells's *Outline of History* to be "the first attempt to present history on a world-wide scale." Huxley, *Memories*, 1:155.

20. Wells et al., *The Science of Life*, 3.

21. Ibid., 3–4.

22. H. G. Wells to Julian Huxley, December 22, 1927, in Huxley, *Memories*, 1:158.

23. Huxley, *Memories*, 1:156.

24. Wells et al., *The Science of Life*, 578.

25. Ibid., 6, 578, 581.

26. Outline notes for *The Science of Life*, Box 62, Folder 1, HP.

27. Charles Elton, *Animal Ecology* (London: Sidgwick and Jackson, 1927).

28. Wells et al., *The Science of Life*, 597, 600–01, 606.

29. Ibid., 613.

30. Ibid., 615–17, 878.

31. Patrick Geddes, "Introduction," in Radhakamal Mukerjee, *The Foundations of Indian Economics* (London: Longmans, Green, 1916), ix–xvii. Ernst Haeckel corresponded with the young Julian Huxley in 1913, Box 5, Folder 10, HP.

32. Radhakamal Mukerjee, "Faiths and Influences," in *The Frontiers of Social Science: In Honour of Radhakamal Mukerjee*, ed. Baljit Singh (London: Macmillan, 1957), 6.

33. Mukerjee, *The Foundations of Indian Economics*, i.

34. Radhakamal Mukerjee, *Regional Sociology* (New York: Century, 1926); Radhakamal Mukerjee, *The Regional Balance of Man: An Ecological Theory of Population* (Madras: University of Madras, 1938).

35. Radhakamal Mukerjee, "Preface to the First Edition," *Man and His Habitation: A Study in Social Ecology*, 2d ed. (1939; Bombay: Popular Prakashanm, 1968), xi.

36.

37. Warder C. Allee and Gertrude Evans, "Certain Effects of Numbers Present on the Early Development of the Purple Sea-Urchin," *Ecology* 18, no. 3 (1937): 337–45.

38. Radhakamal Mukerjee, "The Criterion of Optimum Population," *American Journal of Sociology* 38, no. 5 (1933): 689–91.

39. Ibid., 688–98. See also Mukerjee, *The Political Economy of Population* (London: Longmans, Green, 1942), 160–97.

40. Mukerjee, *The Political Economy of Population*, 191–92, 190.

41. Radhakamal Mukerjee, *Migrant Asia* (Rome: Failli, 1936), 85.

42. Mukerjee, "The Criterion of Optimum Population," 688–98; Mukerjee, *The Political Economy of Population*, 160–97.

43. Ellsworth Huntington, *The Character of Races: As Influenced by Physical Environment, Natural Selection, and Historical Development* (New York: Scribner, 1924).

44. Ellsworth Huntington, *The Human Habitat* (New York: Van Nostrand, 1927).

45. Elazar Barkan, *The Retreat of Scientific Racism* (New York: Cambridge University Press, 1992).

46. Mukerjee, *The Political Economy of Population*, 394, 376.

47. Mukerjee, *Migrant Asia*, 60.

48. A. M. Carr-Saunders, *World Population: Past Growth and Present Trends* (Oxford: Clarendon, 1936), 170.

49. Alexander Carr-Saunders read the work of S. J. Holmes, "The Influence of Season and Climate on the Mortality of the White and Colored Population from Tuberculosis and the Acute Respiratory Infections," *American Journal of Medical Sciences* 195, no. 4 (1938): 501–10, in B/3/5, Alexander Carr-Saunders Papers, LSE.

50. International Institute of Intellectual Cooperation, *Peaceful Change: Procedures, Population, Raw Materials, Colonies. Proceedings of the Tenth International Studies Conference 28 June–3 July 1937* (Paris: IIIC, 1938), 140.

51. Enid Charles, *The Practice of Birth Control: An Analysis of the Birth-Control Experiences of Nine-Hundred Women* (London: Williams & Norgate, 1932).

52. John Boyd Orr to Julian Huxley, December 12, 1936, Box 12, Folder 3, HP. For Hogben's difficult balancing of his declared politics and his academic appointments, see Lancelot Hogben to Julian Huxley, July 16, 1926, August 12, 1926, Box 9, Folder 2, HP.

53. Enid Charles, *The Twilight of Parenthood: A Biological Study of the Decline of Population Growth* (London: Watts, 1934), 7.

54. Ibid., 161–69, 157.

55. Ibid., 190, 7.

56. Arthur G. Tansley, "The Use and Abuse of Vegetational Concepts and Terms," *Ecology* 16, no. 3 (1935): 284–307. See also Golley, *A History of the Ecosystem Concept in Ecology*, 8–34.

57. Alexander von Humboldt, *Kosmos: Entwurf einer physischen Weltbeschreibung*, 5 vols (Tübingen: Cotta, 1845–1862). A full set of volumes was first translated as *Cosmos: A Sketch of a Physical Description of the Universe,* 5 vols (New York: Harper, 1856–1869). See also Gerry Kearns, *Geopolitics and Empire: The Legacy of Halford Mackinder* (Oxford: Oxford University Press, 2009), 65; P. L. Farber, *Finding Order in Nature: The Naturalist Tradition from Linnaeus to E.O. Wilson* (Baltimore, Md.: Johns Hopkins University Press, 2000); Denis Cosgrove, *Geography and Vision: Seeing, Imagining, and Representing the World* (London: Tauris, 2008), 34–43.

58. Benjamin Franklin, "Observations Concerning the Increase of Mankind, Peopling of Countries, etc," in *Political, Miscellaneous, and Philosophical Pieces* (1751: London: J. Johnson, 1779). See also Matthew Connelly, *Fatal Misconception: The Struggle to Control World Population* (Cambridge, Mass.: Belknap, 2008), 6; Joyce E. Chaplin, *Benjamin Franklin's Political Arithmetic: A Materialist View of Humanity* (Washington, D.C.: Smithsonian Institute Libraries, 2006), 24–25. For Franklin's astronomy and meteorology see Joyce E. Chaplin, *The First Scientific American: Benjamin Franklin and the Pursuit of Genius* (New York: Basic, 2006), 280–82, 303–4.

59. Vladimir Vernadsky, *La Biosphère* (Paris: Félix Alcan, 1929), cited in Vaclav Smil, *The Earth's Biosphere: Evolution, Dynamics, and Change* (Cambridge, Mass.: MIT Press, 2002), 6. See also Andrei V. Lapo, "Vladimir Vernadsky (1863–1945), Founder of the Biosphere Concept," *International Microbiology* 4, no. 1 (2001): 47–49. I am grateful to David Christian for conversations on Vernadsky.

60. Halford J. Mackinder, "The Human Habitat," *Scottish Geographical Magazine* 47, no. 6 (1931): 323.

61. See Kearns, *Geopolitics and Empire*, 143.

62. Mackinder, "The Human Habitat," 328.

63. Alfred Russel Wallace, *Man's Place in the Universe* (London: Chapman and Hall, 1903); Alfred Russel Wallace, *Is Mars Habitable?* (London: Macmillan, 1907).

64. Wallace, *Is Mars Habitable?* 105.

65. Wells et al., *The Science of Life*, 6, 7–8.

66. George Handley Knibbs, *The Shadow of the World's Future* (London: Ernest Benn, 1928), 19–20, 124.

67. George Knibbs wrote *The Place of Astronomy in Liberal Education* (Sydney: British Astronomical Association, 1898).

68. "It is likely that all manifestations of life are correlated with the energies received from the centre of our solar system" (Knibbs, *The Shadow of the World's Future*, 13).

69. Ibid., 9, 114.

70. G. H. Knibbs, "The Mathematical Theory of Population, of Its Character and Fluctuations, and of the Factors which Influence Them," Appendix A, *Census of the Commonwealth of Australia, 1911 Volume 1: Statistician's Report*, (Melbourne: Government Printer, 1917), 456.

71. Knibbs, *The Shadow of the World's Future*, 80.

72. Ibid., 69.

73. Knibbs, "Mathematical Theory of Population," 456.

74. Knibbs, *The Shadow of the World's Future*, 46, 116.

75. George H. Knibbs, "Science and Its Service to Man," *Science* 58, no. 1510 (1923): 453.

76. Libby Robin and Will Steffen, "History for the Anthropocene," *History Compass* 5, no. 5 (2007): 1694–95.

77. Typescript of Second Conference on Population Association of America, May 7, 1931, Bureau of Social Hygiene, Series 3, Box 9, Folder 191, RAC; Dennis Hodgson, "The Ideological Origins of the Population Association of America," *Population and Development Review* 17, no. 1 (1991): 2.

78. Alfred J. Lotka, *Elements of Physical Biology* (Baltimore, Md.: Williams and Wilkins, 1925).

79. Kingsland, *Modeling Nature*, 6, 25. See also Golley, *A History of the Ecosystem Concept*, 57–58.

80. Kingsland, *Modeling Nature*, 46.

81. Julian Huxley, "A Journey in Relativity," *North American Review*, July 1923, Box 97, Folder 3, HP.

82. G. H. L. F. Pitt-Rivers, Report of the International Union of the Scientific Investigation of Population Problems, 1937, SA/EUG/D110, Eugenics Society Papers, Wellcome Library. London.

83. G. H. L. F. Pitt-Rivers, "The Urgency of Population Study from the Bio-Anthropological Approach," *Population* 1, no. 1 (1933): 21.

84. Jan C. Smuts, *Holism and Evolution* (New York: Macmillan, 1926); D. C. Phillips, *Holistic Thought in Social Sciences* (Stanford, Calif.: Stanford University Press, 1976); Christopher Lawrence and George Wiesz, "Medical Holism: The Context," in *Greater Than the Parts,* ed. Lawrence and Weisz, 2.

85. Jan Smuts to Julian Huxley, June 24, 1926, Box 9, Folder 1, HP.

86. See James Bohman and Matthias Lutz-Bachmann, eds., *Perpetual Peace: Essays on Kant's Cosmopolitan Ideal* (Cambridge, Mass.: MIT Press, 1997), 25–57. See also K. A. Appiah, *Cosmopolitanism: Ethics in a World of Strangers* (New York: Norton, 2006).

87. John Maynard Keynes, Typescript, "Population" [1913–1914], JMK/SS/1/27, KP [emphasis in original]. See also John Toye, *Keynes on Population* (Oxford: Oxford University Press, 2000), 66.

88. Knibbs, *The Shadow of the World's Future*, 115.

89. Derek Heater, *World Citizenship and Government: Cosmopolitan Ideas in the History of Western Political Thought* (Basingstoke, U.K.: Palgrave, 1996).

90. Mitman, *The State of Nature*, 52–62.

91. Immanuel Kant, *Perpetual Peace* (1795: London: Grotius Society, 1927). See also Chenxi Tang, *The Geographic Imagination of Modernity: Geography, Literature, and Philosophy in German Romanticism* (Stanford, Calif.: Stanford University Press, 2008).

92. Mukerjee, "Faiths and Influences," 16.

93. Anker, *Imperial Ecology*, 185–95. See also Saul Dubow, "Smuts, the United Nations and the Rhetoric of Race and Rights," *Journal of Contemporary History* 43, no. 1 (2008): 43–72.

94. Taraknath Das, "Human Rights and the United Nations," *Annals of the American Academy of Political and Social Science* 252 (1947): 53–62.

95. "Books and Pamphlets by H. G. Wells," in *The Wellsian*, 1960, Box 113, Folder 1, HP. See also H. G. Wells, *The Rights of Man, or, What Are We Fighting For?* (Harmondsworth, U.K.: Penguin, 1940).

96. H. G. Wells, *Anticipations of the Reaction of Mechanical and Scientific Progress upon Human Life and Thought* (1901), cited in John S. Partington, *Building Cosmopolis: The Political Thought of H. G. Wells* (London: Ashgate, 2003), 65.

97. H. G. Wells, *A Modern Utopia* (1905), cited in Partington, *Building Cosmopolis*, 67.

98. Quoted in Partington, *Building Cosmopolis*, 79.

99. *The Wellsian: Journal of the H. G. Wells Society* 1, no. 1 (1906): 6.

100. Partington, *Building Cosmopolis*, 26; W. Warren Wagar, *The City of Man: Prophecies of World Civilization in Twentieth Century Thought* (Baltimore, Md.: Penguin, 1967), 17. See also Heater, *World Citizenship and Government*, 127–35.

101. SA/FPA/A14/173, Family Planning Archives, Wellcome Library; For H. G. Wells's activity around birth control, see Workers' Birth Control Group, "Memorandum on the Question of Birth Control Presented to the Minister of Health on May 9, 1924" (Chelsea: Workers Birth Control Group, 1924); Sir Vepa Ramesam, "Presidential Address," Indian Population Problems: *Proceedings of 2nd All-India Population and 1st Family Hygiene Conferences* ed. G.S. Ghurye (Bombay: Karnatak 1938), 10; Wells lent his name to Marie Stopes's Society. See Society for Constructive Birth Control and Racial Progress, *Constructive Birth Control* (London: Society for Constructive Birth Control, 1921), 1.

102. "The Future of Man: A Talk with H. G. Wells," *Pall Mall Gazette*, October 10, 1903, 2.

103. For H. G. Wells and Margaret Sanger, see Ellen Chesler, *Woman of Valor: Margaret Sanger and the Birth Control Movement in America* (New York: Anchor, 1993), 191, 315.

104. Cited in Margaret Sanger, ed., *International Aspects of Birth Control* (New York: American Birth Control League, 1925), 216.

105. H. G. Wells, *The Open Conspiracy and Other Writings* (London: Waterlow, 1933), 72. Originally published in 1928 with the subtitle "Blue Prints for World Revolution." See also Heater, *World Citizenship and Government*, 130–33.

106. Wells, *The Open Conspiracy*, 27–28.

107. Anker, *Imperial Ecology*, 200.

108. Wells, *The Open Conspiracy*, 34–42.

109. H. G. Wells, cited in Partington, *Building Cosmopolis*, 88–89.

110. Charles, *The Twilight of Parenthood*, 7.

111. Huxley, *UNESCO*, 8, 12. For analysis of Huxley's ideas on evolution, see William B. Provine, "Progress in Evolution and Meaning in Life," in *Julian Huxley: Biologist and Statesman of Science*, ed. C. Kenneth Waters and Albert van Helden (Houston, Tex.: Rice University Press, 1992), 154–80.

112. Huxley, *UNESCO*, 45.

113. Ibid., 13. See Glenda Sluga, "UNESCO and the (One) World of Julian Huxley," *Journal of World History* 21, no. 3 (2010): 393–418.

7. Soil and Food

1. W. H. Beveridge, "Population and Unemployment," *Economic Journal* 33, no. 132 (1923): 460.

2. Adolf Hitler, *Mein Kampf*, trans. Ralph Manheim (1925, 1926; repr. London: Pimlico, 2005), 607.

3. A. B. Wolfe, "Is There a Biological Law of Human Population Growth?" *Quarterly Journal of Economics* 41, no. 4 (1927): 558; Raymond Pearl, *The Nation's Food: A Statistical Study of a Physiological and Social Problem* (Philadelphia: Saunders, 1920).

4. P. K. Whelpton, "Population Trends," *Proceedings of the International Congress on Population and World Resources in Relation to the Family* (London: Family Planning Association, 1948), 55.

5. Alexander Carr-Saunders to Julian Huxley, September 13, 1922, Box 7, Folder 2, HP.

6. John Boyd Orr, "Better Neighbors in a Changing World," Typescript speech to the National Farm Institute, Des Moines, Iowa, February 15, 1946, Folder 2, Boyd Orr Papers, NLS.

7. William Crookes, "Address of the President before the British Association for the Advancement of Science," *Science, New Series* 8, no. 200 (1898): 562–68.

8. American John Hyde responded in 1899 with "America and the Wheat Problem" and Ralph Tichborne Hinckes published *The World's Wheat Problem. Will the Guarantee Help to Solve It?* (Hereford, U.K.: Hereford Times 1920); J. Hay Thorburn, *Our Wheat Supply: The Problem and the Solution* (Eltham, U.K.: Digby, 1927); Vladimir P. Timoshenko, *Agricultural Russia and the Wheat Problem* (Stanford, Calif.: Food Research Institute, 1932).

9. Hugh Dalton, "Review of *Shadow of the World's Future*," *Economica* no. 25 (1929): 83–84.

10. O. E. Baker, "The Potential Supply of Wheat," *Economic Geography* 1, no. 1 (1925): 15.

11. Ibid.

12. Ibid., 49. O. E. Baker, "The Progress of Population," in *Problems of the Pacific, 1927*, ed. J. B. Condliffe (Chicago: University of Chicago Press, 1928), 318–23.

13. Vaclav Smil, "Population Growth and Nitrogen: An Exploration of a Critical Existential Link," *Population and Development Review* 17, no. 4 (1991): 570, 580. See also John H. Perkins, *Geopolitics and the Green Revolution: Wheat, Genes, and the Cold War* (New York: Oxford University Press, 1997), 211–18.

14. William Crookes, *The Wheat Problem*, 3d ed. (London: Longmans, Green, 1917), 38–40; J. Knox, *The Fixation of Atmospheric Nitrogen* (London: Gurney and Jackson, 1914).

15. Crookes, *The Wheat Problem*, 37–38 (emphasis in original). George Knibbs thought nitrogen readily available. *The Shadow of the World's Future* (London: Ernest Benn, 1928), 41–42.

16. L. F. Haber, *The Chemical Industry, 1900–1930* (Oxford: Clarendon, 1971); L. F. Haber, *The Chemical Industry During the Nineteenth Century* (Oxford: Clarendon, 1958); G. J. Leigh, *The World's Greatest Fix: A History of Nitrogen and Agriculture* (Oxford: Oxford University Press, 2004). Vaclav Smil isolates the Haber–Bosch synthesis of ammonia as the most important invention of the twentieth century. Smil, "Population Growth and Nitrogen," 569.

17. Vaclav Smil, *Enriching the Earth: Fritz Haber, Carl Bosch and the Transformation of World Food Production* (Cambridge, Mass.: MIT Press, 2001), 14–20; Leigh, *The World's Greatest Fix*, 113–18.

18. For example "Mixed sample from the plain of Silene. . . Organic matter contained ammonia .16 per cent, equal to .13 per cent of nitrogen" (J. W. Gregory, *Report on the Work of the Commission Sent out by the Jewish Territorial Organization* [London: Jewish Territorial Organization, 1909], 33).

19. Justus von Liebig, *Chemistry in Its Application to Agriculture and Physiology* (London: Wiley and Putnam, 1847). Cited in Smil, *Enriching the Earth*, 8.

20. Their field experiments were largely concerned to increase yields of wheat by examining the source and function of nitrogen in plants, and fertilizers more generally. Bone dust alone, long used in traditional farming, was shown by John Bennet Lawes not to increase yield; it only did so when treated with sulfuric acid. In 1840, Lawes issued a patent for superphosphate, in which he added sulfuric acid to mineral phosphates, that is, treating bone or mineral phosphates with sulfuric acid. Adding sulfuric acid to bones was not new; applying it to mineral phosphates was.

21. A. D. Hall, *The Book of the Rothamsted Experiments* (London: John Murray, 1905).

22. Joseph Morgan Hodge, *Triumph of the Expert: Agrarian Doctrines of Development and the Legacies of British Colonialism* (Athens: Ohio University Press, 2007), 92.

23. J. R. McNeill, "Population and the Natural Environment: Trends and Challenges," *Population and Development Review* 32, no. 1 (2006): 186.

24. Daniel Hall, *The Improvement of Native Agriculture in Relation to Population and Public Health* (Oxford: Oxford University Press, 1936), 6, 4.

25. He also introduced discussion of an organism that could fix nitrogen "when living free in the soil," that is, without a plant host—the bacterium *Azobacter chroococcum*. This bacterium's one need was not a plant host, but lime in the soil.

26. Hall, *The Improvement of Native Agriculture*, 7. Henrietta Moore and Megan Vaughan note that many Anglophone observers of Africa regarded abandonment of fields, and especially the burning of trees in the citemene system, almost automatically as wasteful, but some were aware of the production thereby of phosphate, potash, and calcium. Henrietta L. Moore and Megan Vaughan, *Cutting Down Trees: Gender, Nutrition, and Agricultural Change in the Northern Province of Zambia, 1890–1990* (Portsmouth, N.H.: Heinemann, 1994), 21, 26–28.

27. P. K. Wattal, *The Population Problem in India: A Census Study* (Bombay: Bennett, Coleman, 1916).

28. P. K. Wattal, "The Urgent Need of Birth Control in India," Typescript, Birth Control in Asia Conference, 1933, pp. 19–22, Box 194, SPLC.

29. Harold Cox, *The Problem of Population* (London: Jonathan Cape, 1922), 33.

30. Vaclav Smil writes that return of animal and human waste reached its greatest extent and highest intensity in parts of east Asia, North Africa, and western Europe (China, Japan, the Nile delta, and Dutch farms). Smil, "Population Growth and Nitrogen," 578.

31. See also Bernard. A. Keen and Daniel Hall, eds., *The Physicist in Agriculture: With Special Reference to Soil Problems* (London: Chemical Society, 1925); *Agriculture in the Twentieth Century: Essays on Research, Practice and Organization to be Presented to Sir Daniel Hall* (Oxford: Clarendon, 1939).

32. Hall, *The Improvement of Native Agriculture*, 18.

33. Dana Simmons, "Waste Not, Want Not: Excrement and Economy in Nineteenth-Century France," *Representations* 96, no. 1 (2006): 73–75, 82.

34. Hall, *The Book of the Rothamsted Experiments*, 245–46, 254–58.

35. Cox, *The Problem of Population*, 35.

36. Hall, *The Improvement of Native Agriculture*, 19.

37. Fairfield Osborn, *Our Plundered Planet* (London: Faber & Faber, 1948), 67.

38. Daniel Hall, *Fertilizers and Manures* (1909; New York: Dutton, 1912), 243.

39. Hall, *Improvement of Native Agriculture*, 18, 19.

40. Christopher Hamlin, "William Dibdin and the Idea of Biological Sewage Treatment," *Technology and Culture* 29, no. 2 (1988): 189–218.

41. Radhakamal Mukerjee, *The Political Economy of Population* (London: Longmans, Green, 1942), 121.

42. Hall, *The Improvement of Native Agriculture*, i, 1.

43. Moore and Vaughan, *Cutting Down Trees*, 35–36. See also Helen Tilley, *Africa as a Living Laboratory: Empire, Development, and the Problem of Scientific Knowledge, 1870–1950* (Chicago: University of Chicago Press, 2011), 115–68.

44. O. W. Willcox, *Nations Can Live at Home* (London: Allen & Unwin, 1935), 43.

45. N. J. Ridgewood, "Introduction," in Willcox, *Nations Can Live at Home.*

46. Willcox, *Nations Can Live at Home*, 26, 40, 43.

47. Draft Plan for Regional Demographic Studies, July 18, 1939, Economics Section 10, Box R 446, LNAG.

48. Knibbs, *The Shadow of the World's Future*, 45.

49. Warren S. Thompson, *Danger Spots in World Population* (New York: Knopf, 1929), 215.

50. Paulet, Comment, *Proceedings of the World Population Conference, ed. Margaret Sanger* (London: Edward Arnold, 1927), 292.

51. H. G. Wells, Julian Huxley, and G. P. Wells, *The Science of Life* (London: Cassell, 1931), 616–17.

52. See Richard H. Grove, *Ecology, Climate and Empire: Colonialism and Global Environmental History, 1400–1940* (Cambridge: White Horse, 1997); Richard Drayton,

Nature's Government: Science, Imperial Britain, and the "Improvement" of the World (New Haven, Conn.: Yale University Press, 2000).

53. Mukerjee, *The Political Economy of Population*, 128. See also Tarlok Singh, "Dr Radhakamal Mukerjee on Planning and the Rural Economy," in *The Economic Problems of Modern India*, ed. Gurmukh Ram Madan (Bombay: Allied, 1995), 4.

54. Mukerjee, *The Political Economy of Population*, 129, 139–40, 158.

55. Vaclav Smil writes that soil erosion was and is the "leading global cause of nitrogen loss." "Population Growth and Nitrogen," 576.

56. A. E. Parkins and J. R. Whitaker, eds., "Preface to First Edition," *Our Natural Resources and Their Conservation*, 2d ed. (New York: John Wiley, 1939), ix.

57. Paul B. Sears, *Deserts on the March* (Norman: University of Oklahoma Press, 1935).

58. G. V. Jacks and R. O. Whyte, *Vanishing Lands: A World Survey of Soil Erosion* (New York: Doubleday, Doran, 1939).

59. G. V. Jacks, "Soil Conservation as a Problem of Human Ecology," *Man* 47 (1947): 11.

60. John Boyd Orr, *Soil Fertility: The Wasting Basis of Human Society* (London: Pilot, 1948), 9.

61. Ibid., 9–10.

62. Thomas Robert Malthus, *An Essay on the Principle of Population*, 6th ed., 2 vols. (London: John Murray, 1826), 1:12. On famine and population in India, see Amartya Sen, *Poverty and Famines: An Essay on Entitlement and Deprivation* (Oxford: Clarendon, 1982); Sarah Hodges, "Governmentality, Population and the Reproductive Family in Modern India," *Economic and Political Weekly* 39, no. 11 (2004): 1157–63. On the reporting of late nineteenth-century Indian famines in Britain, see James Vernon, *Hunger: A Modern History* (Cambridge, Mass.: Belknap, 2007), 34–40; David Nally, *Human Encumbrances: Political Violence and the Great Irish Famine* (Notre Dame, Ind.: University of Notre Dame Press, 2011).

63. Annie Besant, *The Law of Population: Its Consequences and Its Bearing upon Human Conduct and Morals* (Chicago: Wilson, 1880).

64. *Report of the Indian Famine Commission Part 1 Famine Relief* (London: HMSO, 1880); *Report of the Indian Famine Commission and Papers Relating Thereto* (London: HMSO, 1901). See also Hari Shanker Srivastava, *The History of Indian Famines and Development of Famine Policy, 1858–1918* (Agra, India: Mehra, 1968); Tim Dyson, ed., *India's Historical Demography: Studies in Famine, Disease and Society* (London: Curzon, 1989). See also, Mike Davis, *Late Victorian Holocausts: El Niño Famines and the Making of the Third World* (London: Verso, 2002), 141–75.

65. For example, ILO economist Rajani Kanta Das told the 1927 Geneva conference that famine was the first and main indicator of Indian overpopulation R. K. Das, *1927 Proceedings*, 117. Arup Maharatna, looking closely at Indian registration data over four Indian famines since the 1870s, shows that in famine years (including the year following) there was a rise in the death rate, a fall in birthrate, and in the year following famine, a larger fall in birthrate. However, his research also showed a "compensatory" rise in fertility in the immediate postfamine period, due largely to a change in age composition of the female population. Arup Maharatna, *The Demography of Famines: An Indian Historical Perspective* (Delhi: Oxford University Press, 1996), 265–66.

66. Alexander Loveday, *The History and Economics of Indian Famines* (London: G. Bell, 1914).

67. See Hodges, "Governmentality," 1157–63.

68. *The Times* [London], November 9, 1920, quoted in Cox, *The Problem of Population*, 96. There were similar reports of a 1927 famine on the Shantung Peninsula; the *New York Times* reported on the selling of girls, mass suicide of children, and cannibalism. *New York Times*, December 24 and 28, 1927; November 16, 1968; May 1, 1929.

69. "There are too many children in the world now. They are being broken in factories and they are dying of hunger. More of them are to die. In this hour of crisis and peril, women alone can save the world. They can save it by refusing for five years to bring a child into being." (Margaret Sanger, "A Birth Strike to Avert World Famine," *Birth Control Review* 4, no. 1 [1920]: 1).

70. Pearl, *The Nation's Food*, 9.

71. Donald F. Jones, *Biographical Memoir of Edward Murray East* (Washington, D.C.: National Academy of Sciences, 1945), 229.

72. "Food and Population in the Pacific," in *Problems of the Pacific, 1929*, ed. J. B. Condliffe, (Chicago: University of Chicago Press, 1930), 52, 56.

73. "For the last ten years or so her food consumption has exceeded her home supply and she has had to rely increasingly on imports from her colonies and abroad" (ibid., 56).

74. T. Nagai, "This Complex Problem of Existence: Japanese Opinions on World Population," *News Bulletin (Institute of Pacific Relations)* (April 1928): 2.

75. Edward East, "Food and Population," *1927 Proceedings*, 89.

76. Henri Brenier, Comment, *1927 Proceedings*, 92–93.

77. Edward M. East, *Mankind at the Crossroads* 1923: (New York: Scribner, 1924), 71.

78. Knibbs, *The Shadow of the World's Future*, 38.

79. Crookes, "Address," 562.

80. Crookes, *The Wheat Problem*, 29. For "the Great Wheat Speech" see William H. Brock, *William Crookes (1832–1919) and the Commercialization of Science* (London: Ashgate, 2008), 374–88.

81. "Japan's Outbreak: Its Cause and Cure," *New Generation* 17, no. 1 (1938): 1.

82. Nick Cullather, "The Foreign Policy of the Calorie," *American Historical Review* 112, no. 2 (2007): 357–58.

83. Jeffrey M. Pilcher, *¡Que vivan los tamales! Food and the Making of Mexican Identity* (Albuquerque: University of New Mexico Press, 1998).

84. M. K. Gandhi, *Diet and Diet Reform* (Ahmedabad: Navajivan, 1949). This was a republication of M. K. Gandhi, "Polished v. Unpolished," *Harijan*, October 26, 1934. In the essay "Waste to Wealth," Gandhi advocated the conversion of manure, urine, and other waste. See Joseph S. Alter, *Gandhi's Body: Sex, Diet, and the Politics of Nationalism* (Philadelphia: University of Pennsylvania Press, 2000). In China too, there was much discussion of the possibility of increasing food supply by leaving rice unpolished, thereby also preventing beriberi caused by deficiency of vitamin B1 that was removed with milling, a significant discovery of nutrition scientists in the 1920s. See Chih-yi Chang, "China's Population Problem—A Chinese View," *Pacific Affairs* 22, no. 4 (1949): 348–49.

85. Sunil S. Amrith, "Food and Welfare in India, c. 1900–1950," *Comparative Studies in Society and History* 50, no. 4 (2008): 1032–33.

86. Gandhi, *Diet and Diet Reform*, vii.

87. Writing in 1990, Vaclav Smil suggests a possible but unlikely way of diminishing global dependence on synthetic nitrogen fixation: substantially lower meat consumption in rich countries, "a shift dispensing with cultivation of feedgrains now dominating Western farming." Smil, "Population Growth and Nitrogen," 590.

88. Knibbs, *The Shadow of the World's Future*, 26.

89. George Knibbs, "The Worlds Possible Population and Its Population Problem," n.d., A8510/1 175/4, Knibbs Papers.

90. "The Consumption of Nutrients by Domestic Animals in the Form of Feeds and Fodders," in Pearl, *The Nation's Food*, 261–68.

91. O. E. Baker, "Land Utilization in the United States: Geographical Aspects of the Problem," *Geographical Review* 13, no. 1 (1923): xiii, 15. See also Wolfe, "Is there a Biological Law of Population Growth?" 570.

92. Benoy Kumar Sarkar, *The Sociology of Population: with Special Reference to Optimum, Standard of Living and Progress* (Calcutta: Ray-Chowdhury, 1936), 44–45.

93. Radhakamal Mukerjee, "Faiths and Influences," in *The Frontiers of Social Science: In Honour of Radhakamal Mukerjee*, ed. Baljit Singh (London: Macmillan, 1957), 4–7.

94. Radhakamal Mukerjee, *Migrant Asia* (Rome: Failli, 1936), 167.

95. Ibid., 168. Per capita consumption in Australia and Argentina per year was 250 lb., he wrote, 160 in the United States, and 120 in the United Kingdom.

96. Vilhjalmur Stefansson, *My Life with the Eskimo* (London: Macmillan, 1913).

97. Clarence W. Lieb, "The Effects on Human Beings of a Twelve Months' Exclusive Meat Diet: Based on Intensive Clinical and Laboratory Studies on Two Arctic Explorers Living under Average Conditions in a New York Climate" *Journal of the American Medical Association* 93, no. 1 (1929): 20–22.

98. John Harvey Kellogg, "Stefansson's Tallow-Eating Stunt," News clipping, 1929, in MSS 98, Folder 1, Stefansson Papers, Dartmouth College.

99. "White Men Can Live on Meats: Doctors Experiment with Explorers," News clipping August 30, 1929, Box 15, Folder 1, Stefansson Papers.

100. Stefansson, "When Is the World Going to Starve to Death?" 1925, Box 13, Folder 31, pp. 7–8, Stefansson Papers.

101. Vilhjalmur Stefansson, "The Population Question and the World's Potential Food Supply," in *Economic Problems: A Book of Selected Readings*, ed. Fred Rogers Fairchild and Ralph Theodore Compton (New York: Macmillan, 1928), 418–35; Vilhjalmur Stefansson "World's Meat Supply," 1935, Box 14, Folder 12, Stefansson Papers.

102. Vilhjalmur Stefansson, *Not By Bread Alone* (New York: Macmillan, 1946).

103. James L. Hargrove, "History of the Calorie in Nutrition," *Journal of Nutrition* 136, no. 12 (2006): 2957–61. In 1920, Raymond Pearl defined "a small calory [*sic*] as the amount of heat necessary to raise 1 gram of water 1 degree centigrade" (*The Nation's Food*, 29).

104. "Military rather than hygienic necessity made the calorie an international standard measure of food" (Cullather, "The Foreign Policy of the Calorie," 347).

105. Cullather, "The Foreign Policy of the Calorie," 340–42; Vernon, *Hunger*, 84–91.

106. Walter Bruno Gratzer, *Terrors of the Table: The Curious History of Nutrition* (New York: Oxford University Press, 2007); Vernon, *Hunger*, 91–96.

107. John Boyd Orr, *Food and the People* (London: Pilot, 1943), 7. See also John Boyd Orr, "The Scientific Basis of Nutrition in Relation to the Prevention and Treatment of Disease," n.d., Folder 1.49, Boyd Orr Papers, NLS.

108. Rocco Santoliquido, Discussion, *1927 Proceedings*, 105.

109. Pearl, *The Nation's Food*, 26.

110. G. R. C. "Russell Henry Chittenden," *Journal of Nutrition* 28, no. 1 (1944): 5. See also Russell Chittenden, *Physiological Economy in Nutrition* (New York: Stokes, 1904).

111. Mixed Committee of the League of Nations, *Final Report on the Relation of Nutrition to Health, Agriculture and Economic Policy* (Geneva: League of Nations, 1937).

112. "Daily Allowances Recommended by the National Research Council, USA 1941," in Boyd Orr, *Food and the People*, 56. By this point, protein; calcium; iron; and vitamins A, B, and C were also tabled with respect to each category.

113. H. K. Stiebling, *Food Budgets for Nutrition and Production Programs*, Miscellaneous Publication No. 183 (Washington, D.C.: USDA, 1933).

114. Cited in Hall, *The Improvement of Native Agriculture*, 17.

115. "Consumption of Human Foods—United States and Germany (Based on Energy Values), in O. E. Baker, "The Potential Supply of Wheat," *Economic Geography* 1, no. 1 (1925): 18, graph.

116. Cullather, "The Foreign Policy of the Calorie," 356–67.

117. Megan Vaughan, *The Story of an African Famine: Gender and Famine in Twentieth-Century Malawi* (Cambridge: Cambridge University Press, 1987), 5. See also, Thomas T. Poleman, "World Food: Myth and Reality," *World Development* 5, no. 5 (1977): 383–94.

118. Radhakamal Mukerjee, *Food Planning for Four Hundred Millions* (London: Macmillan, 1938), 4, viii.

119. B. P. Adarkar and I. B. Adarkar, "The Problem of Nutrition in India," in the 2nd All-India Population and 1st Family Hygiene Conference, Bombay 1938, Programme and Papers, Coll. Misc. 639/1/29, Eileen Palmer Collection, LSE. See also Sunil S. Amrith, *Decolonizing International Health: India and Southeast Asia, 1930–65* (Basingstoke, U.K.: Palgrave, 2006), 26–29.

120. Adarkar and Adarkar, "The Problem of Nutrition in India."

121. See also Robert McCarrison's elaboration of "diet and physique of Indian races," in *Nutrition and National Health* (London: Faber & Faber, 1944), and Vernon's discussion in *Hunger*, 105–8.

122. Adarkar and Adarkar, "The Problem of Nutrition in India," 3.

123. Boyd Orr, *Food and the People*, 5, 9, 12.

124. Adarkar and Adarkar, "The Problem of Nutrition in India."

125. Boyd Orr, *Food and the People*, 7.

126. Anne Hardy, "Beriberi, Vitamin B1 and World Food Policy, 1925–1970," *Medical History* 39, no. 1 (1995): 65.

127. R. Passmore, "Wallace Ruddell Aykroyd," *British Journal of Nutrition* 43, no. 2 (1980): 245. See also David Arnold, "The Discovery of Malnutrition and Diet in Colonial India,"

Indian Economic and Social History Review 31, no. 1 (1994): 1–26; N. Gangulee, *Health and Nutrition in India*, with a foreword by John Boyd Orr (London: Faber & Faber, 1939).

128. W. R. Aykroyd, B. G. Krishnan, R. Passmore and A. R. Sundararajan, *The Rice Problem in India*, Indian Medical Research Memoirs no. 32 (Calcutta: Thacker, Spink, 1940).

129. See also James Vernon, "The Ethics of Hunger and the Assembly of Society: The Techno-Politics of the School Meal in Modern Britain," *American Historical Review* 110, no. 3 (2005): 693–725. For the politics of British nutrition science and policy, see Charles Webster, "Healthy or Hungry Thirties?" *History Workshop Journal* 13 (1982): 110–29; D. F. Smith, ed., *Nutrition in Britain: Science, Scientists and Politics in the Twentieth Century* (London: Routledge, 1997); D. F. Smith and J. Phillips, eds., *Food, Science, Policy and Regulation in the Twentieth Century* (London: Routledge, 2000).

130. Cynthia Brantley, "Kikuyu-Maasai Nutrition and Colonial Science: The Orr and Gilks Study in Late 1920s Kenya Revisited," *International Journal of African Historical Studies* 30, no. 1 (1997): 49–86. John Boyd Orr's interest was initially in the mineral deficiencies in local pastures and diseases in cattle. Only latterly did it shift to human diet and health. See Brantley, "Kikuyu-Maasai Nutrition and Colonial Science," 54–55.

131. Boyd Orr, *Soil Fertility*, 4.

132. John Boyd Orr, *Food: The Foundation of World Unity* (London: National Peace Council, 1948).

133. For the latter, the intake per head of protein is 10 percent higher than prewar and of the main minerals and vitamins about 25 to 30 percent higher. Boyd Orr, *Soil Fertility*, 6–7.

134. Boyd Orr recounts this in "The Hot Springs Conference," 1943, Folder 3, Boyd Orr Papers, NLS.

135. Stanley Bruce, Speech on the Economic and Financial Issues, October 6, 1936, Second Committee of the League of Nations, M104/4/ NAA.

136. "Nutrition and World Agriculture," Speech by Mr. S. M. Bruce, Australian Delegate, before the Second Committee, League of Nations Assembly, September 19, 1935. See also Records of the Sixteenth Ordinary Session of the Assembly, League of Nations Official Journal, Special Supplement, no. 138 (1935): 52.

137. Etienne Burnet and W. R. Aykroyd, "Nutrition and Public Health," *Quarterly Bulletin of the Health Organisation, League of Nations* 4, no. 2 (1935): 323.

138. Boyd Orr, *Soil Fertility*, 11; H. D. Kay, "John Boyd Orr," *Biographical Memoirs of Fellows of the Royal Society* 18, no. 1 (1972): 60.

139. League of Nations Health Organisation, *Report on the Physiological Bases of Nutrition* (Geneva, 1935); League of Nations, "Interim Report of the Mixed Committee on the Problem of Nutrition," League of Nations Papers, Section II (Economic and Financial) Paper B.3, 1936; League of Nations, *Nutrition in Various Countries*, Section II (Economic and Financial) Paper B.5, 1936; Gandhi was only one of many across the world who noted the significance of a universal minimum standard of nutrition and publicized the League's reports locally. See "Findings of the International Commission of Experts Appointed by the Health Committee of the League of Nations," *Harijan*, April 25, 1936, in Gandhi, *Diet and Diet Reform*, 101.

140. John Boyd Orr, "Review of 'Survey of National Nutrition Policies,'" League of Nations, 1938, Folder 1.6, Boyd Orr Papers.

141. Hall, *The Improvement of Native Agriculture*, 17.

142. "Proportion of Energy Yielding to Protective Foods in the Diets of Different Nations," in Boyd Orr, *Food and the People*, 44.

8. Sex

1. John Maynard Keynes to Margaret Sanger, February 4, 1927, Box 191, SPLC.

2. Raymond Pearl, *The Natural History of Population* (Oxford: Oxford University Press, 1939); Edmund Ramsden, "Carving up Population Science: Eugenics, Demography and the Controversy over the 'Biological Law' of Population Growth," *Social Studies of Science* 32, no. 5/6 (2002): 887.

3. Dennis Hodgson, for example, in his subtle explanation of the groups making up the early Population Association of America, writes: "The fact that the birth control movement mainly attracted educated native women made it suspect to the nearly all-male world of academics concerned with population questions . . . Biological Malthusians came to view feminism as part of the problem." It is true that feminism was perceived often to be part of the problem, but birth control was not. Dennis Hodgson, "The Ideological Origins of the Population Association of America," *Population and Development Review* 17, no. 1 (1991): 6.

4. Edward Alsworth Ross, *Standing Room Only?* (London: Chapman and Hall, 1928), 230–31.

5. Anna Davin ventured a slightly different version of this argument in her now classic article "Imperialism and Motherhood: Population and Power," *History Workshop* 5 (1978): 9–65.

6. O. E. Baker, "The Potential Supply of Wheat," *Economic Geography* 1, no. 1 (1925): 51–52, 16.

7. George Handley Knibbs, *The Shadow of the World's Future* (London: Ernest Benn, 1928).

8. Harold Wright, *Population* (London: Nisbet, 1923), 170.

9. Harold Cox, *The Problem of Population* (London: Jonathan Cape, 1922), 13.

10. S. P. Srivastava, "Birth Control," in *Indian Population Problems: Report and Proceedings of the 2nd All-India Population and 1st Family Hygiene Conference*, ed. G.S. Ghurye (Bombay: Karnatak, 1938), 38.

11. Radhakamal Mukerjee, *Food Planning for Four Hundred Millions* (London: Macmillan, 1938).

12. C. R. Drysdale, "International Malthusianism," *The Malthusian* 21, no. 7 (1897): 52.

13. C. P. Blacker, *Birth Control and the State: A Plea and a Forecast* (London: Kegan Paul, Trench, Trubner, 1926), 85.

14. Julian Huxley, *Memories*, 2 vols. (London: Allen & Unwin, 1970), 1:150. Huxley joined Marie Stopes's organization in early May 1924. See Marie Stopes to Julian Huxley,

May 12, 1924, Box 8, Folder 1, HP; Julian Huxley, "The Problem of Birth Control: A Vigorous Viewpoint." *The Humanist*, October 1925, news clipping, Box 97, Folder 4, HP.

15. Julian Huxley, "A Survey of Data Respecting Animal Populations," *Population* 1, no. 1 (1933): 32.

16. W. E. Agar, "Birth Control as a World Problem," Typescript lecture, 1934, MS 1/4/9, W. E. Agar Papers, Australian Academy of the Sciences, Canberra.

17. Sripati Chandrasekhar, *Demographic Disarmament for India: A Plea for Family Planning* (Bombay: Family Planning Association of India, 1951).

18. Warren S. Thompson, *Danger Spots in World Population* (New York: Knopf, 1929), 331, 332, xi, 333.

19. Margaret Sanger, *The Pivot of Civilization* (New York: Brentano's, 1922), 126.

20. "When the Earth Is 'Full Up' Shall We Starve to Death?" *Birth Control News*, in Coll. Misc. 639/1/25, Eileen Palmer Collection, LSE. For another example, see H. C. Dekker, "Overcrowding the World," *Birth Control Review* 2, no. 1 (1918): 6.

21. Margaret Sanger, "Birth Control in China and Japan," Typescript speech, October 30, 1922, Box 199, SPLC.

22. Margaret Sanger, "Birth Control," Typescript, July 6, 1938, Box 199, SPLC.

23. Sanger, "Birth Control in China and Japan."

24. Anna Chou, Birth Control International Information Centre, Newsletter, November 5, 1935, in PP/EPR/f. 1/1, Eileen Palmer Collection, Wellcome Library.

25. *World-Wide Birth Control An Appeal*, Pamphlet from Birth Control International Information Centre, n.d., PP/EPR/F.1/1, Eileen Palmer Collection, Wellcome Library.

26. Anne Kennedy to Mary Boyd, September 7, 1927, Box 191, SPLC.

27. Edith How-Martyn, "Confidential Report Visit to Geneva," Coll. Misc. 0639/1/11, Eileen Palmer Collection, LSE.

28. *The Covenant of the League of Nations* (Geneva: League of Nations, 1919).

29. Viola Kaufman to Secretary-General, League of Nations, November 7, 1930, 11A 23738/305, Social Section, Box R3013, LNAG.

30. "Infant Welfare," 1932, Health Section, Box R6003, LNAG.

31. For actual practices of contraception in Britain, see Hera Cook, *The Long Sexual Revolution: English Women, Sex and Contraception, 1800–75* (Oxford: Oxford University Press, 2004); Kate Fisher, *Birth Control, Sex and Marriage in Britain, 1918–1960* (Oxford: Oxford University Press, 2006); Simon Szreter and Kate Fisher, *Sex Before the Sexual Revolution: Intimate Life in England 1918–1963* (Cambridge: Cambridge University Press, 2010).

32. Programme of the International Birth Control Conference, Zurich, 1930, SA/FPA/NK/73, Family Planning Association Papers, Wellcome Library, London.

33. In 1934, e.g., the guests included Julius Lewin, South Africa; Mrs. Kultaratne, Ceylon; Mr. Allard, Iraq; Dr. Amegather, Gold Coast; Mrs. Ayusawa, Japan. BCIIC Newsletter no. 2, October 1934, in PP/EPR/F.1/1, Eileen Palmer Collection, Wellcome Library.

34. Conference on Birth Control in Asia, PP/EPR/F.1/2, Eileen Palmer Collection, Wellcome Library.

35. Alexander Carr-Saunders, "Population Problems in the East," in *Birth Control in Asia*, ed. Michael Fielding (London: BCIIC, 1935), 12.

36. Edith How-Martyn, Handwritten note, "First Report of the Birth Control International Information Center," 1927, PP/EPR/F1/1, Eileen Palmer Collection, Wellcome Library.

37. Margaret Sanger, "Birth Control," Typescript, July 6 1938, Box 199, SPLC

38. See Susanne Klausen and Alison Bashford, "Fertility Control: Eugenics, Neo-Malthusianism, and Feminism," in *The Oxford Handbook of the History of Eugenics*, ed. Alison Bashford and Philippa Levine (New York: Oxford University Press, 2010), 98–115.

39. David Arnold, "Official Attitudes to Population, Birth Control, and Reproductive Health in India 1921–1946," in *Reproductive Health in India: History, Politics, Controversies*, ed. Sarah Hodges (New Delhi: Orient Longman, 2006), 25–26.

40. B. R. Kher, "Opening Address," in *Indian Population Problems: Report on and Proceedings of the 2nd All-India Population Conference and the 1st Family Hygiene Conference* (Bombay: Karnatak, 1938), 6.

41. Mukerjee, *Food Planning for Four Hundred Millions*, 219.

42. The Madras neo-Malthusians reprinted George Knibb's books on overpopulation.

43. *Madras Birth Control Bulletin* 6, no. 1 (1936), Coll. Misc. 639/1/25, Eileen Palmer Collection, LSE.

44. See Sanjam Ahluwalia, *Reproductive Restraints: Birth Control in India, 1877–1947* (Urbana: University of Illinois Press, 2008), 30–32; Sarah Hodges, "South Asia's Eugenic Past," in *The Oxford Handbook of the History of Eugenics*, ed. Bashford and Levine, 228–42; Narayan Sitaram Phadke, *Sex Problem in India: Being a Plea for a Eugenic Movement in India* (Bombay: Taraporevala, 1927); Barbara N. Ramusack, "Embattled Advocates: The Debate over Birth Control in India, 1920–40," *Journal of Women's History* 1, no. 2 (1989): 34–64.

45. Brochure, Sholapur Eugenics Education Society, n.d.; A. P. Pillay to C. B. S. Hodson, May 23, 1930, SA/EUG/E10, Eugenics Society Papers, Wellcome Library.

46. BCIIC newsletter, February 1937, PP/EPR/F1/1, Eileen Palmer Collection, Wellcome Library; Typescript Report of Tour of Edith How–Martyn, March 28, 1938, PP/EPR/C.2/1, Eileen Palmer Collection, Wellcome Library.

47. "Birth Control Exhibit Is Seized," *Herald Tribune*, August 27, 1936, News clipping, Coll. Misc. 639/1/42, Eileen Palmer Collection, LSE.

48. Singapore Advice Pamphlet on Family Planning, n.d., Coll. Misc. 639/1/25, Eileen Palmer Collection, LSE.

49. "Broadcast by Edith How–Martyn, Third India BC Tour, Bombay, 1937," Typescript, PP/EPR/C.2/1, p. 5, Eileen Palmer Collection, Wellcome Library.

50. Susan Kingsley Kent, *Sex and Suffrage in Britain, 1860–1914* (Princeton, N.J.: Princeton University Press, 1987).

51. See Ramusack, "Embattled Advocates," 50–52.

52. M. K. Gandhi, *Self Restraint versus Self Indulgence* (Ahmedabad: Navajivan, 1947); Hari G. Govil, comment in *The Sixth Intenational Neo-Malthusian and Birth Control Conference: Volume 1, International Aspects of Birth Control*, ed. Margaret Sanger (New York: American Birth Control League, 1925), 185.

53. See M. K. Gandhi, *Birth Control: The Right Way and the Wrong Way* (Ahmedabad: Navajivan, 1959).

54. Margaret Sanger, "For the Independent Woman," Typescript, n.d., Box 199, SPLC.

55. "Gandhi and Mrs. Sanger Debate Birth Control," *Asia Magazine* (November 1936): 1–12.

56. Margaret Sanger, "Does Mr Gandhi Know Women? What He Told Me at Wardha," *Madras Birth Control Bulletin* 6, no. 1 (1936): 2–3.

57. Quoted in editorial, "Mrs. Sanger's Talks with Gandhiji," *Madras Birth Control Bulletin* 6, no. 1 (1936): 9–11.

58. Blacker, *Birth Control and the State*, 22–26, 40–41, 45–46.

59. For Malthus's history with respect to gender, sexual practices, and emancipatory politics, see Gail Bederman, "Sex, Scandal, Satire, and Population in 1798: Revisiting Malthus's First Essay," *Journal of British Studies* 47, no. 4 (2008): 768–95.

60. Thomas Robert Malthus, *An Essay on the Principle of Population,* 1st ed (London: J. Johnson, 1798), 41–42.

61. John Stuart Mill, *Principles of Political Economy,* 2 vols. (Boston: Little and Brown, 1848), 1:426, 1:446.

62. Thomas Robert Malthus, *An Essay on the Principle of Population*, 2d ed. (London: J. Johnson, 1803), ix.

63. John Toye, "Keynes on Population and Economic Growth," *Cambridge Journal of Economics* 27, no. 1 (1997): 8.

64. M. Godelier, "Malthus and Ethnography," in *Malthus: Past and Present*, ed. J. Dupâquier, A. Fauce-Chamoux, and E. Grebenik (London: Academic, 1983), 131. See also J. C. Caldwell, "Malthus and the Third World." www.acolasecretariat.org.au/papers.htm (accessed 10 May 2013).

65. Malthus, *Essay* (1803), 20–22.

66. E. A. Wrigley, *Population and History* (New York: McGraw-Hill, 1969), 42–44. See also Alison Bashford, "Malthus and Colonial History," *Journal of Australian Studies* 36, no. 1 (2012): 99–110.

67. Malthus, *Essay* (1798), 364.

68. Alexander Carr-Saunders drew from accounts such as those of Mrs. Edgeworth David, *Funafuti: Or, Three Months on a Coral Island, an Unscientific Account of a Scientific Expedition* (London: John Murray, 1899), 195; W. H. R. Rivers, *The History of Melanesian Society* (Cambridge: Cambridge University Press, 1914); A. C. Haddon, *Reports of the Cambridge Anthropological Expedition to the Torres Straits, 1898*, 6 vols. (Cambridge: Cambridge University Press, 1901–35).

69. Discussed in chapter 2.

70. A. M. Carr-Saunders, *Population* (London: Oxford University Press, 1925), 63, 15–16. See W. H. R. Rivers, ed., *Essays on the Depopulation of Melanesia* (Cambridge: The University Press, 1922). For assessment, see Margaret Jolly, "Other Mothers: Maternal 'Insouciance' and the Depopulation Debate in Fiji and Vanuatu, 1890–1930," in *Maternities and Modernities: Colonial and Postcolonial Experiences in Asia and the Pacific*, ed. Kalpana Ram and Margaret Jolly (Cambridge: Cambridge University Press, 1998), 177–212.

71. Following Carr-Saunders, the International Union for the Scientific Investigation of Population Problems incorporated infanticide and abortion as demographic dynamics.

72. A. M. Carr-Saunders, *World Population: Past Growth and Present Trends* (Oxford: Clarendon Press, 1936), 106.

73. Carr-Saunders, *Population*, 18–19.

74. Similarly, Japanese specialist E. F. Penrose normalized infanticide in world history terms "practiced in some degree in most, if not all, parts of the world in historic times." E. F. Penrose, *Population Theories and Their Application with Special Reference to Japan* (Stanford, Calif.: Food Research Institute, 1934), 113.

75. Carr-Saunders, *Population*, 71.

76. At least that was the story Carr-Saunders passed on to his Oxford colleague Julian Huxley, topped with gossip about Stopes's recently divorced husband and their biologist colleague Ruggles Gates. Alexander Carr-Saunders to Julian Huxley, September 13, 1922, Box 7, Folder 2, HP.

77. See Rosanna Ledbetter, *A History of the Malthusian League, 1877–1927* (Columbus: Ohio University Press, 1976), 65–66.

78. Malthus, *Essay* (1798), 61.

79. Annie Besant, *The Law of Population* (Chicago: Wilson, 1880), 19.

80. C. R. Drysdale, *The Population Question According to T. R. Malthus and J. S. Mill* (London: Strandring, 1892), 30–31.

81. Ibid., 31.

82. Ross, *Standing Room Only?*, 248–49.

83. Margaret Sanger, Open forum lecture outline, January 3, 1926, Box 199, SPLC.

84. Radhakamal Mukerjee, "The Criterion of Optimum Population," *American Journal of Sociology* 38, no. 5 (1933): 698.

85. Carr-Saunders, *Population*, 65–66.

86. A. M. Carr-Saunders, "An Outline of Population History," *Population* 1, no. 1 (1933): 27.

87. George Standring, "The Need for Birth control in India," n.d., PP/EPR/F.1/2, Eileen Palmer Collection, Wellcome Library.

88. Ross, *Standing Room Only?*, 208.

89. "The sacrifices—of freedom and happiness, of health, nay, even of life—which unlimited child-bearing imposes upon the mother, so far exceed the burden a large family imposes upon the father that it is not too much to say that family limitation is chiefly women's concern" (Ross, *Standing Room Only?*, 143).

90. Ibid., 230–31, 241 (emphasis in original).

91. Ibid., 252, 271, 144.

92. Prabbu Dutt Shastri, "The Outlook in India," in *International Aspects of Birth Control*, ed. Sanger, 122.

93. Ross, *Standing Room Only?*, 355.

94. Though according to Havelock Ellis, Charles Darwin thought that Malthus failed to lay enough stress on infanticide. Havelock Ellis, "The Evolutionary Meaning of Birth Control," in *The Sixth International Neo-Malthusian and Birth Control Conference: Volume 3, Medical and Eugenics Aspects of Birth Control*, ed. Margaret Sanger (New York: American Birth Control League, 1926), 5.

95. Herbert Spencer, *The Principles of Biology* (rev. ed. 1899), cited in Warren Thompson, *Population: A Study in Malthusianism* (New York: Columbia University, 1915), 161.

96. See Ross, *Standing Room Only?*, 217.

97. J. M. Keynes, "Preface," in Wright, *Population*, vii.

98. Michel Foucault never fully developed birth control or population control as a major manifestation of modern governance of the self and governance of the state—"governmentality." See Graham Burchell, Colin Gordon and Peter Miller eds, *The Foucault Effect* (Chicago: University of Chicago Press, 1991).

99. Mukerjee, "The Criterion of Optimum Population," 698, 696.

100. H. G. Wells, in Sanger, *The Pivot of Civilization*, preface (n.p.).

101. Sanger, "Birth Control in China and Japan," 18–19.

102. Enid Charles, *The Menace of Under-Population: A Biological Study of the Decline of Population Growth* (London: Watts, 1936), 200.

103. Carr-Saunders, *World Population*, 243–59.

104. See Frank Lorimer and Frederick Osborn, *Dynamics of Population: Social and Biological Significance of Changing Birth Rates in the United States* (New York: Macmillan, 1934). See also the major study commissioned by Harold Ickes, secretary of the interior and chair of the National Resources Committee, which included Secretary of War Harry H. Woodring: *The Problems of a Changing Population* (Washington D.C.: U.S. National Resources Committee, 1938). Frank Lorimer was director of the technical staff on this report, and Warren Thompson was a member of its Population Problems Committee.

105. J. M. Keynes, Galton Lecture, reported in *The Birth Control News* 15, no. 10 (1937), located in B/3/5, Alexander Carr-Saunders Papers, LSE.

106. Indeed, Robert René Kuczynski had said as early as 1929, at the Harris Foundation Lectures, "The population of Western and Northern Europeans, North America, and Australia combined, no longer reproduce themselves." Quoted in John Coatman, "Migration in the Twentieth Century," *Population* 1 no. 1 (1933): 61.

107. D. V. Glass and C. P. Blacker, *Population and Fertility* (London: Population Investigation Committee, 1939), 19.

108. D. V. Glass, "The Population Problem," in BCIIC Newsletter, February 1937, PP/EPR/F1/1, Eileen Palmer Collection, Wellcome Library. This formed part of the thesis of Glass's *The Struggle for Population*. Thomas Horder, Julian Huxley, Carlos Paton Blacker, and Helena Wright had already worked closely together as part of the National Birth Control Association's medical conferences. See Minutes, July 13, 1933, National Birth Control Association, SA/EUG/D.19, Eugenics Society Papers, Wellcome Library.

109. Glass and Blacker, *Population and Fertility*, 100–1.

110. C. P. Blacker, "Stages in Population Growth," *Eugenics Review* 39, no. 1 (1947): 88–97.

111. For example, Vanessa Baird, *The No-Nonsense Guide to World Population* (Oxford: New Internationalist, 2011).

112. Knibbs, *The Shadow of the World's Future*, 110.

113. Blacker, *Birth Control and the State*, 95.

114. Alexander Carr-Saunders, "Eugenics in the Light of Population Trends" (1935), reprinted in *Eugenics Review* 60, no. 1 (1968): 48, 50.

115. Carr-Saunders, *Population*, 57.

116. Simon Szreter, "The Idea of Demographic Transition and the Study of Fertility Change: A Critical Intellectual History," *Population and Development Review* 19, no. 4 (1993): 659–701.

117. Kingsley Davis, "The World Demographic Transition," *Annals of the American Academy of Political and Social Science* 237, no. 1 (1945): 1–11.

9. The Species

1. Edward Alsworth Ross, *Standing Room Only?* (London: Chapman and Hall, 1928), 355.

2. George Handley Knibbs, *The Shadow of the World's Future* (London: Ernest Benn, 1928), 108, 114.

3. See, for example, Johanna Schoen, *Choice and Coercion: Birth Control, Sterilization, and Abortion in Public Health and Welfare* (Chapel Hill: University of North Carolina Press, 2005); Laura Briggs, *Reproducing Empire: Race, Sex, Science, and US Imperialism in Puerto Rico* (Berkeley: University of California Press, 2003).

4. With Edmund Ramsden, I see eugenics playing an important and continuing role within the establishment of demography, but also perceive a closer alignment than he allows between emerging professional demography, Malthusianism, and birth control activism. Edmund Ramsden, "Social Demography and Eugenics in the Interwar United States," *Population and Development Review* 29, no. 4 (2003): 551.

5. William Allen Pusey, *Medicine's Responsibilities in the Birth Control Movement* (privately printed paper: Sixth Neo-Malthusian and Birth Control Conference, 1925), 3.

6. Diane B. Paul, "Eugenics and the Left," *Journal of the History of Ideas* 45, no. 4 (1984): 567–90.

7. Francis Galton, *Memories of My Life* (London: Methuen, 1908), 323.

8. See Rosanna Ledbetter, *A History of the Malthusian League, 1877–1927* (Columbus: Ohio State University Press, 1976), 204–5; C. R. Drysdale, *The Population Question According to T. R. Malthus and J. S. Mill* (London: Standring, 1892), 36.

9. International Neo-Malthusian Conference, *The Malthusian* 34, no. 8 (1910): 75.

10. C. V. Drysdale, *Neo-Malthusianism and Eugenics* (London: Bele, 1912).

11. John Maynard Keynes, Notes for Speech to the Malthusian League, July 26, 1927, JMK/PS/3/114, KP. See also Donald Markwell, *John Maynard Keynes and International Relations: Economic Paths To War and Peace* (Oxford: Oxford University Press, 2006), 167.

12. See, e.g., John Toye's discussion of Keynes's "race prejudice," *Keynes on Population* (Oxford: Oxford University Press, 2000), 4–5.

13. Drysdale, *Neo-Malthusianism and Eugenics*, 3–4.

14. Radhakamal Mukerjee, "The Criterion of Optimum Population," *American Journal of Sociology* 38, no. 5 (1933): 698.

15. G. H. Knibbs, "The New Malthusianism in the Light of Actual World Problems of Population," *Scientia* 40 (1926): 383, 387.

16. Knibbs, *The Shadow of the World's Future*, 103, 109–111, 113.

17. Margaret Sanger, "Plans for Peace," *Birth Control Review* 16, no. 4 (1932): 106.

18. William H. Schneider, *Quality and Quantity: The Quest for Biological Regeneration in Twentieth-Century France* (Cambridge: Cambridge University Press 1990), 37.

19. C. S. Hsueh and George Y. Auhang, "A Plea for International Birth Control," *The Peoples Tribune*, n.d., Coll. Misc. 639/F1, Eileen Palmer Collection, LSE.

20. Typescript of Proceedings, Second Conference on Population Association of America, New York City, May 7, 1931, Bureau of Social Hygiene 1931, Series 3, Box 9, Folder 191, RAC.

21. Minutes of Conference on Coordination, Population Association of America, February 15, 1934, Council on Population Policy, Frederick Osborn Papers, APS, Philadelphia.

22. Henry Pratt Fairchild (1940), cited in Dennis Hodgson and Susan Cotts Watkins, "Feminists and Neo-Malthusians: Past and Present Alliances," *Population and Development Review* 23, no. 3 (1997): 478.

23. Mission Statement of the Population Association of America, cited in Dennis Hodgson, "The Ideological Origins of the Population Association of America," *Population and Development Review* 17, no. 1 (1991): 9.

24. Hodgson, "Ideological Origins," 9.

25. *Conference on Birth Control in Asia*, Pamphlet, 1933, SA/EUG/D.14, Eugenics Society Papers, Wellcome Library, London.

26. Ramsden, "Social Demography and Eugenics," 557–60.

27. Frederick Osborn, "Development of a Eugenic Philosophy," *American Sociological Review* 2, no. 3 (1937): 389–97.

28. On Julian Huxley's advocacy of sterilization, see Garland E. Allen, "Julian Huxley and the Eugenical View of Human Evolution," in *Julian Huxley: Biologist and Statesman of Science*, ed. C. Kenneth Waters and Albert Van Helden (Houston: Rice University Press, 1992), 212–16.

29. John Macnicol, "Eugenics and the Campaign for Voluntary Sterilization in Britain Between the Wars," *Social History of Medicine* 2, no. 2 (1989): 147–69.

30. H. G. Wells, Julian Huxley, and G. P. Wells, *The Science of Life* (London: Cassell, 1931), 875.

31. Margaret Sanger, "Birth Control and Racial Betterment," *Birth Control Review* 3 (1919): 11.

32. Raymond Pearl, "The Biology of Superiority," *American Mercury* 12, no. 47 (1927): 257–66; Elazar Barkan, *The Retreat of Scientific Racism* (Cambridge: Cambridge University Press, 1992); Edmund Ramsden, "Carving Up Population Science: Eugenics, Demography and the Controversy over the 'Biological Law' of Population Growth," *Social Studies of Science* 32, no. 5/6 (2002): 857–99.

33. Warren Thompson, *Population: A Study in Malthusianism* (New York: Columbia University, 1915), 165.

34. Warren S. Thompson, *Danger Spots in World Population* (New York: Knopf, 1929), 327.

35. Warren S. Thompson, "Eugenics as Viewed by a Sociologist," *Monthly Labor Review* 18, no. 2 (1924): 16, 20.

36. Daniel Kevles, *In the Name of Eugenics: Genetics and the Uses of Human Heredity* (Cambridge Mass.: Harvard University Press, 1985), 97; See also Barkan, *The Retreat of Scientific Racism*, 194–203.

37. Prescott F. Hall, "Immigration Restriction and World Eugenics," *Journal of Heredity* 10, no. 3 (1919): 125–27; Hodgson, "Ideological Origins," 10.

38. Hall, "Immigration Restriction and World Eugenics," 126. See also Matthew Connelly, "To Inherit the Earth: Imagining World Population, from the Yellow Peril to the Population Bomb," *Journal of Global History* 1, no. 3 (2006): 299–319.

39. Charles Davenport, Comment, *Proceedings of the World Population Conference*, ed. Margaret Sanger (London: Edward Arnold, 1927), 275 [hereafter *1927 Proceedings*].

40. Margaret Sanger, "Address of Welcome," in *Sixth International Neo-Malthusian and Birth Control Conference: Volume 1. International Aspects of Birth Control*, ed. Margaret Sanger (New York: American Birth Control League, 1925), 4–5, 24.

41. Immigration Act of 1917 (U.S.), H.R. 10384, sect. 3, 64th Cong. (1917).

42. Immigration Act of 1910 (Canada), Edward VII, chap. 27, Section 3(a).

43. Griffith Taylor, *Environment and Race* (London: Oxford University Press, 1927), 341.

44. K. S. Inui, Comment, *1927 Proceedings*, 272.

45. Marilyn Lake and Henry Reynolds, *Drawing the Global Colour Line: White Men's Countries and the International Challenge of Racial Equality* (Cambridge: Cambridge University Press, 2008), 284–309.

46. Gregory, "International Migration and Its Control," *1927 Proceedings*, 303–4.

47. Hall, "Immigration Restriction and World Eugenics," 125–27.

48. Charles Davenport to Dame Rachel Crowdy, November 23, 1920, Box R 642, International Eugenics, 12 9789/7260 (1920), LNAG.

49. Nitobe Inazō, Handwritten note, December 24, 1920, Box R 642, International Eugenics, 12 9789/7260 (1920), LNAG.

50. Sybil Neville-Rolfe to Dame Rachel Crowdy, September 28, 1920, Box R 642, International Eugenics, 12 9789/7260 (1920), LNAG.

51. Typescript, Eugenics Society evidence to the Royal Commission on Population, 1944, B/3/10, Alexander Carr-Saunders Papers, LSE.

52. Karl Ittmann, "Demography as Policy Science in the British Empire, 1918–1969," *Journal of Policy History* 15, no. 4 (2003): 417–48; Karl Ittmann, "The Colonial Office and the Population Question in the British Empire, 1918–62," *The Journal of Imperial and Commonwealth History* 27, no. 3 (1999): 55–81. See also Karl Ittmann, Dennis D. Cordell, and Gregory H. Maddox, eds., *The Demographics of Empire: The Colonial Order and the Creation of Knowledge* (Athens: Ohio University Press, 2010).

53. Dame Rachel Crowdy to Sybil Neville-Rolfe, October 5, 1920, Box R 642, International Eugenics, 12 9789/7260 (1920), LNAG.

54. D. F. Ramos to Director, Department of Hygiene, September 21, 1926, Infant Welfare Enquiry Dossier respecting Cuba, Box R 975, Health Section, LNAG.

55. Report on the work of the Health Committee 8th Session, October 1926, Box R 912, Health Section, LNAG.

56. Todd Hedrick, "Race, Difference, and Anthropology in Kant's Cosmopolitanism," *Journal of the History of Philosophy* 46, no. 2 (2008): 245–68. See also Saul Dubow,

"Smuts, the United Nations and the Rhetoric of Race and Rights," *Journal of Contemporary History* 43, no. 1 (2008): 45–74.

57. F. James Dawson, *Aggression and Population* (London: Rockliff, 1946), 73.

58. Barkan, *The Retreat of Scientific Racism*; Warwick Anderson, "Ambiguities of Race: Science on the Reproductive Frontier of Australia and the Pacific Between the Wars," *Australian Historical Studies*, 40, no. 2 (2009): 143–60.

59. Leonard Darwin, "Presidential Address," *Proceedings of the First International Eugenics Congress, 1912* (London: Eugenics Education Society, 1913), 1.

60. Sir John Macdonell, *Eugenics Review* (1916), cited in C. E. A. Bedwell, "Eugenics in International Affairs," *Eugenics Review* 14, no. 3 (1922): 188.

61. Edward Isaacson, *The New Morality* (New York: Moffat, Yard, 1913), 173.

62. Thompson, *Danger Spots*, 179.

63. G. H. L. F. Pitt-Rivers, "The Urgency of Population Study for the Bio-Anthropological Approach," *Population* 1, no. 1 (1933): 21.

64. Radhakamal Mukerjee, *Migrant Asia* (Rome: Failli, 1936), 84–85.

65. Knibbs, "The New Malthusianism," 383.

66. George H. Knibbs, "The Fundamental Elements of the Problems of Population and Migration," *Eugenics Review* 19, no. 4 (1928): 276, 285–89.

67. "No country was prepared willingly to absorb the derelict population of another country to the detriment of itself or to the prejudice of its future development" (George Knibbs, "Overcrowding the World: the Problem of Distribution," *Argus*, September 3, 1927).

68. Thompson, "Eugenics as Viewed by a Sociologist," 22.

69. George Knibbs to Alfred Mjøen, September 30, 1924, A8510/1 175/4, Knibbs Papers, NLA.

70. Alexander Carr-Saunders to Julian Huxley, October 10, 1926, Box 9, Folder 2, HP; Alexander Carr-Saunders to Julian Huxley, February 24, 1925, Box 8, Folder 4, HP.

71. A. M. Carr-Saunders, *Eugenics* (London: Thornton Butterworth, 1926), 201, 228, 170–71, 166–168.

72. Ibid., 89.

73. J. H. Bennett, "Introduction," in *Natural Selection, Heredity and Eugenics*, ed. J. H. Bennett (Oxford: Clarendon, 1983), 5.

74. R. A. Fisher, "The Correlation between Relatives on the Supposition of Mendelian Inheritance," *Philosophical Transactions of the Royal Society of Edinburgh* 52 (1918): 399–433; R. A. Fisher, *The Genetical Theory of Natural Selection* (Oxford: Clarendon, 1930).

75. Bennett, "Introduction," 22, 12–13.

76. For example, R. A. Fisher, "On the Existence of Daily Changes in the Bacterial Numbers in American Soil," *Soil Science* 23 (1927): 253–59; R. A. Fisher, Comment, *1927 Proceedings*, 43–46; R. A. Fisher to Julian Huxley, May 22, 1924, Box 8, Folder 1, HP.

77. Bennett, "Introduction," 16.

78. R. A. Fisher to Julian Huxley, April 2, 1942, Box 16, Folder 2, HP.

79. Bennett, "Introduction," 25. For Robert René Kuczynski, see Enid Charles, *The Twilight of Parenthood: A Biological Study of the Decline of Population Growth* (London: Watts, 1934), 45–76.

80. Bennett, "Introduction," 33; Fisher, *The Genetical Theory of Natural Selection*,170–256. See also Barkan, *The Retreat of Scientific Racism*, 220–27.

81. A. M. Carr-Saunders, "Eugenics in the Light of Population Trends" (1935), reprinted in *Eugenics Review* 60, no. 1 (1968): 55.

82. Ramsden, "Social Demography and Eugenics," 560–62.

83. "The Eugenics Outlook for the Future," October 7, 1937, MS 1/4/14, W. E. Agar Papers, National Academy of Sciences, Canberra.

84. Cecil Cook, Statement, Commonwealth of Australia, *Conference on Aboriginal Welfare* (Canberra: Government Printer, 1937), 14.

85. See "Draft Memorandum on Refugee Problems," B/3/9, Alexander Carr-Saunders Papers, LSE. See also Gavin Schaffer, *Racial Science and British Society, 1930–62* (Basingstoke, U.K.: Palgrave, 2008), 32–35.

86. Julian S. Huxley, A. C. Haddon, and A. M. Carr-Saunders, *We Europeans: A Survey of "Racial" Problems* (1935; Harmondsworth, U.K.: Penguin, 1939), 7. See Allen, "Julian Huxley," 206–07.

87. A. C. Haddon to Julian Huxley, August 9, 1935, and September 15, 1935, Box 11, Folder 9, HP.

88. Huxley et al., *We Europeans*, 20.

89. Ibid., 27, 86, 24–27, 84–85.

90. Ibid., 54, 88–89, 90.

91. Ibid., 92.

92. A. C. Haddon to Julian Huxley, June 18, 1936, Box 12, Folder 2, HP.

93. Barkan, *The Retreat of Scientific Racism*.

94. United Nations Educational, Scientific and Cultural Organization, *The Race Concept* (Paris: UNESCO, 1952), 27. See also Perrin A. Selcer, "Beyond the Cephalic Index: Negotiating Politics to Produce Unesco's Scientific Statements on Race," *Current Anthropology* 55, no. 5 (2012): 173–84.

95. Other signatories included Joseph Needham, Gunnar Dahlberg, and Theodore Dobzhansky.

96. For example, John P. Jackson and Nadine M. Weidman, *Race, Racism, and Science: Social Impact and Interaction* (Santa Barbara, Calif.: ABC Clio, 2004); William B. Provine, "Geneticists and Race," *American Zoologist* 26 (1986): 857–87; Barkan, *The Retreat of Scientific Racism*, 280, 342.

97. F. A. E. Crew et al., "Social Biology and Population Improvement," *Nature* 144, no. 3646 (1939): 521.

98. Huxley et al., *We Europeans*, 63.

10. Food and Freedom

1. Amy L. S. Staples, *The Birth of Development: How the World Bank, Food and Agriculture Organization, and World Health Organization Changed the World, 1945–1965* (Kent, Ohio: Kent State University Press, 2006); Paul Greenough, *Prosperity and Misery in Modern Bengal: The Famine of 1943–1944* (New York: Oxford University Press, 1982); Amartya

Sen, *Poverty and Famines: An Essay on Entitlement and Deprivation (*Oxford: Clarendon, 1982), 52–85; Sugata Bose, "Starvation Amidst Plenty: The Making of Famine in Bengal, Honan and Tonkin, 1942–45," *Modern Asian Studies* 24, no. 4 (1990): 699–727.

2. Suzanne Pepper, *Civil War in China: The Political Struggle, 1945–1949* (Oxford: Rowman and Littlefield, 1999), 71–72.

3. John Boyd Orr, *Soil Fertility: The Wasting Basis of Human Society* (London: Pilot, 1948), 11.

4. Nick Cullather, *The Hungry World: America's Cold War Battle Against Poverty in Asia* (Cambridge, Mass.: Harvard University Press, 2010).

5. John Perkins, *Geopolitics and the Green Revolution: Wheat, Genes, and the Cold War* (New York: Oxford University Press, 1997).

6. John Sharpless, "World Population Growth, Family Planning and American Foreign Policy," *Journal of Policy History* 7, no. 1 (1995): 72–102.

7. Kingsley Davis, "Population and Power in the Free World," in *Population Theory and Policy*, ed. Joseph Spengler and Otis Dudley Duncan (Glencoe, Ill.: Free Press, 1956), 342–56. Cited also in Mohan Rao, *From Population Control to Reproductive Health: Malthusian Arithmetic* (New Delhi: Sage, 2004), 105.

8. Ellsworth Bunker, Will. L. Clayton, and Hugh Moore to JDR3, November 26, 1954, Population Interests, Series 1, Subseries 5, Box 81, Folder 676, JDR3 Papers, RFam.A, RAC.

9. Alfred Sauvy, "Trois Mondes, Une Planète," *L'Observateur*, August 14, 1952, 14.

10. Kingsley Davis, "The World Demographic Transition," *Annals of the American Academy of Political and Social Science* 237, no. 1 (1945): 1–11. This was a special issue edited by Davis, titled "World Population in Transition."

11. Josué de Castro, *The Geopolitics of Hunger* (New York: Monthly Review, 1977). This was first published as *The Geography of Hunger: Hunger in Brazil* (1946), and then as *The Geography of Hunger (*1952). See Kirit S. Parikh, "Chronic Hunger in the World: Impact of International Policies," in *The Political Economy of Hunger*, 3 vols., ed. Jean Drèze and Amartya Sen (Oxford: Clarendon, 1990), 1:114–46.

12. Sripati Chandrasekhar, *Hungry People and Empty Lands: An Essay on Population Problems and International Tensions* (London: Allen & Unwin, 1954), 249.

13. Patricia Clavin and Jens-Wilhelm Wessels, "Transnationalism and the League of Nations: Understanding the Work of its Economic and Financial Organisation," *Contemporary European History* 14, no. 4 (2005): 465–92.

14. Frank Notestein et al., *The Future Population of Europe and the Soviet Union: Population Projections, 1940–1970* (Princeton: Economic, Financial and Transit Department of the League of Nations, 1944).

15. See Dennis Hodgson, "The Ideological Origins of the Population Association of America," *Population and Development Review* 17, no. 1 (1991): 1–34; Richard Symonds and Michael Carder, *The United Nations and the Population Question, 1945–1970* (London: Chatto & Windus, 1973), 40–51.

16. The Princeton group had undertaken work from 1938 on the "Study of Social and Psychological Factors Affecting Fertility," and the social sciences of fertility motivation

began to take off. See Edmund Ramsden, "Social Demography and Eugenics in the Interwar United States," *Population and Development Review* 29, no. 4 (2003): 573–77.

17. See the section "Problems of Policy in Relation to Areas of Heavy Population Pressure," in *Demographic Studies of Selected Areas of Rapid Growth: Proceedings of the Round Table on Population Problems* (New York: Milbank Memorial Fund, 1944).

18. Frank W. Notestein, "Population: The Long View," in *Food for the World*, ed. Theodore W. Schultz (Chicago: University of Chicago Press, 1945), 36.

19. Symonds and Carder, *The United Nations and the Population Question*, 69–71.

20. Julian Huxley, "Statement by the Representative of UNESCO on the Desirability of Convening a UN Conference on World Population Problem," May 13, 1948, GX 6/1/7, United Nations Archive, Geneva.

21. Report of the Program Commission, General Conference First Session, 1946, pp. 222, 235, UNESCO/C/30, UNESCO Archive, Paris.

22. Resolution 4.315, Records of the General Conference of UNESCO, 3rd Session 1948, Volume II, Resolutions, p. 24, UNESCO Archive.

23. UNESCO, "Studies of Social Tensions, 1947–1952," January 19, 1951, UNESCO/SS/2, UNESCO Archive.

24. Conrad Taeuber, "Demographic Activities of FAO," Conférence de Genève 1949, Union internationale pour l'étude scientifique de la population, Typescript, Box 3, Folder 13, Alfred J. Lotka Papers, Stanley G. Rudd Manuscript Library, Princeton University.

25. H. D. Kay, "John Boyd Orr," *Biographical Memoirs of Fellows of the Royal Society* 18 (1972): 62.

26. See Amy Staples, "To Win the Peace: The Food and Agriculture Organization, Sir John Boyd Orr, and the World Food Board Proposals," *Peace & Change* 28, no. 4 (2003): 495–523.

27. John Boyd Orr, "The Hot Springs Conference," 1943, Folder 3, Boyd Orr Papers, NLS.

28. John Boyd Orr and David Lubbock, *The White Man's Dilemma* (1953; London: Allen & Unwin, 1964), 52.

29. R. Passmore, "Wallace Ruddell Aykroyd," *British Journal of Nutrition* 43, no. 2 (1980): 248.

30. Boyd Orr, *Soil Fertility*, 12.

31. John Boyd Orr, "The Road Through Plenty to Peace," Typescript, c. 1945, Folder 2, Boyd Orr Papers.

32. Ibid.

33. "Overpopulation a Major World Problem," 1955; "Families and Famine," News clippings, 1955, SA/FPA/A17/10, Family Planning Association Papers, Wellcome Library, London.

34. Typescript, "Notes on Population," September 1959, Box 75, Folder 4, HP.

35. Julian Huxley, "The Impending Crisis," n.d, Box 75, Folder 5, HP.

36. Julian Huxley, *Memories*, 2 vols. (London: Allen & Unwin, 1970), 2: 54.

37. Julian Huxley, "Statement by the Representative of UNESCO on the Desirability of Convening a UN Conference on World Population Problem."

38. For example, Symonds and Carder, *The United Nations and the Population Question*, xii. See also R. S. Deese, "The New Ecology of Power: Julian and Aldous Huxley in the Cold War Era," in *Environmental Histories of the Cold War*, ed. J. R. McNeill and Corinna Unger (Cambridge: Cambridge University Press, 2010), 279–300. Deese's study of Huxley rightly focuses on his "ecological humanism," though underappreciates the extent to which that emerges from Huxley's own Malthusianism and the post–World War I generation's linking of population and ecology at a global level.

39. See Symonds and Carder, *The United Nations and the Population Question*, 54–57.

40. Margaret Mead, *The Family's Food* (London: Bureau of Current Affairs, 1949); Alva Myrdal and P. Vincent, *Are We Too Many? UNESCO, Food and People* (London: Bureau of Current Affairs, 1949).

41. Charles E. Kellogg, *Need We Go Hungry?* (London: Bureau of Current Affairs, 1950), 9.

42. Ibid., 46, 45.

43. E. J. Russell to Julian Huxley, October 20, 1941, Box 15, Folder 5, HP.

44. E. J. Russell, *The Fertility of Soil* (Cambridge: The University Press, 1913); E. J. Russell, *Manuring for Higher Crop Production* (Cambridge: The University Press, 1916); E. J. Russell, *Artificial Fertilizers in Modern Agriculture* (London: Ministry of Agriculture and Fisheries, 1931).

45. Preparatory Document for Conference on Population Problems, National Academy of Sciences, 1952, Series 1, Subseries 5, Box 85, Folder 718, p. 35, JDR3 Papers, RFam.A, RAC.

46. Huxley, *Memories*, 1:170.

47. *Enough to Eat* (Merton Park Studios, UK, 1936); John Boyd Orr, *Food, Health and Income* (London: Macmillan, 1936); John Boyd Orr, *Food and the People* (London: Pilot, 1943).

48. See Joseph Morgan Hodge, *Triumph of the Expert: Agrarian Doctrines of Development and the Legacies of British Colonialism* (Athens: Ohio University Press, 2007), 256–57.

49. Sir John Boyd Orr, *Proceedings of the International Congress on Population and World Resources in Relation to the Family* (London: Family Planning Association, 1948), 9 [hereafter *1948 Proceedings*].

50. "Group Discussions on the Social Implications of Science: Food and People," November 22, 1948, UNESCO/NS/SIS/1, UNESCO Archive.

51. *The Social Implications of Science*, July 30, 1949, 1. Food and People Pamphlets, UNESCO/NS/SIS/4, UNESCO Archive.

52. Davis, "The World Demographic Transition," 1–11; Kingsley Davis, "The Theory of Change and Response in Modern Demographic History," *Population Index* 29, no. 4 (1963): 345–66.

53. Kingsley Davis and Julius Isaac, *People on the Move* (London: Bureau of Current Affairs, 1950).

54. Megan Vaughan, *The Story of an African Famine: Gender and Famine in Twentieth-Century Malawi* (Cambridge: Cambridge University Press, 1987); Henrietta L. Moore and Megan Vaughan, *Cutting Down Trees: Gender, Nutrition, and Agricultural Change in the Northern Province of Zambia, 1890–1990* (Portsmouth, N.H.: Heinemann, 1994).

55. Davis and Isaac, *People on the Move*, 16, 18–19.

56. Aldous Huxley, *Brave New World* (London: Chatto & Windus, 1932); *Eyeless in Gaza* (London: Chatto & Windus, 1936); *Ends and Means* (London: Chatto & Windus, 1937); *Science, Liberty and Peace* (London: Chatto & Windus, 1947).

57. Aldous Huxley, *The Double Crisis* (Paris: UNESCO, 1949), 1.

58. Boyd Orr, *Soil Fertility*, 8–9.

59. Aldous Huxley, *The Double Crisis*, 1.

60. See also McNeill and Unger, eds., *Environmental Histories of the Cold War*.

61. Fairfield Osborn, *Our Plundered Planet* (London: Faber & Faber, 1948); William Vogt, *Road to Survival* (New York: Sloane, 1948); Paul B. Sears, *The Living Landscape* (New York: Ecological Association of America, 1949),

62. Thomas Robertson, *The Malthusian Moment: Global Population and the Birth of American Environmentalism* (New Brunswick, N.J.: Rutgers University Press, 2012).

63. Sears, *The Living Landscape*, 91. For excellent context, see Thomas Robertson, "'This Is the American Earth': American Empire, the Cold War, and American Environmentalism," *Diplomatic History* 32, no. 4 (2008): 561–84.

64. Sears, *The Living Landscape*, 91.

65. F. James Dawson, *Aggression and Population* (London: Rockliff, 1946), 28.

66. E. G. Hutchinson, "On Living in the Biosphere," *Scientific Monthly* 67, no. 6 (1948): 393–94.

67. See Robertson, *The Malthusian Moment*, 72–73.

68. Pierre Desrochers and Christine Hoffbauer, "The Post War Intellectual Roots of the Population Bomb: Fairfield Osborn's *Our Plundered Planet* and William Vogt's *Road to Survival* in Retrospect," *Electronic Journal of Sustainable Development* 1, no. 3 (2009): 73–97.

69. Osborn, *Our Plundered Planet*, 43, 176.

70. Rachel Carson, *Silent Spring* (Boston: Houghton Mifflin, 1962).

71. Osborn, *Our Plundered Planet*, 74–75.

72. Ibid., 18, 23, 54–55, 73–75. For Fairfield Osborn and William Vogt, and world population organizations, see Matthew Connelly, *Fatal Misconception: The Struggle to Control World Population* (Cambridge Mass.: Belknap, 2008), 128–29.

73. Fairfield Osborn, *The Limits of the Earth* (Boston: Little, Brown, 1953), 6.

74. Osborn, *Our Plundered Planet*, 94.

75. Osborn, "Our Reproductive Potential," Typescript Lecture, n.d., GC/142/22, Wellcome Library.

76. Transcript, Conference on Population Problems, Williamsburg, June 20, 1952, Series 1, Subseries 5, Box 85, Folder 720, JDR3 Papers, RFam.A, RAC.

77. Memo, John McLean to JDR3, July 18, 1952, "Population Interests," Series 1, Subseries 5, Box 81, Folder 674, JDR3 Papers, RFam.A, RAC.

78. See also Perkins, *Geopolitics and the Green Revolution*, 136–38; Maureen A. McCormick, "Of Bird, Guano, and Men: William Vogt's 'Road to Survival,'" Ph.D. thesis, University of Oklahoma, 2005.

79. Transcript, Conference on Population Problems, Williamsburg, June 20, 1952, Series 1, Subseries 5, Box 85, Folder 720, JDR3 Papers, RFam.A, RAC.

80. Samuel W. Anderson to JDR3, September 18, 1952, Series 1, Box 80, Folder 668, JDR3 Papers, RFam.A, RAC.

81. Accepting an award in 1950, Margaret Sanger said to her audience: "I will not attempt to point out to you the wastefulness of the resources of our plundered planet but I would rather refer you to the books by Fairfield Osborn and William Vogt . . . There is an old saying that a hungry man is an angry man and this is no less true of famished populations." Address by Margaret Sanger, October 25, 1950, Box 199, SPLC.

82. William Vogt to Sripati Chandrasekhar, March 10, 1949, Box 1, Folder 7, CP.

83. William Vogt to Sripati Chandrasekhar, January 27, 1954, Box 1, Folder 14, CP.

84. William Vogt, "Preface," in Sripati Chandrasekhar, *Hungry People and Empty Lands.*

85. William Vogt to Sripati Chandrasekhar, January 27, 1954, Box 1, Folder 14, CP.

86. Thomas Horder, Comment, *1948 Proceedings*, 4.

87. Ibid., 5–6.

88. Sanger, Comment, *1948 Proceedings*, 85–86.

89. Pascal Whelpton, "Population Trends," *1948 Proceedings*, 49, 48–56.

90. Frank Lorimer, "Essential Standards of Living," *1948 Proceedings*, 28, 40; Blacker, Comment, *1948 Proceedings*, 57.

91. Report by Chandrasekhar to Head, Department of Social Science, August 31, 1948, 312.620.91, UNESCO Archive. See also Symonds and Carder, *The United Nations and the Population Question*, 53–55.

92. Sripati Chandrasekhar, "UNESCO and Population Problems," *1948 Proceedings*, 156.

93. Sripati Chandrasekhar, "Indian Demographer Looks at World Population Problems," *UNESCO Courier* 2 (1 February 1949): 10.

94. Joseph Needham, Supplementary Report to Director General, August 31, 1948, 312.620.91 A 06 (41–1) "49," UNESCO Archive.

95. Report by Sripati Chandrasekhar to Head, Department of Social Sciences, UNESCO, August 31, 1948, 312.620.91, UNESCO Archive.

96. John Boyd Orr, Comment, *1948 Proceedings*, 12–14.

97. Radhakamal Mukerjee, "Faiths and Influences," in *The Frontiers of Social Science: In Honour of Radhakamal Mukerjee*, ed. Baljit Singh (London: Macmillan, 1957), 11–12.

98. In the food and people project, the "importance of the agricultural development of the arid zone" was evident. UNESCO, Report on Activities of UNESCO related to Arid Zone Research and Development, August 31, 1951, UNESCO/NS/87, UNESCO Archive.

99. Lord Boyd Orr, "Overpopulated World," 1958, SA/FPA/A17/103, Folder 2, Family Planning Papers, Wellcome Library.

100. Kate Frankenthal, Comment, *1948 Proceedings*, 19–20.

101. JDR3, Memo, September 9, 1954, Series 1, Subseries 5, Box 81, Folder 676, JDR3 Papers, RFam.A, RAC.

102. JDR3, "Memorandum: Objective and Focus of My Asian Interest," 1953, Series 1, Subseries 5, Box 81, Folder 675, JDR3 Papers, RFam.A, RAC.

103. Memorandum, June 18, 1948; Resolution, Rockefeller Fund, 48097, August 6, 1948; Memo "Notestein–Population–Orient," February 11, 1948; all in "Letters re Far East," RG 1.2, Series 600, Box 1, Folder 4, RFA, RAC.

104. Rockefeller Foundation, ECAFE Progress Report, April 9, 1952, RG 2, Series 100, Box 4, Folder 17, RFA, RAC.

105. "The Population Problem: A Tentative Analysis," Memorandum to JDR3, February 1,1952, in Series 1, Subseries 5, Box 80, Folder 667, JDR3 Papers, RFam.A, RAC.

106. Connelly, *Fatal Misconception*, 156–57; Robertson, *The Malthusian Moment*, 66–72.

107. Donald McLean to JDR3, January 10, 1952, Series 1, Subseries 5, Box 81, Folder 674, JDR3 Papers, RFam.A, RAC. Notestein himself recommended they include someone knowledgeable on the general subject of soil. Memo to Donald McLean, April 3, 1952, Series 1, Subseries 5, Box 81, Folder 674, JDR3 Papers, RFam.A, RAC.

108. Transcript, Conference on Population Problems, Williamsburg, June 20, 1952, Series 1, Subseries 5, Box 85, Folder 720, JDR3 Papers, RFam.A, RAC.

109. Connelly, *Fatal Misconception*, 156.

110. Warren Weaver, "Population and Food," Typescript, 1952, Series 1, Box 80, Folder 667, JDR3 Papers, RFam.A, RAC.

111. Warren Weaver, Transcript, Conference on Population Problems, Williamsburg, June 21, 1952, Series 1, Subseries 5, Box 85, Folder 723, JDR3 Papers, RFam.A, RAC.

112. Weaver, "Population and Food."

113. Ibid.

114. Julian Huxley, "World Population," *Scientific American*, 1956. Offprint in Box 100, Folder 7, HP; Nick Cullather, "The Foreign Policy of the Calorie," *American Historical Review* 112, no. 2 (2007): 363.

115. Weaver, "Population and Food."

116. Warren Weaver, Transcript, Conference on Population Problems, Williamsburg, June 21 1952, Series 1, Sub Series 5, Box 85, Folder 723, JDR3 Papers, RFam.A, RAC.

117. William Godwin, *Of Population* (London: Longman, 1820), 500–1.

118. "Bread from the Breezes: Tapping the Atmosphere for Food," News clipping, January 28, 1922, A8510 175/2, Knibbs Papers, NAA.

119. Vilhjalmur Stefansson, "World Meat Supply," Stef Ms 98, Box 14, Folder 12, Stefansson Papers, Dartmouth College.

120. Boyd Orr "Overpopulated World"; Boyd Orr, *1948 Proceedings*, 12.

121. See de Castro, *The Geopolitics of Hunger*, 447–84, 485–502.

122. Huxley, *The Double Crisis*, 3–4.

123. Weaver, "Population and Food," 38.

124. Sterling B. Hendricks, Transcript, Conference on Population Problems, Williamsburg, June 21, 1952, Series 1, Sub Series 5, Box 85, Folder 723, p. 73, JDR3 Papers, RFam.A, RAC.

125. Davis and Isaac, *People on the Move*, 76.

126. Warren Weaver, cited in memorandum to JDR3, "The Population Problem: A Tentative Analysis," February 1, 1952, in Series 1, Subseries 5, Box 80, Folder 667, JDR3 Papers, RFam.A, RAC.

127. Thomas Robert Malthus, *An Essay on the Principle of Population*, 1st ed. (London: J. Johnson, 1798), 12.

128. Hibbert R. Roberts, "Meals for Millions Foundation: Limits and Potential for International Aid," *International Journal of Comparative Sociology* 21, no. 3/4 (1980): 182–95.

129. Norman Cousins, "Food for a Better World," *Saturday Review* 39 (July 21, 1956): 20; John F. Kennedy, "United States to Cooperate with FAO in Freedom from Hunger Campaign," *Department of State Bulletin* 45, no. 1173 (December 18, 1961): 1020–21; "FAO Launches Worldwide Freedom-from-Hunger Campaign," *United Nations Review* 7 (August 1960): 13. See also, D. John Shaw, *World Food Security: A History since 1945* (Basingstoke, U.K.: Palgrave, 2007).

130. For the "freedoms" and the Atlantic Charter, see A. W. Brian Simpson, *Human Rights and the End of Empire* (Oxford: Oxford University Press, 2001), 180–81.

131. See Alec. L. Parrott, "Social Security: Does the Wartime Dream Have to Become a Peacetime Nightmare," *International Labour Review* 131, no. 3 (1992): 375.

132. John Boyd Orr, "The World of Plenty: Restriction to Expansion," Folder 1.92. Boyd Orr Papers.

133. Boyd Orr, *Food and the People*, 5.

134. Boyd Orr, *Soil Fertility*, 14, 10.

135. See also Uday S. Mehta, "Indian Constitutionalism: The Articulation of Political Vision," in *From the Colonial to the Postcolonial: India and Pakistan in Transition*, ed. Dipesh Chakrabarty, Rochonoa Majumdar, and Andrew Sartori (New Delhi: Oxford University Press, 1994).

136. Julian Huxley, "What Are People For?" Lasker Award Address, 1959, Box 75, Folder 9, HP.

137. Imre Ferenczi, "Freedom from Want and International Population Policy," *American Sociological Review* 8, no. 5 (1943): 537, 539.

138. H. G. Wells, *The New World Order* (London: Secker and Warburg, 1940), 107.

139. Chandrasekhar, *Hungry People and Empty Lands*, 49.

140. "The Third Freedom," September 1954, Box 1, Folder 17, CP; B. C. Horton to Sripati Chandrasekhar, August 16, 1954; BBC payment document, September 22, 1954, both in Box 1, Folder 16, CP.

141. Chandrasekhar, *Hungry People and Empty Lands*, 256, 42.

142. Ibid., 49,190, 233, 257–58, 258.

143. "If the Commonwealth redistributes its people in the way best suited to develop its immense resources, then in one or two generations it can become the most powerful force on earth. This would immensely benefit, not only Britain and the Commonwealth nations overseas, but the whole world." (Migration Council, draft by Dudley Barker, August 25, 1951, SA/EUG/D/125, Eugenics Society Papers, Wellcome Library.)

144. A. James Hammerton and Alistair Thomson, *Ten Pound Poms: Australia's Invisible Migrants* (Manchester: Manchester University Press, 2005); Freda Hawkins, *Canada and Immigration: Public Policy and Public Concern* (Montreal: McGill-Queens University Press, 1988).

145. Clifford Heathcote-Smith, "The Crowded Island: Towards a Greater Britain Overseas," 1951, News clipping, SA/EUG/D/126, Eugenics Society Papers, Wellcome Library.

146. "Commonwealth Development Need for a New Approach," Migration Council, SA/EUG/D/124, Eugenics Society Papers, Wellcome Library.

147. Radhakamal Mukerjee, *Races, Lands, and Food: A Program for World Subsistence* (New York: Dryden, 1946), 7–8.

148. Harold L. Ickes, "Introduction," in Mukerjee, *Races, Lands, and Food*, 2.

149. Mark Mazower, *No Enchanted Palace: The End of Empire and the Ideological Origins of the United Nations* (Princeton, N.J.: Princeton University Press, 2009), 111–16.

150. Immigration and Nationality Act of 1965, H.R. 2580, 89th Cong. (1965).

151. Ken Rivett to Sripati Chandrasekhar, March 13, 1964, Box 14, Folder 14, CP.

152. J. Hardjono, "The Indonesian Transmigration Program in Historical Perspective," *International Migration* 26, no. 4 (1988): 427–39; Jonathan Rigg, "Land Settlement in Southeast Asia: The Indonesian Transmigration Program," in *Southeast Asia: A Region in Transition* (London: Unwin Hyman, 1991), 80–108.

153. W. A. Douglas Jackson, "The Virgin and Idle Lands of Western Siberia and Northern Kazakhstan: A Geographical Appraisal," *Geographical Review* 46, no. 1 (1956): 1–19.

154. Sripati Chandrasekhar to Ken Rivett, April 3, 1965, Box 14, Folder 14, CP.

155. Sripati Chandrasekhar, *Asia's Population Problems: With a Discussion of Population and Immigration in Australia* (London: Allen & Unwin, 1967).

156. Sripati Chandrasekhar to John C. Caldwell, January 22, 1964, Box 14, Folder 14, CP.

157. Visit to Australia by Dr. Chandrasekhar, Indian Minister for Health, 1967, A463, 1967/2918, NAA.

158. Sripati Chandrasekhar, *From India to Australia: A Brief History of Immigration: The Dismantling of the "White Australia" Policy: Problems and Prospects of Assimilation* (La Jolla, Calif.: Population Review, 1992).

159. Norris E. Dodd, "Address to the UN General Assembly, 1948," *UNESCO Courier* (February 1949): 10.

160. Boyd Orr and Lubbock, *The White Man's Dilemma*, 12.

161. Ibid., 1.

162. "There is an immense gap between the 'haves' and 'have-nots,' the privileged and the underprivileged" (Julian Huxley, "The Impending Crisis," n.d., Box 75, Folder 5, HP).

163. Boyd Orr and Lubbock, *The White Man's Dilemma*, 73.

164. Lothrop Stoddard, *The Rising Tide of Color Against White World-Supremacy* (London: Chapman and Hall, 1923), 149.

165. Boyd Orr and Lubbock, *The White Man's Dilemma*, 18.

166. Ibid., 15.

167. Sunil S. Amrith, "Food and Welfare in India, c. 1900–1950," *Comparative Studies in Society and History* 50, no. 4 (2008): 1032.

168. Boyd Orr and Lubbock, *The White Man's Dilemma*, 73.

169. John Boyd Orr, "Better Neighbors in a Changing World," February 15, 1946, Folder 2, Boyd Orr Papers.

170. John Boyd Orr, "Science and World Unity," 1940, Folder 1, Boyd Orr Papers.

171. Boyd Orr, "Better Neighbors in a Changing World."

172. John Boyd Orr, "The Choice Ahead—One World or None," December 14, 1946, RG1.1, Series D1, Food and Agriculture Organization Archives, Rome.

173. See Kristen L. Ahlberg, *Transplanting the Great Society: Lyndon Johnson and Food for Peace* (Columbia: University of Missouri Press, 2008).

11. Life and Death

1. Guy Irving Burch and Elmer Pendell, *Human Breeding and Survival: Population Roads to Peace or War* (Penguin: New York, 1947), 12.

2. Giorgio Agamben, *Homo Sacer: Sovereign Power and Bare Life* (Stanford, Calif.: Stanford University Press, 1998); Roberto Esposito, "Thanatopolitics," in *Bios: Biopolitics and Philosophy*, trans. Timothy Campbell (Minneapolis: University of Minnesota Press, 2008), 110–45.

3. For example, Dennis Hodgson, "Orthodoxy and Revisionism in American Demography," *Population and Development Review* 14, no. 4 (1988): 541–69.

4. Kingsley Davis, "The World Demographic Transition," *Annals of the American Academy of Political and Social Science* 237, no. 1 (1945): 1.

5. Ibid., 6.

6. Davis took his data from A. M. Carr-Saunders, *World Population: Past Growth and Present Trends* (Oxford: Clarendon Press, 1936), and from the League of Nations: *Statistical Year-book for 1941–42* (Geneva: Economic Financial and Transit Department, League of Nations, 1943).

7. Davis, "The World Demographic Transition," 4.

8. Ibid., 7, 8.

9. Ibid., 8.

10. Hera Cook, *The Long Sexual Revolution: English Women, Sex, and Contraception 1800–1975* (Oxford: Oxford University Press, 2004), 1–39.

11. Davis, "The World Demographic Transition," 5, 11.

12. Kingsley Davis, *The Population of India and Pakistan* (Princeton, N.J.: Princeton University Press, 1951). See also David M. Heer, "Forging a Distinguished Career," in *Kingsley Davis: A Biography and Selections from His Writings*, ed. David M. Heer (New Brunswick, N.J.: Transaction, 2005), 47.

13. Simon Szreter, "The Idea of Demographic Transition and the Study of Fertility Change: A Critical Intellectual History," *Population and Development Review* 19, no. 4 (1993): 659–701.

14. Ibid., 149, 150.

15. Ta Chen, *Population in Modern China* (Chicago: University of Chicago Press, 1946), 72.

16. Richard Bailey to Carlos Paton Blacker, April 16, 1957, PP/CPB/B.18/2, Blacker Papers, Wellcome Library.

17. Deborah Oakley, "American-Japanese Interaction in the Development of Population Policy in Japan, 1945–52," *Population and Development Review* 4, no. 4 (1978): 619.

18. Matthew Connelly also details the Japanese postwar context in *Fatal Misconception: The Struggle to Control World Population* (Cambridge, Mass.: Belknap, 2008), 134–41.

19. A member of the House of Councillors and former minister of education Mr Tanaka "said that the Japanese population problem should be considered as an international problem and Japanese given the right to emigrate to sparsely settled countries." Roger F. Evans to O. R. McCoy, May 18, 1949, RG 2, Series 609, Box 464, Folder 3113, RFA, RAC.

20. Cited in Sripati Chandrasekhar, *Hungry People and Empty Lands: An Essay on Population Problems and International Tensions* (London: Allen & Unwin, 1954), 126.

21. Warren S. Thompson, *Population and Peace in the Pacific* (Chicago: University of Chicago Press, 1946). By Pacific, he meant lands and people west from Hawaii, including Japan, China, Korea, the Philippines, the southwest Asian peninsula, Australia and the Pacific Islands, and India as well. Warren S. Thompson, *Danger Spots in World Population* (New York: Knopf, 1929).

22. Oakley, "American-Japanese Interaction," 634.

23. See correspondence in RG2, Series 609, Box 464, Folder 3113, RFA, RAC. See also Takeda Hiroko, *The Political Economy of Reproduction in Postwar Japan: Between Nation-State and Everyday Life* (London: Routledge Curzon, 2005).

24. Fairfield Osborn memo, October 13, 1953, RG5, Subseries 1, Box 81, Folder 675, RFam.A, RAC.

25. Warren Thompson, Transcript, Conference on Population Problems, Williamsburg, June 21, 1952, RG5, Series 1, Subseries 5 Box 85, Folder 723, p. 88, RFam.A, RAC.

26. Evans to McCoy, May 18, 1949, RFA.

27. Connelly, *Fatal Misconception*, 140.

28. Christiana Norgren, *Abortion Before Birth Control: The Politics of Reproduction in Post-War Japan* (Princeton, N.J.: Princeton University Press, 2001).

29. Abstract, Eugenic Protection Law No. 156 of July 13, 1948, *International Digest of Health Legislation* 16, no. 4 (1965): 690–99.

30. Koya Yoshio, "The Program for Family Planning in Japan," 1952, Report presented to Planned Parenthood Conference, Bombay, 1952. Offprint in RG 5, Series 1, Subseries 5, Box 80, Folder 671, RFam.A, RAC. For a full account of these studies, and their relevance to both eugenics and family-planning establishments, see C. P. Blacker, "Dr. Yoshio Koya: A Memorable Story," *Eugenics Review* 55, no. 3 (1963): 153–57.

31. See also Irene B. Taeuber and Marshall Balfour, "Means of Fertility Control in Japan," November 13, 1951, RG5, Series 1, Box 80, Folder 671, RFam.A, RAC.

32. Koya Yoshio, "The Program for Family Planning in Japan," 4.

33. Sarah Kovner, *Occupying Power: Sex Workers and Servicemen in Postwar Japan* (Stanford, Calif.: Stanford University Press, 2012).

34. Koya Yoshio, "The Program for Family Planning in Japan," 2.

35. Ibid., 2–3. See also Taeuber and Balfour, "Means of Fertility Control in Japan."

36. C. P. Blacker, "Review of Warren Thompson, *Population and Peace in the Pacific*," *Eugenics Review* 39, no. 1 (1947): 30.

37. Dorothy Thomas, Transcript, Conference on Population Problems, Williamsburg, June 21 1952, RG 5, Series 1, Subseries 5, Box 85, Folder 721, JDR3 Papers, RFam.A, RAC.

38. Ibid., pp. 10–22.

39. R. Jureidini and K. White, "Life Insurance, the Medical Examination and Cultural Values," *Journal of Historical Sociology* 13, no. 2 (2000): 190–214.

40. United Nations Population Division, *World Population Prospects*, Infant Mortality Rates (both sexes) 1950–2100 http://esa.un.org/unpd/wpp/Excel-Data/mortality.htm. [accessed 10 May 2013]. The *Epidemiological and Vital Statistics Reports* (Geneva: World Health Organisation, 1967), recorded lower infant mortality rates in general, but with

similar declines. India, for example, on its calculation, declined from 131 deaths/1,000 live births at Independence to 95 in 1960 and 65 in 1965. Sripati Chandrasekhar used WHO Reports in *Infant Mortality, Population Growth, and Family Planning in India* (Chapel Hill: University of North Carolina Press, 1972).

41. Population Bulletin, March 1958, cited in "Family Planning International Campaign, Detailed Statement of Purpose," n.d., Box 112, Folder 6, HP.

42. Offprint, review of Political and Economic Planning Report "Land and People," 1955, RG5, Subseries 5, Box 80, Folder 669, RFam.A, RAC.

43. Julian Huxley, "Population—or the Battle of Quantity and Quality," 1946, Box 86, Folder 2, HP.

44. "When death control raises human reproductive power to dangerous levels, birth control must provide a brake. It is the function of eugenics to apply this brake to the best advantage of the individual and the community." ("Death Control," Eugenics Society's Statement of Objects, Suggestions from C. P. Blacker, June 21, 1957, PP/CPB/B.18/2, Blacker Papers, Wellcome Library).

45. Shoichi Arai, "The Family Planning Movement in Japan," 1953, Box 199, SPLC.

46. Memorandum to JDR3, February 1, 1952, Series 1, Subseries 5, Box 80, Folder 667, JDR3 Papers, RFam.A, RAC

47. Marshall Balfour, Transcript, Conference on Population Problems, Williamsburg, June 20, 1952, Series 1, Subseries 5, Box 85, Folder 720, JDR3 Papers, RFam.A, RAC.

48. Chester Barnard to Cardinal Spellman, June 21, 1948, RG1.2, Series 600, Box 1, Folder 4, RFA, RAC.

49. For example, "Edith How Martyn's Lectures," p. 10, Coll. Misc. 639/1/25, Eileen Palmer Collection, LSE.

50. BCIIC Newsletter, no. 8, August 1937, PP/EPR/F1/1, Eileen Palmer Collection, Wellcome Library.

51. *World-Wide Birth Control: An Appeal*, BCIIC Pamphlet, n.d., PP/EPR/F.1/1, Eileen Palmer Collection, Wellcome Library.

52. Okli Ampofo, Comment, *Proceedings of the International Conference on Population and World Resources* in Relation to the Family (London: Family Planning Association, 1948), 83, 136–37 [hereafter *1948 Proceedings*].

53. Elizabeth Watkins, *On the Pill: A Social History of Oral Contraceptives, 1950–1970* (Baltimore, Md.: Johns Hopkins University Press, 1998), 50. See also Lara Marks, *Sexual Chemistry: A History of the Contraceptive Pill* (New Haven, Conn.: Yale University Press, 2001), 6–7.

54. Brock Chisholm to Paul Henshaw, January 21, 1952, GH 12 Birth Control, WHO Archives, Geneva.

55. "Three Billion People," PP/CPB/C.4/1, Blacker Papers, Wellcome Library. This talk included C. P. Blacker, Dhanvanthi Rama Rau, Margaret Sanger, Theodore W. Schultz, William Vogt, Ellen Watamull, and R. Richard Wohl.

56. William Vogt to Brock Chisholm, November 23, 1951, GH 12 Birth Control, WHO Archives, Geneva.

57. "The Prime Minister Speaks," *Planned Parenthood: Monthly Bulletin of the Family Planning Association of India* 6, no. 8–10 (1959): 3.

58. Julian Huxley, *Memories*, 2 vols. (London: Allen & Unwin, 1970), 1:152. Emphasis in original.

59. For example, the Ford Foundation sponsored motivation research within its Motivation Action Research Committee "to provide better understanding of the basic factors in the acceptance of family planning," Typescript extract from the *Hindu Weekly Review*, September 26, 1960, CPB/C4/5, Blacker Papers, Wellcome Library.

60. "Building Stronger Families, Happier Children, International Planned Parenthood Federation," pamphlet n.d., Box 115, Folder 4, HP.

61. Donald McLean, Memo, September 9, 1952, Series 1, Subseries 5, Box 81, Folder 674, JDR3 Papers, RFam.A, RAC.

62. Rufus E. Miles Jr., "Balancing Population and Resources—the Greatest Challenge of Social Engineering," March 15, 1952, Series 1, Subseries 5, Box 80, Folder 667, JDR3 Papers, RFam.A, RAC.

63. Transcript, Conference on Population Problems, Williamsburg, June 21, 1952, Series 1, Subseries 5, Box 85, Folder 721, JDR3 Papers, RFam.A, RAC.

64. Ibid., Folder 723.

65. Margaret Sanger, "Population Planning—Program of Birth Control Viewed as Contributing to World Peace," *New York Times*, January 3, 1960.

66. "Three Billion People," PP/CPB/C.4/1, Blacker Papers, Wellcome Library.

67. *Planned Parenthood News*, no. 25 (1959): 1–3.

68. Ibid., 1; Planned Parenthood 39th Annual Luncheon, seating arrangements, Box 115, Folder 4, HP.

69. Indian Envoy, "Rockefeller Urges US Action on Population Control," *Planned Parenthood News*, no. 27 (1960): 1.

70. Chandrasekhar, *Hungry People and Empty Lands*, 246.

12. Universal Rights?

1. Betsy Hartmann, *Reproductive Rights and Wrongs: The Global Politics of Population Control and Contraceptive Choice* (Boston: South End, 1995).

2. Ian Dowbiggin, *The Sterilization Movement and Global Fertility in the Twentieth Century* (Oxford: Oxford University Press, 2008), 6; Garland Allen, "From Eugenics to Population Control," *Science for the People* 12, no. 4 (1980): 22–28. This is an element of a wider critique of postwar population control. For another statement of apparently underground eugenics, see John P. Jackson and Nadine M. Weidman, *Race, Racism, and Science: Social Impact and Interaction* (Santa Barbara, Calif.: ABC Clio, 2004), 167.

3. Dowbiggin, *The Sterilization Movement*. Mohan Rao, to take another example, has suggested that "in the 1940s and 1950s, eugenic ideas were considered embarrassing, close as they were to those of the architect of the Nazi holocaust. Population control, however, came to occupy a respectable position" (Mohan Rao, *From Population Control to Reproductive Health: Malthusian Arithmetic* [New Delhi: Sage, 2004], 102). For the politics surrounding sterilization and feminism in the United States see Linda Gordon, *The Moral Property of Women: A History of Birth Control Politics in America* (Urbana:

University of Illinois Press, 2007). For a nuanced account of Puerto Rican sterilization, see Iris Lopez, *Matters of Choice: Puerto Rican Women's Struggle for Reproductive Freedom* (New Brunswick, N.J.: Rutgers University Press, 2008).

4. As I have argued elsewhere, it was not really until well into the 1970s that the term "eugenics" became almost entirely undefended (but never totally without some defendants). See Alison Bashford, "Where Did Eugenics Go?" in *The Oxford Handbook of the History of Eugenics*, ed. Alison Bashford and Philippa Levine (Oxford: Oxford University Press, 2010), 539–58.

5. See Edmund Ramsden, "Social Demography and Eugenics in the Interwar United States," *Population and Development Review* 29, no. 4 (2003): 550.

6. Matthew Connelly, *Fatal Misconception: The Struggle to Control World Population* (Cambridge, Mass.: Belknap, 2008), 163.

7. "The population control movement is eugenics put into action" (Terry Melanson, "Crypto–Eugenics Population Control: Policy and Propaganda," August 9, 2009; http://www.conspiracyarchive.com [accessed March 28, 2011]).

8. Nikolas Rose, *Powers of Freedom: Reframing Political Thought* (Cambridge: Cambridge University Press, 1999), 138.

9. "War Time Eugenic Measures in Germany," August 10, 1947, PP/CPB/H.1/9, Blacker Papers, Wellcome Library.

10. C. P. Blacker, "'Eugenic Experiments Conducted by the Nazis on Human Subjects," *Eugenics Review* 44, no. 1 (1952): 9.

11. Julian Huxley, *UNESCO: Its Purpose and its Philosophy* (Washington D.C.: Public Affairs, 1947), 38. See Paul Weindling, "Julian Huxley and the Continuity of Eugenics in Twentieth-Century Britain," *Journal of Modern European History* 10, no. 4 (2012): 480–99.

12. Julian Huxley, ed. *The Humanist Frame* (London: Allen & Unwin, 1961), 23.

13. Huxley, *UNESCO*, 15.

14. Julian Huxley, Typescript Notes, "What Are People For?" Lasker Award Address, 1959, Box 75, Folder 9, HP.

15. Julian Huxley, "The Impending Crisis," n.d., Box 75, Folder 5, HP; Huxley, "Human variety is being threatened," 1961, Box 77, Folder 11, HP.

16. F. A. E. Crew et al., "Social Biology and Population Improvement," *Nature* 144, no. 3646 (1939): 521–22.

17. Stefan Kühl, *The Nazi Connection: Eugenics, American Racism, and German National Socialism* (Oxford: Oxford University Press, 2002), 5–6.

18. Barry Mehler, "Foundation for Fascism: The New Eugenics Movement in the United States," *Patterns of Prejudice* 23, no. 4 (1989): 17–25; William H. Tucker, *The Funding of Scientific Racism: Wickliffe Draper and the Pioneer Fund* (Urbana: University of Illinois Press, 2007).

19. Reginald Ruggles Gates to Fraser Roberts, September 27, 1954, SA/EUG/D104, Eugenics Society Papers, Wellcome Library, London.

20. Fraser Roberts to Reginald Ruggles Gates, October 5, 1954, SA/EUG/D104, Eugenics Society Papers, Wellcome Library.

21. Colin Bertram correspondence and reports, 1961–1963, SA/EUG/104, Eugenics Society Papers, Wellcome Library.

22. Gavin Schaffer, *Racial Science and British Society, 1930–62* (Basingstoke, U.K.: Palgrave, 2008), 151–2.

23. [Carlos Paton Blacker], "The Eugenic Aspects of West Indian Immigration: some general notes," n.d., SA/EUG/D103, Eugenics Society Papers, Wellcome Library. Although this document is not named, it is in my view clearly a commentary by Blacker.

24. Huxley, "The Impending Crisis," 1961.

25. Huxley, Lasker Award Address, 1959. This is a phrase and an idea that circulated between Huxley and Colin Bertram. See G. C. L. Bertram, "What Are People For?" *The Humanist Frame*, ed. Huxley, 373–84.

26. See, e.g., Untitled typescript, notes, and manuscripts, "Adventures of the Mind, 1958–59," Box 63, Folder 7, HP.

27. Huxley, "Eugenics in Evolutionary Perspective," Box 79, Folder 4, HP.

28. "Dr. R. A. Fisher, Director of the Galton Laboratory, England, made an analysis of the subject and demonstrated that if such sterilization could be consistently carried out for one complete generation, sound mathematical estimates based on reasonable assumptions would show in the next generation" (Human Betterment Foundation, *Human Sterilization Today*, n.d. B/3/10, Alexander Carr–Saunders Papers, LSE).

29. John Macnicol, "Eugenics and the Campaign for Voluntary Sterilization in Britain Between the Wars," *Social History of Medicine* 2, no. 2 (1989): 161–62.

30. Hugh Moore, "Can We Avoid War?" n.d, Box 1, Folder 6, Hugh Moore Fund Collection, Rudd Manuscript Library, Princeton University.

31. League of Nations Association Correspondence, 1937–44, Hugh Moore Fund Collection, Box 24, Folder 23–24; United Nations Association Congressional Speaking Tour, 1943, Box 25, Folder 11; See also, World Peace Foundation, 1946, Box 27, Folder 10.

32. Hugh Moore Fund, *The Population Bomb* (New York: Hugh Moore Fund, 1954).

33. Hugh Moore to JDR3, December 1959, Box 80, Folder 670, JDR3 Papers, RFam.A, RAC.

34. *The Population Bomb,* 14.

35. Rao, *From Population Control to Reproductive Health*, 33, 38–40.

36. Brock Chisholm, Endorsing letter, jacket, *The Population Bomb*.

37. Julian Huxley, *Memories*, 2 vols. (London: Allen & Unwin, 1970), 1:151.

38. Guy Irving Burch and Elmer Pendell, *Human Breeding and Survival: Population Roads to Peace or War* (New York: Penguin, 1947), 100.

39. Jacqueline Kasun, *The War Against Population: The Economics and Ideology of Population Control* (San Francisco: Ignatius, 1988), 161.

40. See, e.g., Yolanda Eraso, "Biotypology, Endocrinology, and Sterilization: The Practice of Eugenics in the Treatment of Argentinean Women During the 1930s," *Bulletin of the History of Medicine* 81, no. 4 (2007): 817. For connections between abortion and sterilization debates, see Marius Turda, "'To End the Degeneration of a Nation': Debates on Eugenic Sterilization in Inter-War Romania," *Medical History* 53, no. 1 (2009): 77–104.

41. Johanna Schoen, *Choice and Coercion: Birth Control, Sterilization, and Abortion in Public Health and Welfare* (Chapel Hill: University of North Carolina Press, 2005), 3.

42. A. M. Carr-Saunders, "Eugenics in the Light of Population Trends" (1935), reprinted in *Eugenics Review* 60, no. 1 (1968): 55.

43. Typescript, Draft Resolution about Sterilization for the [Eugenic Society] Council, June 20, 1961, Box 112, Folder 5, HP.

44. "Knowledge of and Attitudes Towards Eugenics, A Mail Survey Among Scientists and Classicists," October 1962. Report, Box 112, Folder 5, HP.

45. "Five Year Plan Committee Report" *Planned Parenthood* (India) 7, no. 10–11 (1960): 2.

46. C. P. Blacker to J. P. Brander, February 17, 1960, CPB/C.4/4, Blacker Papers, Wellcome Library.

47. Rao, *From Population Control to Reproductive Health*, 38–53.

48. Sripati Chandrasekhar, "Population Problems of India and Pakistan," *UNESCO Courier* 2 (1949): 9.

49. Ken Rivett to Sripati Chandrasekhar, July 7, 1964, Box 14, Folder 14, CP.

50. See correspondence between Sripati Chandrasekhar and Douglas Ensminger, Ford Representative in India between 1955 and 1969, Box 16, Folder 39, CP.

51. Sripati Chandrasekhar to Japanese Ambassador (Washington, D.C.), October 12, 1965, Box 14, Folder 1, CP.

52. Chancellor H. Whiting, M.D., to Sripati Chandrasekhar, February 18, 1947, Box 1, Folder 6, CP.

53. C. P. Blacker to Sripati Chandrasekhar, December 4, 1954, Box 1, CP.

54. Connelly, *Fatal Misconception*, 115–54.

55. Sarah Hodges, "Indian Eugenics in an Age of Reform," in *Reproductive Health in India: History, Politics, Controversies*, ed. Sarah Hodges (New Delhi: Orient Longman, 2006), 132.

56. Radhakamal Mukerjee, *Food Planning for Four Hundred Millions* (London: Macmillan, 1938), 221.

57. National Planning Committee Report, 1938, quoted in Benjamin Zachariah, "Uses of Scientific Argument: The Case of 'Development' in India, c. 1930–1950," *Economic and Political Weekly* 36, no. 39 (2001): 3695. See also Rao, *From Population Control to Reproductive Health*, 109.

58. Connelly, *Fatal Misconception*, 141–46.

59. Sanjam Ahluwalia, *Reproductive Restraints: Birth Control in India, 1877–1947* (Urbana: University of Illinois Press, 2008), 54–84.

60. Sripati Chandrasekhar, *Hungry People and Empty Lands: An Essay on Population Problems and International Tensions* (London: Allen & Unwin, 1954), 247, 253.

61. "It is not generally known that the Mahatma approved sterilization (vasectomy for the husband) as early as 1935 when male sterilization was not widely known" (Typescript, draft chapter, Gandhi and Birth Control, Box 61, Folder 19, CP).

62. International Planned Parenthood Federation, Brochure, n.d, Box 112, Folder 6, HP.

63. Chandrasekhar, *Hungry People and Empty Lands*, 251.

64. V. H. Wallace, "A World Population Policy as a Factor in Maintaining Peace," in *Paths to Peace: A Study of War, Its Causes and Prevention*, ed. V. H. Wallace (Melbourne: Melbourne University Press, 1957), 292.

65. United Nations, Universal Declaration of Human Rights (1948), Preamble.

66. United Nations, Universal Declaration of Human Rights (1948), Article 16.

67. Convention on the Prevention and Punishment of the Crime of Genocide (1948), Article II(d).

68. Convention on Civil and Political Rights (1966), Article 23.

69. "U Thant Urges World Wide Use of Birth Control," *Lodi News Sentinel*, December 10, 1966. See also Stanley P. Johnson, *World Population and the United Nations: Challenge and Response* (Cambridge: Cambridge University Press, 1987), xxi.

70. See Kingsley Davis, "Zero Population Growth: The Ends and the Means," in *Kingsley Davis: A Biography and Selections from His Writings*, ed. David M. Heer (New Brunswick, N.J.: Transaction, 2005), 537.

71. Richard Symonds and Michael Carder, *The United Nations and the Population Question, 1945–1970* (London: Chatto & Windus, 1973), 157, 163.

72. Convention on the Elimination of all Forms of Discrimination against Women (1979), Article 12, Article 16.

73. C. P. Blacker, "Eugenics in the Middle Nineteen-Sixties," June 20, 1966, PP/CPB/H.1/54, Blacker Papers, Wellcome Library.

74. Dugald Baird, "A Fifth Freedom?" *Eugenics Review* 58, no. 4 (1966): 195. Originally published in *British Medical Journal* 2, no. 5471 (1965): 1141–48.

75. Ibid., 195–204.

76. Ibid., 204.

77. Hartmann, *Reproductive Rights and Wrongs*; Sarah Hodges, "Malthus Is Forever: The Global Market for Population Control," *Global Social Policy* 10, no. 1 (2010): 120–26.

78. Frederick Osborn, "Development of a Eugenic Philosophy," *American Sociological Review* 2, no. 3 (1937): 395.

79. Douglas Martin, "Sripati Chandrasekhar, Indian Demographer, Dies at 83," *New York Times*, June 23, 2001, A11.

80. Chandrasekhar, *Hungry People and Empty Lands*, 243.

81. Susan Greenhalgh and Edwin A. Winckler, *Governing China's Population: From Leninist to Neoliberal Biopolitics* (Stanford, Calif.: Stanford University Press, 2005).

82. Kingsley Davis, "Population Policy: Will Current Programs Succeed?" *Science* 158, no. 3802 (1967): 730–39. See also Kingsley Davis, "Zero Population Growth: The Goal and the Means," in *The No-Growth Society*, ed. Mancur Olson and Hans H. Landsberg (New York: Norton, 1973).

83. World Population Growth Rates, 1950–2050, U.S. Census Bureau International Database, June 2011. www.census.gov/population/international/data/idb/worldgrgraph.php (accessed February 10, 2012).

84. *Z.P.G.* (produced by Frank de Felitta, 1972).

85. "Mr. Symes on the Cause of Poverty," *The Malthusian* 72 (1885): 582.

Conclusion

1. Thomas Robert Malthus, *An Essay on the Principle of Population*, 2d ed. (London: J. Johnson, 1803), 8.

2. Julian Huxley, "What Are People For?" Lasker Award Address, 1959, Box 75, Folder 9, HP.

3. Libby Robin and Will Steffen, "History for the Anthropocene," *History Compass* 5, no. 5 (2007): 1694–95.

4. Aldous Huxley, *Brave New World Revisited* (London: Chatto & Windus, 1959), 15, 18.

5. See "Food and People," Box 75, Folder 2, HP.

6. Huxley, "What Are People for?"

7. V. H. Wallace, "A World Population Policy as a Factor in Maintaining Peace," in *Paths to Peace: A Study of War, Its Causes and Prevention*, ed. V. H. Wallace (Melbourne: Melbourne University Press, 1957), 284, 288, 289–90.

8. C. P. Blacker, "Population," *Eugenics Review* 39, no. 1 (1947): 27.

9. Radhakamal Mukerjee, "The Commonwealth and Rights of Mankind," in *The Oneness of Mankind* (New Delhi: Radha, 1965), 57–67.

10. Mukerjee, "The Philosophical Unity of Mankind," in *The Oneness of Mankind*, 92–93.

11. Sripati Chandrasekhar, *"A dirty filthy book": The writings of Charles Knowlton and Annie Besant on Reproductive Physiology and Birth Control* (Berkeley: University of California Press, 1981).

12. Charles Galton Darwin, *The Next Million Years* (New York: Doubleday, 1953), 26, 141, 207.

13. Lord Simon of Wythenshawe, "Some Aspects of World Population and Food Resources," *Eugenics Review* 46, no. 2 (1954): 100.

14. George Knibbs, "Waste," 1926, A8510, 175/6, Knibbs Papers, NAA.

15. For example, John Glad, *Jewish Eugenics* (Washington, D.C.: Wooden Shore, 2011). And see Kevin McDonald's review in the anti-Semitic *Occidental Observer: White Identity, Interests, and Culture*, May 15, 2011. http://www.theoccidentalobserver.net/2011/05/review-of-john-glads-jewish-eugenics/ (accessed March 16, 2012).

16. Darwin, *The Next Million Years*, 32, 179–80.

17. Kenneth Pomeranz, *The Great Divergence: China, Europe, and the Making of the Modern World Economy* (Princeton, N.J.: Princeton University Press, 2000), 313–15.

18. Georg Borgström, *The Hungry Planet: The Modern World at the Edge of Famine* (New York: Macmillan, 1965), 70–71.

19. Margaret Sanger, "Plans for Peace," *Birth Control Review* 16, no. 4 (1932): 106.

20. Statement by President Lyndon B. Johnson upon signing the Food for Peace Act of 1966, November 12, 1966 http://www.presidency.ucsb.edu/ws/?pid=28025 [accessed 10 May 2013]. See also Kristen L. Ahlberg, *Transplanting the Great Society: Lyndon Johnson and Food for Peace* (Columbia: University of Missouri Press, 2008).

21. Borgström, *The Hungry Planet*, 470.

22. Paul R. Ehrlich, *The Population Bomb: Population Control or Race to Oblivion* (New York: Ballantine, 1968).

23. Humanae Vitae, Encyclical Letter of His Holiness Pope Paul VI, On the Regulation of Birth, 25 July 1968, http://www.vatican.va/holy_father/paul_vi/encyclicals/documents/hf_p-vi_enc_25071968_humanae-vitae_en.html [Accessed 10 May 2013]; Garrett Hardin, "The Tragedy of the Commons," *Science* 162, no. 3859 (1968): 1243–1248.

Archival Collections

Alexander Morris Carr-Saunders Papers, London School of Economics Archives
Alfred J. Lotka Papers, Stanley G. Rudd Manuscript Library, Princeton University
American Birth Control League Papers, Houghton Library, Harvard University
Eileen Palmer Collection, London School of Economics Archives
Eileen Palmer Collection, Wellcome Library, London
Eugenics Society Papers, Wellcome Library, London
Family Planning Association Papers, Wellcome Library, London
Frederick Osborn Papers, American Philosophical Society, Philadelphia
George Knibbs, Semi-Personal papers (Commonwealth Institute of Science and Industry), National Archives of Australia, Canberra
Hugh Moore Fund Collection, Stanley G. Rudd Manuscript Library, Princeton University
Isaiah Bowman Papers, Milton Eisenhower Library Special Collections, Johns Hopkins University
John Boyd Orr Papers, National Library of Scotland, Edinburgh
John Maynard Keynes Papers, Modern Archives Centre, King's College, Cambridge
John Walter Gregory Collection, University of Glasgow
Julian Sorell Huxley Papers, Woodson Research Center, Fondren Library, Rice University, Texas
League of Nations Archive, Palais des Nations, Geneva
Margaret Sanger Papers, Manuscript Division, Library of Congress, Washington, D.C.
Raymond Pearl Papers, American Philosophical Society, Philadelphia

Rockefeller Archives Center, Sleepy Hollow, New York (Rockefeller Family Papers; Rockefeller Foundation Papers).

Sripati Chandrasekhar Papers, Ward M. Canaday Center, University of Toledo Library, Toledo, Ohio

UNESCO Archive, Paris

United Nations Archive, New York City

Vilhjalmur Stefansson Papers, Rauner Special Collections Library, Dartmouth College Hanover, New Hampshire

W. E. Agar Papers, Australian Academy of Sciences, Canberra

World Health Organization Archives, Geneva

Index